Simulation in textile technology

The Textile Institute and Woodhead Publishing

The Textile Institute is a unique organisation in textiles, clothing and footwear. Incorporated in England by a Royal Charter granted in 1925, the Institute has individual and corporate members in over 90 countries. The aim of the Institute is to facilitate learning, recognise achievement, reward excellence and disseminate information within the global textiles, clothing and footwear industries.

Historically, The Textile Institute has published books of interest to its members and the textile industry. To maintain this policy, the Institute has entered into partnership with Woodhead Publishing Limited to ensure that Institute members and the textile industry continue to have access to high calibre titles on textile science and technology.

Most Woodhead titles on textiles are now published in collaboration with The Textile Institute. Through this arrangement, the Institute provides an Editorial Board which advises Woodhead on appropriate titles for future publication and suggests possible editors and authors for these books. Each book published under this arrangement carries the Institute's logo.

Woodhead books published in collaboration with The Textile Institute are offered to Textile Institute members at a substantial discount. These books, together with those published by The Textile Institute that are still in print, are offered on the Woodhead website at: www.woodheadpublishing.com. Textile Institute books still in print are also available directly from the Institute's website at: www.textileinstitutebooks.com.

A list of Woodhead books on textile science and technology, most of which have been published in collaboration with The Textile Institute, can be found towards the end of the contents pages.

Woodhead Publishing Series in Textiles: Number 136

Simulation in textile technology

Theory and applications

Edited by

D. Veit

The Textile Institute

WOODHEAD PUBLISHING

Oxford Cambridge Philadelphia New Delhi

Published by Woodhead Publishing Limited in association with The Textile Institute
Woodhead Publishing Limited,
80 High Street, Sawston, Cambridge CB22 3HJ, UK
www.woodheadpublishing.com
www.woodheadpublishingonline.com

Woodhead Publishing, 1518 Walnut Street, Suite 1100, Philadelphia, PA 19102-3406, USA

Woodhead Publishing India Private Limited, G-2, Vardaan House, 7/28 Ansari Road, Daryaganj, New Delhi – 110002, India
www.woodheadpublishingindia.com

First published 2012, Woodhead Publishing Limited

British Library Cataloguing in Publication Data
A catalogue record for this book is available from the British Library.

Library of Congress Control Number: 2012904961

ISBN 978-0-85709-029-4 (print)
ISBN 978-0-85709-708-8 (online)
ISSN 2042-0803 Woodhead Publishing Series in Textiles (print)
ISSN 2042-0811 Woodhead Publishing Series in Textiles (online)

The publisher's policy is to use permanent paper from mills that operate a sustainable forestry policy, and which has been manufactured from pulp which is processed using acid-free and elemental chlorine-free practices. Furthermore, the publisher ensures that the text paper and cover board used have met acceptable environmental accreditation standards.

Typeset by Replika Press Pvt Ltd, India
Printed by TJ International Ltd, Padstow, Cornwall, UK

Contents

Contributor contact details

(* = main contact)

Editor and chapters 1, 2, 3, 4 and 9

Dr-Ing. Dieter Veit
Institut für Textiltechnik of RWTH Aachen University
Otto-Blumenthal-Str. 1
52074 Aachen
Germany

E-mail: dieter.veit@ita.rwth-aachen.de

Chapter 5

Dr-Ing. Philip Jungbecker and Dr-Ing. Dieter Veit*
Institut für Textiltechnik of RWTH Aachen University
Otto-Blumenthal-Str. 1
52074 Aachen
Germany

E-mail: dieter.veit@ita.rwth-aachen.de

Chapters 6 and 8

Professor Y. K. Kyosev
Faculty of Textile and Clothing Technology
Hochschule Niederrhein – University of Applied Sciences
Webschulstr. 31
41065 Mönchengladbach
Germany

E-mail: yordan.kyosev@hs-niederrhein.de

Chapter 7

Professor Bohuslav Neckář* and Dr Monika Vyšanská
Department of Textile Technologies
Faculty of Textile Engineering
Technical University of Liberec
Studentska 2
Liberec 461 17
Czech Republic

E-mail: bohuslav.neckar@tul.cz; monika.vysanska@tul.cz

Woodhead Publishing Series in Textiles

1 **Watson's textile design and colour Seventh edition**
Edited by Z. Grosicki

2 **Watson's advanced textile design**
Edited by Z. Grosicki

3 **Weaving Second edition**
P. R. Lord and M. H. Mohamed

4 **Handbook of textile fibres Volume 1: Natural fibres**
J. Gordon Cook

5 **Handbook of textile fibres Volume 2: Man-made fibres**
J. Gordon Cook

6 **Recycling textile and plastic waste**
Edited by A. R. Horrocks

7 **New fibers Second edition**
T. Hongu and G. O. Phillips

8 **Atlas of fibre fracture and damage to textiles Second edition**
J. W. S. Hearle, B. Lomas and W. D. Cooke

9 **Ecotextile '98**
Edited by A. R. Horrocks

10 **Physical testing of textiles**
B. P. Saville

11 **Geometric symmetry in patterns and tilings**
C. E. Horne

12 **Handbook of technical textiles**
Edited by A. R. Horrocks and S. C. Anand

13 **Textiles in automotive engineering**
W. Fung and J. M. Hardcastle

14 **Handbook of textile design**
J. Wilson

15 **High-performance fibres**
Edited by J. W. S. Hearle

16 **Knitting technology Third edition**
D. J. Spencer

17 **Medical textiles**
Edited by S. C. Anand

18 **Regenerated cellulose fibres**
Edited by C. Woodings

19 **Silk, mohair, cashmere and other luxury fibres**
Edited by R. R. Franck

20 **Smart fibres, fabrics and clothing**
Edited by X. M. Tao

21 **Yarn texturing technology**
J. W. S. Hearle, L. Hollick and D. K. Wilson

22 **Encyclopedia of textile finishing**
H-K. Rouette

23 **Coated and laminated textiles**
W. Fung

24 **Fancy yarns**
R. H. Gong and R. M. Wright

25 **Wool: Science and technology**
Edited by W. S. Simpson and G. Crawshaw

26 **Dictionary of textile finishing**
H-K. Rouette

27 **Environmental impact of textiles**
K. Slater

28 **Handbook of yarn production**
P. R. Lord

29 **Textile processing with enzymes**
Edited by A. Cavaco-Paulo and G. Gübitz

30 **The China and Hong Kong denim industry**
Y. Li, L. Yao and K. W. Yeung

31 **The World Trade Organization and international denim trading**
Y. Li, Y. Shen, L. Yao and E. Newton

32 **Chemical finishing of textiles**
W. D. Schindler and P. J. Hauser

33 **Clothing appearance and fit**
J. Fan, W. Yu and L. Hunter

34 **Handbook of fibre rope technology**
H. A. McKenna, J. W. S. Hearle and N. O'Hear

35 **Structure and mechanics of woven fabrics**
J. Hu

36 **Synthetic fibres: nylon, polyester, acrylic, polyolefin**
Edited by J. E. McIntyre

37 **Woollen and worsted woven fabric design**
E. G. Gilligan

38 **Analytical electrochemistry in textiles**
P. Westbroek, G. Priniotakis and P. Kiekens

39 **Bast and other plant fibres**
R. R. Franck

40 **Chemical testing of textiles**
Edited by Q. Fan

41 **Design and manufacture of textile composites**
Edited by A. C. Long

42 **Effect of mechanical and physical properties on fabric hand**
Edited by Hassan M. Behery

43 **New millennium fibers**
T. Hongu, M. Takigami and G. O. Phillips

44 **Textiles for protection**
Edited by R. A. Scott

45 **Textiles in sport**
Edited by R. Shishoo

46 **Wearable electronics and photonics**
Edited by X. M. Tao

47 **Biodegradable and sustainable fibres**
Edited by R. S. Blackburn

48 **Medical textiles and biomaterials for healthcare**
Edited by S. C. Anand, M. Miraftab, S. Rajendran and J. F. Kennedy

49 **Total colour management in textiles**
Edited by J. Xin

50 **Recycling in textiles**
Edited by Y. Wang

51 **Clothing biosensory engineering**
Y. Li and A. S. W. Wong

52 **Biomechanical engineering of textiles and clothing**
Edited by Y. Li and D. X-Q. Dai

53 **Digital printing of textiles**
Edited by H. Ujiie

54 **Intelligent textiles and clothing**
Edited by H. R. Mattila

55 **Innovation and technology of women's intimate apparel**
W. Yu, J. Fan, S. C. Harlock and S. P. Ng

56 **Thermal and moisture transport in fibrous materials**
Edited by N. Pan and P. Gibson

57 **Geosynthetics in civil engineering**
Edited by R. W. Sarsby

58 **Handbook of nonwovens**
Edited by S. Russell

59 **Cotton: Science and technology**
Edited by S. Gordon and Y-L. Hsieh

60 **Ecotextiles**
Edited by M. Miraftab and A. R. Horrocks

61 **Composite forming technologies**
Edited by A. C. Long

62 **Plasma technology for textiles**
Edited by R. Shishoo

63 **Smart textiles for medicine and healthcare**
Edited by L. Van Langenhove

64 **Sizing in clothing**
Edited by S. Ashdown

65 **Shape memory polymers and textiles**
J. Hu

66 **Environmental aspects of textile dyeing**
Edited by R. Christie

67 **Nanofibers and nanotechnology in textiles**
Edited by P. Brown and K. Stevens

68 **Physical properties of textile fibres Fourth edition**
W. E. Morton and J. W. S. Hearle

69 **Advances in apparel production**
Edited by C. Fairhurst

70 **Advances in fire retardant materials**
Edited by A. R. Horrocks and D. Price

71 **Polyesters and polyamides**
Edited by B. L. Deopura, R. Alagirusamy, M. Joshi and B. S. Gupta

72 **Advances in wool technology**
Edited by N. A. G. Johnson and I. Russell

73 **Military textiles**
Edited by E. Wilusz

74 **3D fibrous assemblies: Properties, applications and modelling of three-dimensional textile structures**
J. Hu

75 **Medical and healthcare textiles**
Edited by S. C. Anand, J. F. Kennedy, M. Miraftab and S. Rajendran

76 **Fabric testing**
Edited by J. Hu

77 **Biologically inspired textiles**
Edited by A. Abbott and M. Ellison

78 **Friction in textile materials**
Edited by B. S. Gupta

79 **Textile advances in the automotive industry**
Edited by R. Shishoo

80 **Structure and mechanics of textile fibre assemblies**
Edited by P. Schwartz

81 **Engineering textiles: Integrating the design and manufacture of textile products**
Edited by Y. E. El-Mogahzy

82 **Polyolefin fibres: Industrial and medical applications**
Edited by S. C. O. Ugbolue

83 **Smart clothes and wearable technology**
Edited by J. McCann and D. Bryson

84 **Identification of textile fibres**
Edited by M. Houck

85 **Advanced textiles for wound care**
Edited by S. Rajendran

86 **Fatigue failure of textile fibres**
Edited by M. Miraftab

87 **Advances in carpet technology**
Edited by K. Goswami

88 **Handbook of textile fibre structure Volume 1 and Volume 2**
Edited by S. J. Eichhorn, J. W. S. Hearle, M. Jaffe and T. Kikutani

89 **Advances in knitting technology**
Edited by K-F. Au

90 **Smart textile coatings and laminates**
Edited by W. C. Smith

91 **Handbook of tensile properties of textile and technical fibres**
Edited by A. R. Bunsell

92 **Interior textiles: Design and developments**
Edited by T. Rowe

93 **Textiles for cold weather apparel**
Edited by J. T. Williams

94 **Modelling and predicting textile behaviour**
Edited by X. Chen

95 **Textiles, polymers and composites for buildings**
Edited by G. Pohl

96 **Engineering apparel fabrics and garments**
J. Fan and L. Hunter

97 **Surface modification of textiles**
Edited by Q. Wei

98 **Sustainable textiles**
Edited by R. S. Blackburn

99 **Advances in yarn spinning technology**
Edited by C. A. Lawrence

100 **Handbook of medical textiles**
Edited by V. T. Bartels

101 **Technical textile yarns**
Edited by R. Alagirusamy and A. Das

102 **Applications of nonwovens in technical textiles**
Edited by R. A. Chapman

103 **Colour measurement: Principles, advances and industrial applications**
Edited by M. L. Gulrajani

104 **Fibrous and composite materials for civil engineering applications**
Edited by R. Fangueiro

105 **New product development in textiles: Innovation and production**
Edited by L. Horne

106 **Improving comfort in clothing**
Edited by G. Song

107 **Advances in textile biotechnology**
Edited by V. A. Nierstrasz and A. Cavaco-Paulo

108 **Textiles for hygiene and infection control**
Edited by B. McCarthy

109 **Nanofunctional textiles**
Edited by Y. Li

110 **Joining textiles: Principles and applications**
Edited by I. Jones and G. Stylios

111 **Soft computing in textile engineering**
Edited by A. Majumdar

112 **Textile design**
Edited by A. Briggs-Goode and K. Townsend

113 **Biotextiles as medical implants**
Edited by M. King and B. Gupta

114 **Textile thermal bioengineering**
Edited by Y. Li

115 **Woven textile structure**
B. K. Behera and P. K. Hari

116 **Handbook of textile and industrial dyeing. Volume 1: Principles, processes and types of dyes**
Edited by M. Clark

117 **Handbook of textile and industrial dyeing. Volume 2: Applications of dyes**
Edited by M. Clark

118 **Handbook of natural fibres. Volume 1: Types, properties and factors affecting breeding and cultivation**
Edited by R. Kozlowski

119 **Handbook of natural fibres. Volume 2: Processing and applications**
Edited by R. Kozlowski

120 **Functional textiles for improved performance, protection and health**
Edited by N. Pan and G. Sun

121 **Computer technology for textiles and apparel**
Edited by Jinlian Hu

122 **Advances in military textiles and personal equipment**
Edited by E. Sparks

123 **Specialist yarn and fabric structures**
Edited by R. H. Gong

124 **Handbook of sustainable textile production**
M. I. Tobler-Rohr

125 **Woven textiles: Principles, developments and applications**
Edited by K. Gandhi

126 **Textiles and fashion: Materials design and technology**
Edited by R. Sinclair

127 **Industrial cutting of textile materials**
I. Viļumsone-Nemes

128 **Colour design: Theories and applications**
Edited by J. Best

129 **False twist textured yarns**
C. Atkinson

130 **Modelling, simulation and control of the dyeing process**
R. Shamey and X. Zhao

131 **Advanced process control in textile manufacturing**
Edited by A. Majumdar, A. Das, R. Alagirusamy and V. K. Kothari

132 **Understanding and improving the durability of textiles**
Edited by P. A. Annis

133 **Smart textiles for protection**
Edited by R. Chaptman

134 **Functional nanofibers and applications**
Edited by Q. Wei

135 **The global textile and clothing industry**
Edited by R. Shishoo

136 **Simulation in textile technology: Theory and applications**
Edited by D. Veit

137 **Pattern cutting for clothing using CAD: How to use Lectra Modaris pattern cutting software**
M. Stott

Acknowledgements

I would like to express my gratitude towards several of my colleagues at the Institut für Textiltechnik of RWTH Aachen University who supported me while writing and editing this book. Our librarian Christiane Cremer helped me in finding a large number of scientific papers on many subjects covered in this book. Roswitha Jacobs did a great job in creating most of the drawings that illustrate this book without which it could not have been published. Sabine Weber also contributed a range of drawings that appear in several chapters. Inke Eley created most of the diagrams with much more skill than I could. I am also indebted to the other authors of this book, Professor Yordan Kyosev from Hochschule Niederrhein, Moenchengladbach, and my colleague Philip Jungbecker, who co-authored the chapter on computational fluid dynamics at very short notice. I especially acknowledge the contributions by Professor Bohuslav Neckář, Ing. Gabriela Krupincová and Dr Monika Vyšanská of TU Liberec for Chapter 7.

Dieter Viet

Preface

This book is intended, on the one hand, to give an overview of common techniques to simulate textile processes and machines and to explain the theory behind them to the extent which is needed for engineers to understand the respective models and computer programs. On the other hand, the many examples from published papers and from the authors' own experience illustrate the wide range of possibilities that these methods offer. This book is based on the authors' research and their reading of the available literature. The selection of the methods described therefore does not claim to be complete.

Each chapter begins with an introduction to theoretical aspects of that particular method. Selected examples of applications are then explained in varying levels of detail before the chapter concludes with some practical advice on how to apply the method. The references at the end of each chapter cover significant papers on the subject, with an emphasis on recent research. In addition, where available, the references mention books that provide a deeper insight into a particular simulation method for those readers wishing to understand the theory in greater detail. Where appropriate, the references also list websites offering computer programs applying the principles covered in each chapter.

The book is intended for graduate students of textile engineering as well as professionals wishing to get a deeper insight into the basics of simulation in textile technology without having to know too much about the mathematics behind it. A general knowledge of textile processes and machines is thus sufficient to understand the contents of this book. We do hope that this book will be interesting reading. Comments on the contents are welcome in order to further improve the book in its next edition.

Dieter Veit

1
Introduction to modelling and simulation in textile technology

D. VEIT, RWTH Aachen University, Germany

Abstract: This chapter introduces the concept of simulation both with and without computers. It then discusses the role of models, including white, black and grey box models. It concludes by considering expert and other knowledge-based systems and their applications.

Key words: simulation, models, expert systems, knowledge-based systems, textile technology.

1.1 Introduction to simulation

Simulation is normally used to analyse systems that are hard or impossible to describe using systems of explicit equations. In particular, this applies to highly complex dynamic systems such as most textile machines and processes. By carrying out a simulation experiment, knowledge about the real system can be gathered comparatively easily in contrast to trials on the machine or during the actual process. This includes practical experiments as well as a simulation using a computer program.

There are several reasons for the application of simulations. Investigating the real system can be:

- too time-consuming (e.g. changing the setting of a polymer production plant);
- too costly (e.g. crash trials);
- ethically questionable (e.g. medical trials of a new textile implant);
- too dangerous (e.g. reaction of new textile-based building materials on fire).

Systems that do not yet exist provide another field of application of simulations. This comprises the testing of new plant design concepts or, for example, the development of new aerospace or automotive vehicles using carbon textile reinforced composites. In some cases it is not possible to investigate the actual system directly. A simulation can then be used to gain valuable information which is otherwise unavailable. Typical examples are the simulation of the molecular movement in a fluid and the movement of continental plates which takes place over a long time period and thus cannot be observed directly.

1.2 Simulation with and without computers

In general, simulation methods can be divided into those that use a computer and those that do not, as shown in Fig. 1.1. Those simulations without computer can be separated into destructive and nondestructive methods. Simulations that use a computer can be based on technical models (e.g. finite element method (FEM), computational fluid dynamics (CFD)), on examples taken from nature (e.g. artificial neural networks, evolutionary algorithms), or on other methods, such as sociological models.

For each simulation, independent of whether it is carried out using a computer or not, it is absolutely crucial that the results can be reproduced every single time that the simulation is run. The use of a simulation must also make sense economically. Thus, the model on which the simulation is based is normally simplified as much as possible, but still takes into account the relevant factors that have an influence on the results. This often results in a mathematical model which is valid only for a certain range of the respective values. Hence, it is very important to verify the results of a simulation by running actual trials, e.g. on a machine, in order to ensure that the simulation accurately represents the problem at hand. It must be noted, though, that measurements are never truly accurate, hence the measured results can deviate from reality to a certain extent.

1.2.1 Simulation without a computer

A typical example for a simulation without a computer is a car crash test. In this case the circumstances of an accident are simplified using crash test dummies. These are equipped with sensors that register the mechanical impact the crash would inflict on real persons. This can be achieved by making further simplifications. Experimental investigations of fluid flows also fall into this category when they are carried out, for example, in wind tunnels. Making flows in air visible by using another medium, e.g. water,

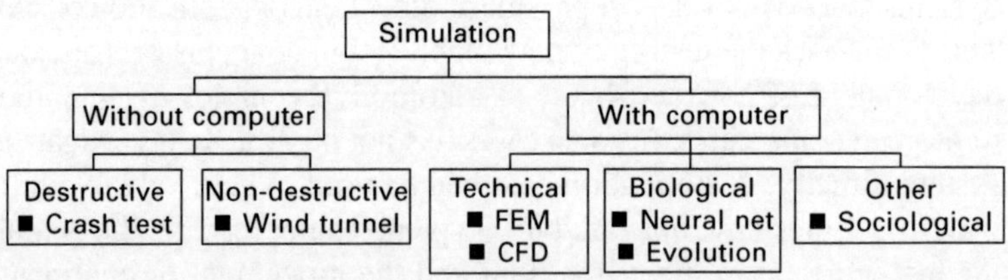

1.1 Simulation methods.

by applying the theory of similarity, is another typical example for this kind of simulation.

1.2.2 Simulation with a computer

Most technical simulations belong to this kind. Typical examples are calculations of the stresses, e.g. in machine components or textile structures, applying FEM, fluid flow simulations using the CFD method and the simulation of machines and whole production lines, e.g. in a textile mill. Biological simulations such as evolutionary algorithms and neural networks also fall into this category.

1.3 Modelling: white, black and grey box models

Prior to carrying out the actual simulation, a model, in most cases based on equations, must be devised which describes the system with all relevant parameters. In order to find these equations, experiments, often applying experimental design, are normally carried out. Alternatively, the equations are derived from theoretical assumptions. This leads to an image of reality which is either representative of the whole range of parameters or a certain number of parameters. A model is hence an abstract image of the system which is representative of the real system. It is not important that the model exactly mirrors reality in all aspects, but is much more important that the model produces sufficiently accurate results to explain the real case. In general, models are divided into white, black and grey box models as shown in Fig. 1.2.

In order to save time and hence money, most simulation models use a simplified description of the real process or machine. The following methods are commonly used:

- Components or parameters that are not of crucial importance are neglected. In order to determine these components or parameters, experiments applying factorial design can be helpful as the results normally clearly show the size of the effects when changing the settings of these components or parameters.

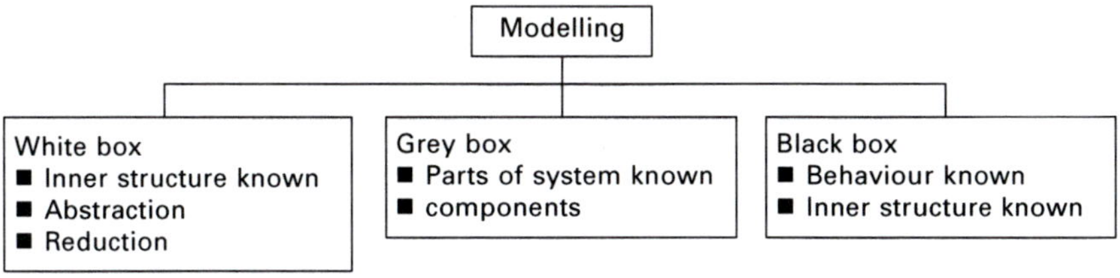

1.2 Different kinds of computer models.

- Unimportant details are not considered.
- The system is divided into its components which are then investigated separately. However, possible interactions are then hard to determine.
- Combining several attributes into classes can reduce the complexity of a simulation considerably. A major drawback of this approach is that valuable information can also be lost. This method should therefore only be applied carefully and if other methods fail.

1.3.1 White box model

This kind of model is suitable if the inner structure of the system is known. This structure is then deliberately abstracted, modified and reduced to the most important influencing parameters. A typical example is the modelling of a weaving machine (see Chapter 9).

1.3.2 Black box model

If the inner structure of a system is unknown but its behaviour or its interaction with other systems can be observed and modelled, this is called a black box model. A typical example is the use of neural networks to simulate textile processes.

1.3.3 Grey box model

In many cases, only parts of a system are fully known and only a few, but not all interactions between its components are established. The respective model is then called a grey box model. In order to save costs, this approach is widely used. A fuzzy model (see Chapter 4) can be regarded as a grey box model in some cases.

1.4 Expert systems and other knowledge-based models

Many simulation models are based on knowledge of experts of knowledge acquired by running practical trials. In order to design such a model, several steps are required.

- Determination and structuring of explicit and implicit knowledge.
- Build-up of a knowledge database and abstracting of relations between parameters.
- Processing of knowledge, e.g. by inference (see Chapter 4).
- Visualization of knowledge.

1.4.1 Expert systems

This kind of system is based on knowledge of experts which hence allows instructions to be deduced that can be used to solve or to assess certain problems. Expert systems are a branch of artificial intelligence dating back to the 1960s. In many cases, they cannot only be used to make decisions based on existing knowledge but also to give recommendations for new, previously unknown situations. When neural networks are applied, this is called 'learning by discovery'. Figure 1.3 gives an overview of currently used methods.

1.4.2 Case-based systems

These systems use a database in which actual problems and suitable solutions are stored. On a case-by-case basis, the system then determines which stored case is closest to the problem in question. This approach often leads to a viable solution but this method also has some drawbacks: in practice, it is often difficult to implement as similar problems can require totally different solutions so that this approach fails if the database is too small. In order to overcome that difficulty, it is therefore advisable to first weight the parameters describing the problem. Neural networks that use the counterpropagation algorithm are a typical example for the latter approach (see Chapter 2). This can greatly improve the accuracy of the results.

1.4.3 Rule-based systems

These systems comprise sets of rules, e.g. 'if X then Y'. Based on these rules, conclusions can be drawn to solve the problem at hand. In order to design the rules, normally experts in the field are asked and their 'knowledge' is then fed into the system by formulating rules. A typical example is fuzzy logic (see Chapter 4).

1.4.4 Decision trees

This kind of model is based on a large database which describes the parameter range. It is especially suitable for classification problems. In a

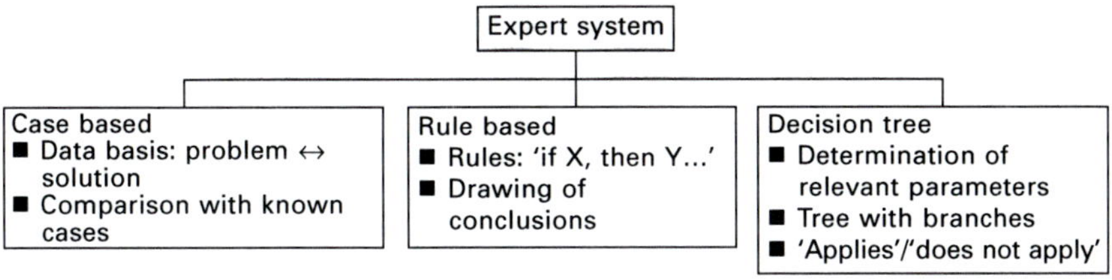

1.3 Expert systems.

first step, those parameters that describe the problem are identified. They are then used to describe the elements to be classified. In a second step, a tree is designed where each branch represents a question that can be answered with 'applies' or 'does not apply'. Following the branches eventually leads to the classification of the unknown object. A typical example for a decision tree is shown in Fig. 1.4. The numbers represent 'applies' (1) and 'does not apply' (0) which leads to a string of 0 and 1 as the classification result. A typical practical example is the classification of trash particles in cotton. The trash particles can be described according to size, shape, contrast to background, etc. These criteria can then be used to determine the kind of trash particle when an image is taken of the cotton material that contains the trash particles to be analysed.

1.5 Applications of expert systems to textile technology

Expert systems are normally used for highly complex problems when human experts are not readily available or the problem itself is too complex to be solved by them. In these cases, model-based expert systems can represent a successful approach to support the investigator. Typical applications of expert systems are:

- interpretation of data sets (e.g. in digital image processing);
- supervision and control of machines and mills;

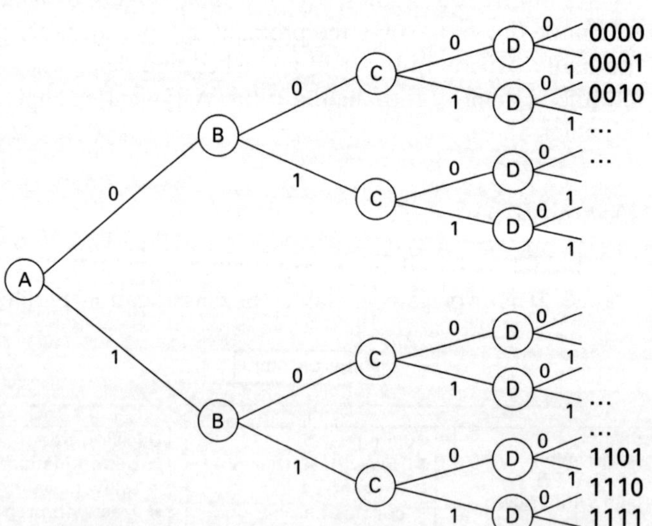

1.4 Decision tree.

- diagnosis of disruptions;
- elimination of disruptions.

The main drawback of expert systems is their limited knowledge due to their restrictions with regard to the number of available data sets. These in turn can contain errors which does affect the accuracy of the prediction. Besides, the range of values is normally confined to an area in the vicinity of the current problem. This can result in severe problems as only solutions that are close to the known range of parameters may be calculated or considered. This normally does not lead to truly innovative solutions, hence drawing to the conclusion that these systems can never replace the human mind and they are never to be trusted blindly (see example at the end of Chapter 2). A typical example for the disastrous consequences of blindly trusting expert systems is the Black Monday in 1987 when unsupervised computer programs caused havoc in the stock market when cascades of wrong decisions that were programmed into the systems led to a complete loss of control by traders.

1.6 References and further reading

There is a wide range of books available that cover different aspects of simulation. Some deal with general topics, others concentrate on certain methods. This subsection briefly describes some recently published books that are recommended reading. This is not intended to represent a complete list but only a selection of books.

Sokolowski and Banks (2009) explains the principles of modelling and simulation for both discrete and continuous systems. The authors of the book cover a wide range of applications not only in engineering but also in computer science and related fields. Practical examples from transportation, medicine and business decision illustrate the concepts of simulation and modelling presented in this book. Hence, it gives a good introduction into the subject although it does not go into too many details regarding the algorithms behind the examples.

Tettamanzi and Tomassini (2010) was written as a workbook for students and gives a comprehensive description of neural networks, fuzzy systems and evolutionary algorithms. It also details the combination of these systems, e.g. evolutionary design of neural networks, neuro-fuzzy systems. It is therefore especially suited for those interested in understanding the basics of soft computing. A concise description of computational fluid dynamics is given in Wendt (2010). The book covers the basics as well as a wide range of applications and is written by experts in the field and hence a good introduction to the subject.

Kleijnen (2010) is an excellent introduction into many methods that can be used to verify simulation results. The focus is laid on design and analysis of simulation experiments. Polynomial regression models as well as simulation optimization and screening design methods are covered and explained with many examples. An excellent book on the design of experiments (DoE) is Antony (2003). Its examples cover a wide range of manufacturing problems and the book is hence especially suitable for engineers. Another book which is designed to serve as a concise workbook with examples covering 2, 3 and up to 31 factor experiments is Barrentine (1999).

A very useful book on the simulation of textiles is Majumdar (2011a). It describes various soft computing techniques from yarn manufacturing through fabric production to textile properties, applications and quality evaluation. Another recent publication by the same author cites more than 300 useful references relating to textile applications of soft computing (Majumdar, 2011b). Chen (2010) presents a wide range of mathematical models, e.g. for fibres, yarns, fabrics and composite structures. Many examples are explained in great detail which gives an excellent overview of the field. Both are therefore perfect complements to this book which covers many of the respective machines and processes. Another book covering different aspects of textile production and the possibilities to simulate the respective processes is Sztandera and Pastore (2010). It outlines the basics of soft computing and details a large number of applications in the textile industry.

Antony, J. (2003), *Design of experiments for engineers and scientists*, Butterworth-Heinemann.

Barentine, L.B. (1999), *An introduction to design of experiments: A simplified approach*, American Society for Quality.

Chen, X. (Ed., 2010), *Modelling and predicting textile behaviour*, Woodhead Publishing.

Kleijnen, J.P.C. (2010), *Design and analysis of simulation experiments*, Springer.

Majumdar, A. (Ed., 2011a), *Soft computing in textile engineering*, Woodhead Publishing.

Majumdar, A. (2011b), *Soft computing in fibrous materials engineering*, CRC Press Inc.

Sokolowski, J.A., Banks, C.M. (Eds) (2009), *Principles of modeling and simulation: A multidisciplinary approach*, Wiley.

Sztandera, L.M., Pastore, C. (Eds), 2010), *Soft computing in textile sciences (studies in fuzzyness and soft computing)*, Physica-Verlag.

Tettamanzi, M.T., Tomassini, M. (2010), *Soft computing: Integrating evolutionary, neural and fuzzy systems*, Springer.

Wendt, J.F. (Ed., 2010), *Computational fluid dynamics: An introduction*, Springer.

2
Neural networks and their application to textile technology

D. VEIT, RWTH Aachen University, Germany

Abstract: Neural networks are adaptive computer programs. The algorithms behind these programs resemble the inner structure of the human brain but they are much less complex. Since the 1990s components of machines and also complete production systems have been optimized and controlled by applying neural nets. The biological background of neural nets is briefly explained, followed by a description of simple computer algorithms applying these principles. Training methods for neural networks, i.e. the backpropagation algorithm, are outlined. The chapter concludes with selected examples of successful applications of neural networks to solve textile-related problems.

Key words: neural network, backpropagation, prediction, textile applications.

2.1 Introduction: biological background

The nervous system enables living beings to react quickly and specifically to stimuli from the environment. It is also used to control and adjust the inner functions of the body. The nervous system can be divided into peripheral and central parts. The main task of the peripheral nervous system is the transmission of signals from the receptor cells (e.g. temperature perception, sense of hearing) to the central nervous system and from there to the effector cells (e.g. muscles). In the central nervous system, the signals from the receptor cells are processed and signals are generated for the effector cells to cause them to react adequately to the respective stimuli. It is estimated that the human brain consists of between 10^{10} and 10^{12} nerve cells. This results in between 10^{14} and 10^{18} possible connections between neurons.

Since the early days of brain research, the location of certain tasks within the brain structure has been one of the main areas of interest. This led to several important findings: by defunctionalizing certain parts of the brain, some tasks can no longer be fulfilled (e.g. speaking, calculating). After a certain period of time though, other parts of the brain compensate by taking over new tasks previously not assigned to them. Apparently, information is not only stored in one particular brain area but in several independent regions that are not necessarily directly connected. The failure of one region does therefore not always lead to a loss of the information which was stored there. These important aspects need to be taken into account when designing a mathematical model to mimic the functionality of the human brain.

2.1.1 Build-up of a neuron

In order to design such a mathematical model, it is essential to first understand the way neurons work and also the interactions between them. Figure 2.1 depicts a nerve cell which consists of a cell body (soma) with the cell nucleus, the axon (nerve filaments) and the dendrites. At the end of a tube-like axon, which can measure up to one metre in length, synapses with dendrites are located that connect to other neurons. The axon is often surrounded by the myelin sheath which acts as an insulating layer to the environment. Thereby, the transmission speed is more than a hundred times higher than that of axons without a myelin sheath.

The dendrites are connected via synapses with the dendrites of other neurons. Neurons can have up to 200 000 dendrites; commonly in the range of 10 000. If the number of dendrites gathering information from other neurons exceeds the number of dendrites passing on information to subsequent neurons, the signal processing is called convergent. If it is the other way around, it is called divergent.

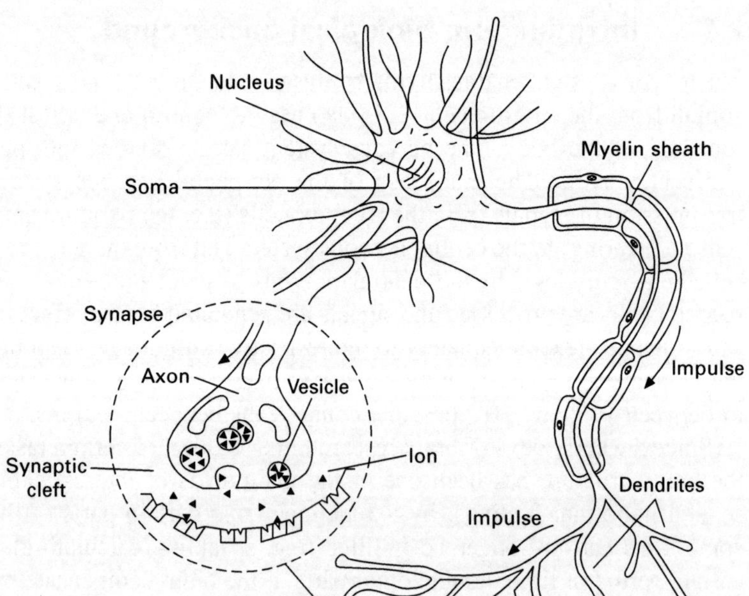

2.1 Schematic representation of a nerve cell (adapted from Schöneburg *et al.*, 1990).

2.1.2 Information transmission between neurons

Figure 2.2 depicts the principal design of a synapse at the end of an axon consisting of the synaptic vesicle containing certain transmitter substances, the synaptic cleft between neurons and the receptors of the subsequent neuron. Neurons transmit information by a combination of electric impulses and by releasing a neurotransmitter that binds to chemical receptors of the connected neurons. In the passive state, each neuron has a so-called equilibrium resting potential of approximately –70 mV. This potential is the result of different ion concentrations inside and outside the cell. Positively charged calcium ions leave the cell due to diffusion forces, whereas the respective anions cannot penetrate the cell wall because of their size. This leads to a negative charge inside the cell. This difference in the electrical potential generates a force counteracting the diffusion force. This in turn leads to an equilibrium where ions neither leave nor enter the cell. The cell has then reached a passive state.

When an electrical impulse reaches a synapsis, the vesicle releases transmitter substances that bind to chemical receptors of the targeted cell. This opens the ion channels of this cell, allowing ions to enter it. That leads to either an increase or a decrease of its electrical potential. If a certain threshold is exceeded, an electrical impulse is generated, the cell 'fires'. If the potential of the receptor cell is increased by the opening of its ion channels, the probability that the threshold is exceeded increases. The preceding neuron is therefore excitatory. In the reverse case, if the potential is decreased owing to an inflow of negatively charged ions, the preceding neuron and therefore this synapsis of the actual neuron are inhibitory, making it more difficult for the neuron to become 'active'.

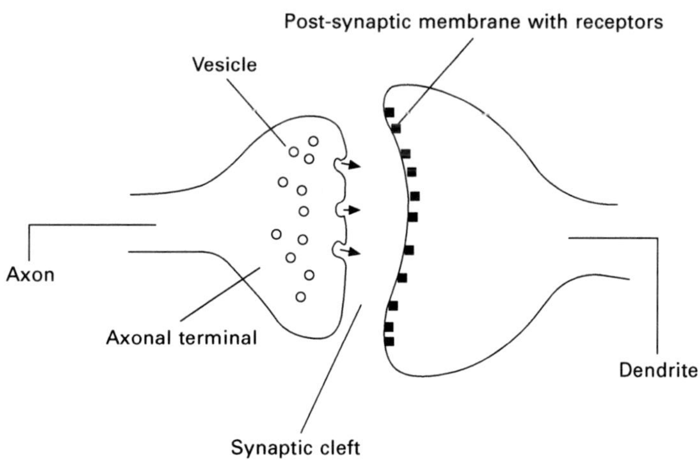

2.2 Design of a synapse (adapted from Schöneburg *et al.*, 1990).

If either several synapses send neurotransmitters simultaneously to the same neuron or if the sum of signals reaching this neuron over a certain period of time exceeds its resting potential, the neuron becomes active and transmits neurotransmitters to all neurons with which it is connected (feed-forward). Each signal is therefore an 'all or nothing' signal, which means it is either logical nil (0) or logical one (1).

Various analyses of these processes have shown that the number of ion channels of the target cell that open up when neurotransmitters reach them differs between the synapses of the target cell. It is therefore possible that the activity of a certain single synapsis causes the cell to 'fire' because the resting potential of the cell is exceeded whereas other synapses have to be active as a group in order to cause the same effect. This can be taken into account mathematically by assigning a weight factor w_i to each synapsis. If the sum of all weight factors of simultaneously active synapses exceeds the threshold value of the target neuron to which they are connected, this neuron becomes active and information is passed on to the subsequent neurons to which it is connected.

The threshold of a neuron can be lowered if this neuron is often active. The neuron is then 'trained' and this causes it to become active more easily. Neurons that are not activated regularly raise their threshold accordingly. This means that the information to be learned should be processed often in order to achieve a memory effect of the neurons. This realization led to a variety of different neural networks and the respective learning rules (e.g. Hebb rule, see Section 2.3.2).

Table 2.1 shows typical values for the neurons of the human brain and for computer-based neural networks. It is apparent that the processing speed of natural neurons is very low, owing to the slow velocity of the chemical transmission between neurons. The processing speed of modern computers is more than 100 000 times higher. On the other hand, the number of connections between neurons is extremely high in the human brain and has not been reached even remotely by a computer. The enormous performance and reaction speed of the human brain are therefore apparently caused by processing information in parallel. The best result when training such a neural network can be achieved if the number of subsequently connected

Table 2.1 Comparison of human brain and computer power

Parameter	Brain	Computer
Circuit time	10^{-3} s	10^{-9} s
Switching operations per neuron/transistor	10^{3} s^{-1}	10^{9} s^{-1}
Overall switching operations per second (theoretical)	10^{14} s^{-1}	10^{18} s^{-1}
Overall switching operations per second (actual)	10^{12} s^{-1}	10^{10} s^{-1}

Source: Zell, 1997.

neurons is small compared with the number of neurons working in parallel. The main reason for this is that subsequent neurons exchange information via slow chemical processes whereas neurons working in parallel process information without the need to exchange ions. It has to be noted, though, that all information that a neural networks is supposed to learn is stored in the synapses (weight factors, see earlier). This means that the number of connections between the neurons determines how much information can be learned by such a system. A multi-layer neural network can hence lead to better results than a single-layer network for an equal number of neurons.

2.2 Models of artificial neural networks

The origins of artificial neural networks (ANN) date back to the 1940s when Warren McCulloch and Walter Pitts presented the first simple systems (McCulloch and Pitts, 1945). They proved that an ANN can learn any arithmetic or logical function. In 1949, Donald Hebb described a learning rule for single neurons which is still being used today and is named after him, Hebb's rule (Hebb, 2002). Frank Rosenblatt and others introduced the first working neuro-computer in 1958 (Rosenblatt, 1958). It could be used to recognize patterns and consisted of neurons of the perceptron type. The basic principle of this system is still applied in many of today's ANNs. In 1969, though, Marvin Minsky and Seymour Papert showed that a large number of important problems could not be represented and thus not be learned by perceptrons (Minsky and Papert, 1969). They assumed that ANNs could not learn certain mathematical problems, which led to an almost total cut-back in public funding of research projects into the development of ANNs and hence to an almost complete standstill in this area for about 20 years. In the mid-1980s, new types of ANNs, e.g. Hopfield nets, backpropagation and Kohonen nets, led to a renaissance of ANNs. Ironically, it was, among others, Minsky, one of the hardest critics of ANNs, who successfully developed new and working neural nets.

The basic idea of ANNs is depicted in Fig. 2.3. Data sets are fed into the input layer, passed on to one or more hidden layers where the information is processed, thus generating 'knowledge' before the output is calculated. In order to understand the functional principle of ANNs that are used today, in the following chapters the basic algorithms of perceptrons and later developed neural networks are explained.

2.2.1 Perceptron

The perceptron was developed by F. Rosenblatt and others in 1958 (Rosenblatt, 1958). It can be used to recognize and thus to classify patterns. A perceptron is a feed-forward neuron, which means that the data flow is

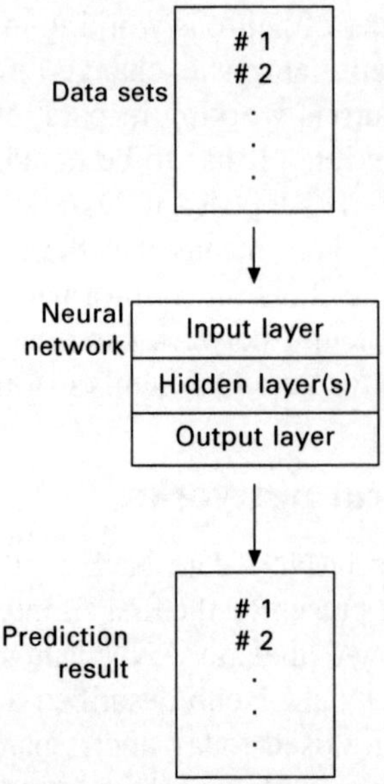

2.3 Principle of artificial neural networks.

unidirectional from input to output. The main configuration of perceptron networks is shown in Fig. 2.4. In the elements of the input layer, each input data item is multiplied with a constant weight factor w_{ij}, and passed on to the neurons of the subsequent intermediate layer. Every neuron of the input layer can propagate its output to either only one or to several neurons of the intermediate layer. Each element of the intermediate layer sums its inputs to the so-called net_j value which is then multiplied with a variable weight and propagated to the neurons of the output layer. Each element of the output layer is connected to all elements of the intermediate layer. Depending on their weighted input net_i which may or may not exceed their respective threshold value, these elements deliver either the value 0 or the value 1. Figure 2.5 shows the design of such an element by the example of output neuron no. 1 in Fig. 2.4.

Learning rule of a perceptron

During the learning process of perceptrons (Fig. 2.6), the desired output z_i for each neuron in the output layer is known. It can thus be constantly compared with the calculated output o_i. The variable weight factors of the neurons in the intermediate layer are then changed according to the following rules:

- If $o_i = z_i$, then do not change the weights as the patterns is already recognized correctly.

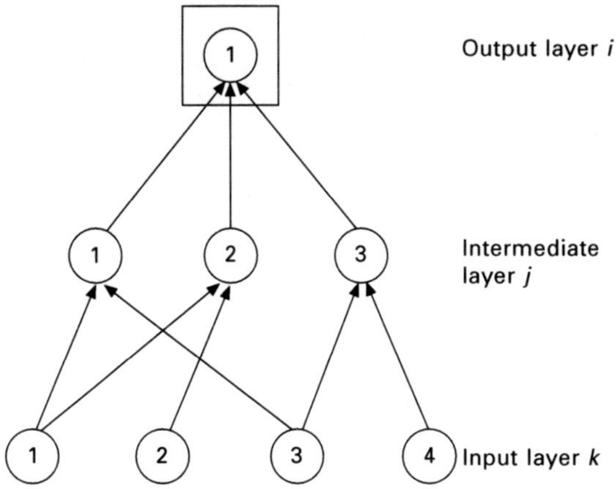

2.4 Design of a perceptron network.

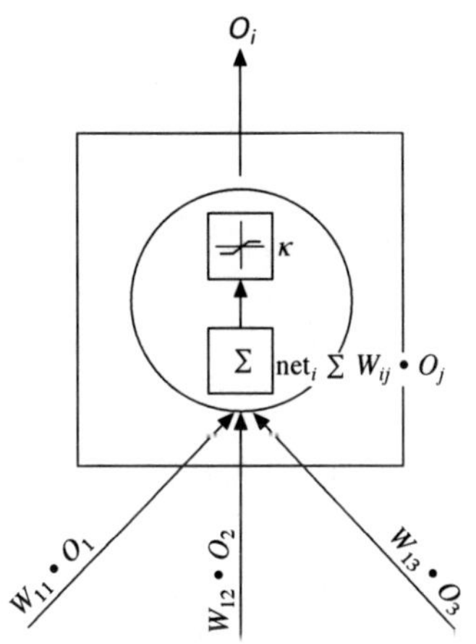

2.5 Design of a perceptron neuron.

- If $z_i = 1$ and $o_i = 0$, then increase the weights of those neurons in the output layer which are connected to active elements in the intermediate layer j, hence $w_{ij}(t + 1) = w_{ij}(t) + \sigma \cdot o_j$. In the next learning step, this increases the chance that $z_i = 1$, so that the pattern is recognized correctly.
- If $z_i = 0$ and $o_i = 1$, then decrease the weights of those neurons in the output layer which are connected to active elements in the intermediate layer j, hence $w_{ij}(t + 1) = w_{ij}(t) - \sigma \cdot o_j$. In the next learning step, this increases the chance that $z_i = 0$, so that the pattern is recognized.

The value σ stands for the learning rate and determines the size of the weight change in each learning step. It normally assumes values in the range of [0; 1]. The value o_j is the output of neuron j in the intermediate layer. The

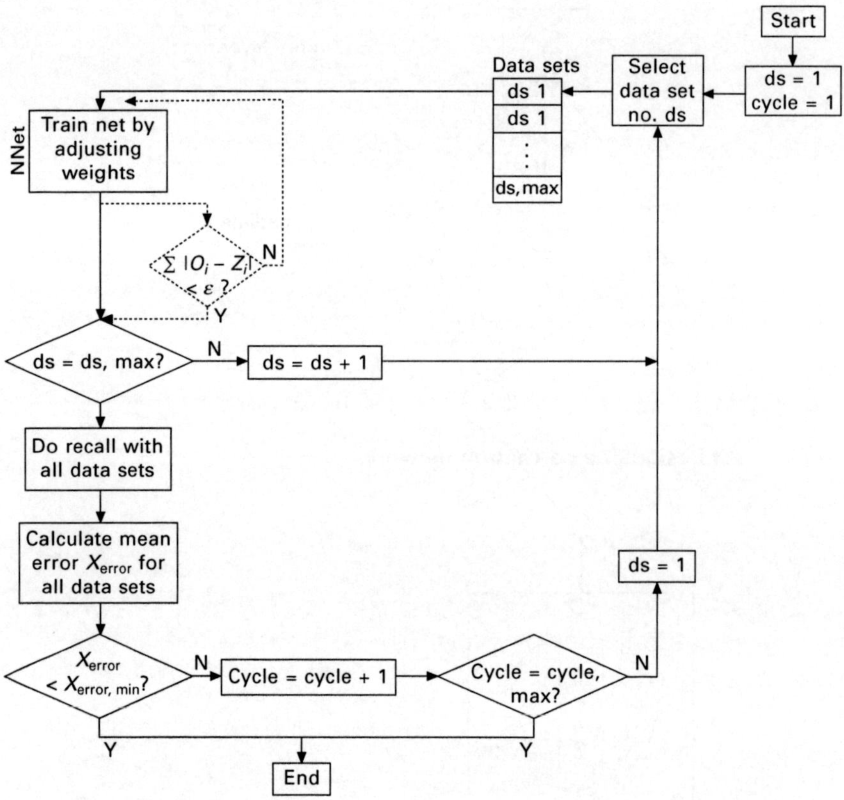

2.6 Training process of a neural network consisting of perceptrons.

factor w_{ij} is the weight value with which o_j is multiplied before the new value is fed into neuron i of the output layer. This type of learning process is called supervised learning.

Perceptron convergence theorem

Rosenblatt postulated that the learning algorithm as described above will converge within a limited period of time to the desired result provided that the neural network can represent the patterns in general. This gives no information on how long it will take for the ANN to learn to recognize all patterns. This depends on the structure of the network and how and in which order the patterns are fed into it. Besides, the initial values of the variable weight factors and also the values of the constant weight factors play an important role. The crucial aspect when applying perceptrons for pattern recognition is the question whether it can actually represent the problem at hand in general.

Problem of linear divisibility

In 1969, Minsky and Pappert showed that perceptrons can learn not to recognize certain patterns. One of their examples was the parity determination and its special case, the XOR problem. This showed directly the weakness of perceptrons. The XOR function is one of 16 boolean operations with two variables. Its truth table is shown in Table 2.2. If the sum net_i of the weighted inputs of the intermediate layer in the output element exceeds a threshold κ, the output element gives either a 0 or a 1. This is shown in Fig. 2.7 by the straight line, the so-called threshold line. It separates the area with $net_i > \kappa$ from the area with $net_i < \kappa$. In order to represent the XOR problem with a single perceptron, the weights w_1 and w_2 must be determined in such a way that the points (0/0) and (1/1) are on one side of the threshold line and the points (0/1) and (1/0) on the other. It is apparent that this is not possible hence the XOR problem cannot be solved with a single perceptron. This type of function is called 'non-linearly separable' for which the perceptron does not converge.

This problem can be overcome by introducing a second perceptron element which is connected with the first one with logical AND, as shown in

Table 2.2 Truth table of the XOR function

Input			Target
x	*y*	→	*z*
0	0	→	0
1	0	→	1
0	1	→	1
1	1	→	0

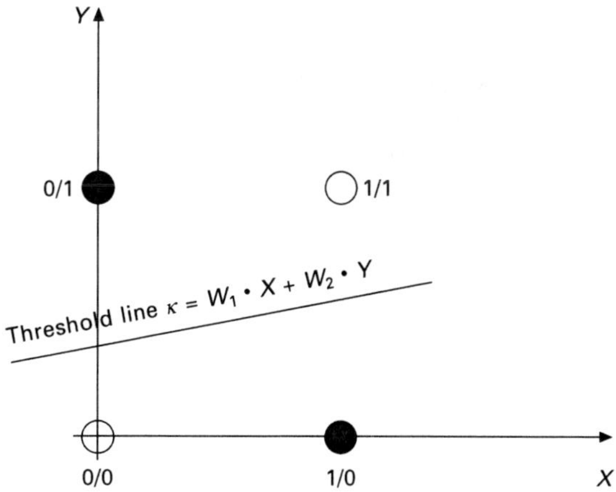

2.7 XOR problem (adapted from Schöneburg *et al.*, 1990).

Fig. 2.8. This leads to two threshold lines. The shaded area there is located in between the two threshold lines of the two perceptrons. It can therefore be separated from the surrounding area and the XOR problem can be solved. In principle, any non-linearly separable problem can be solved by logically connecting a sufficiently large number of perceptron elements. The resulting neural network structure can become very complicated and the required training time can be very long.

2.2.2 Adaline and madaline networks

Adaline and madaline networks are the result of further developments of the perceptron and the predecessor of today's most common ANN, the backpropagation network (see Section 2.3).

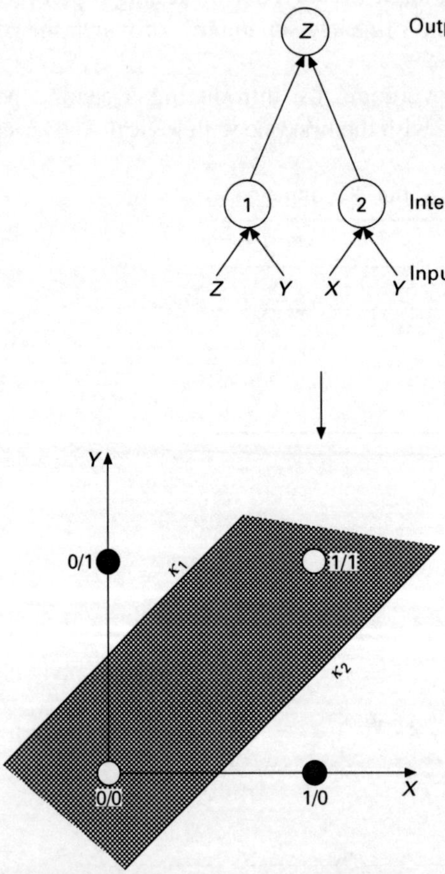

x	y	→	z
0	0	→	0
1	0	→	1
0	1	→	1
1	1	→	0

2.8 Structure of neural network with two perceptrons that are connected with logical AND (top) and solution of XOR problem (bottom).

Adaline network

The **ada**ptive **li**near **ne**uron network was developed in the 1950s and is similar to the perceptron. Figure 2.9 shows its design. It consists of an input layer which is completely connected with the output layer via variable weights. The major difference to perceptron networks is the delta learning rule which is also an integral part of the backpropagation algorithm.

Similar to the perceptron, training is supervised, which means that input and output are known to the network during the training. In contrast to the perceptron algorithm, not the difference between desired output z_i and actual output o_i is calculated. Instead, the difference between the desired output z_i and the activity a_i of the neuron is determined. The actual output o_i is a function of the activity a_i (e.g. output $o_i = 1$, if activity a_i exceeds a threshold value κ). The delta learning rule is then as follows:

$$w_{ij}(t+1) = w_{ij}(t) + \left(\frac{\sigma}{n}\right) \cdot \delta_i \cdot o_j \qquad [2.1]$$

with $\delta_i = z_i - a_i$, n = number of inputs (weights) of neuron i and σ = learning rate.

The prefix of the error and the value of the output of neuron j determine the direction and size of the weight change. A subsequent recall (another feeding into the network of the just trained pattern) leads to an error of 0 if the output function is chosen appropriately. Each adaline element learns (adjusts its weights) until its activity reaches the value of the desired output. Hence elements that already produce the correct output still adjust their weights as long as their activity is close to the threshold value. Thus, the distance between activity and threshold value is increased, which leads to a more stable state of the neural network as slight changes in the input pattern do not affect the output. In addition, once a pattern is trained, the neural network does not 'forget' the pattern as easily when other patterns are fed.

This behaviour resembles the biological basics as described in Section 2.1: a synapse which is activated often lowers its activation threshold, which leads to a quicker reaction. In the mathematical model, instead of lowering the threshold value, the weight factors are increased or decreased, which

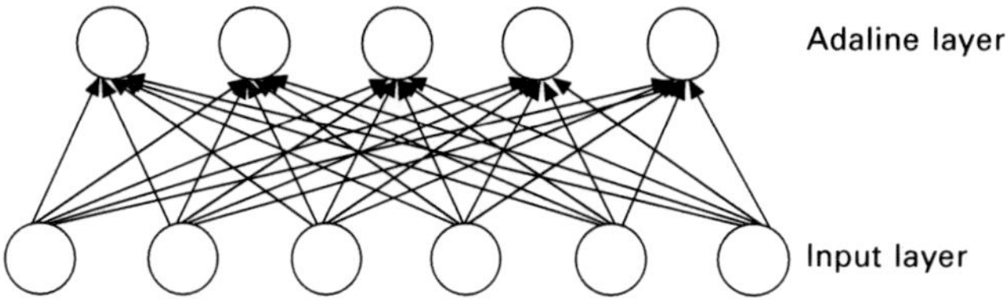

2.9 Design of an adaline network.

leads to the same result. Owing to this learning rule, the adaline network learns much quicker than a perceptron network. But it still faces the problem of non linearly-separable problems which is apparent by:

$$w_i \cdot \text{input}_1 + w_2 \cdot \text{input}_2 + \ldots + w_n \cdot \text{input}_n = output_\text{i} \quad [2.2]$$

where n = number of input parameters and i = number of output neurons.

This equation system can be solved for at best n linearly independent (separable) input vectors. If more than n patterns are trained, only an approximate solution for each pattern can be found. In general, the capacity of adaline networks is in the range of n to $2n$ patterns.

Madaline network

The **m**ultiple **ada**ptive **li**near **ne**uron (madaline) network is an adaline network with an additional layer as shown in Fig. 2.10. Each element in the madaline layer is connected to exactly one madaline element via a constant weight factor of value 1. The output of the madaline element then depends on the difference between the number of active and passive neurons it is connected to and is hence always either +1 or –1. The learning rule of the adaline layer is similar to the learning rule as described in the previous section. If the classification of a madaline element is wrong, the weights of that element which activity is closest to 0 and which output has the wrong prefix are adjusted. Hence, the network can represent the input vectors internally in the hidden adaline layer and classify through its madaline elements. This separation also allows non-linearly separable problems to be solved, which is a prerequisite to tackle more complex tasks.

2.3 The backpropagation algorithm

In 1986, David E. Rumelhart, Geoffrey E. Hinton and Ronald J. Williams rediscovered a learning algorithm developed by Paul Werbos in 1974 (Werbos,

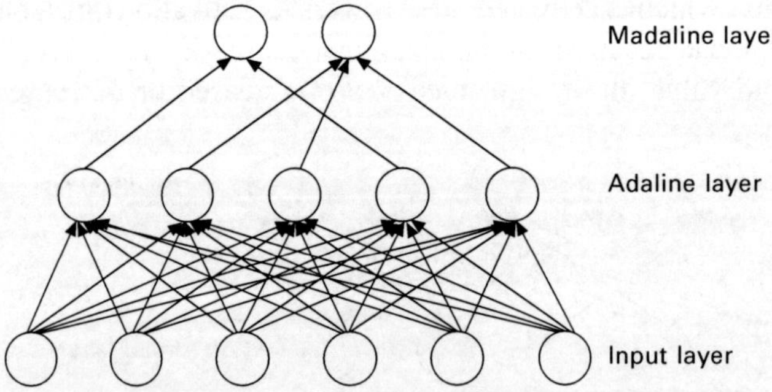

2.10 Structure of a madaline network.

1974), the so-called backpropagation algorithm (Rumelhart *et al.*, 1986). This was an improved madaline network type with a different structure and learning rule. The data are represented internally, the classification takes place in the output layer and learning is supervised. This algorithm makes it possible to solve non-linearly separable problems and, owing to the new learning rule, its learning speed is considerably higher than previous networks of, for example, the madaline type. Besides, much more complex network structures can be realized. Hence, backpropagation networks have a much higher capacity and can learn many more data sets than the nets described above. The direction of the information flow is feed-forward, the error of the output neurons is backpropagated through all layers back to the input neurons. This allows weights of neurons to be changed in the so-called hidden layers which are, in contrast to input and output layer, not visible from the outside. This is a new feature that did not previously exist.

2.3.1 Structure of the net

Figure 2.11 shows the structure of a neural net based on the backpropagation algorithm. In this typical example, each neuron is connected to all neurons of the previous and the subsequent layer. The capacity of the net (the number

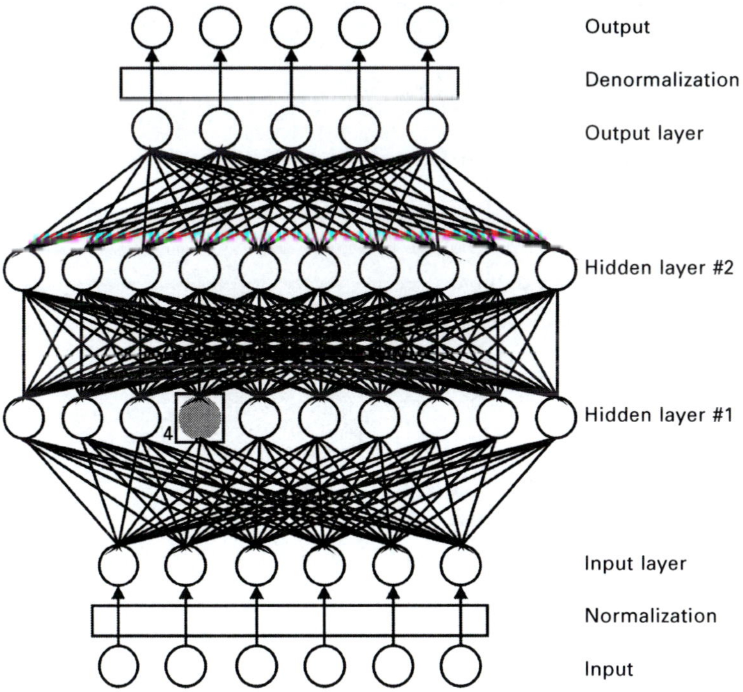

2.11 Structure of a backpropagation network.

of data sets it can learn) is thus at a maximum. In the following, each layer is denoted with a letter, e.g. h, i, j, k, where layer j precedes layer i, hence being closer to the input layer. The weight factor w_{ij} therefore denotes the weight between neuron i in the current layer and the neuron j in the preceeding layer.

The input data are fed to the input layer j and multiplied with a variable weight factor w_{ij}. The sum of all inputs to a neuron of input layer j is then fed to the neurons of the first hidden layer i. The input function net_i of these neurons converts that input into a new value which is fed to the activation function F_i and subsequently to the output function f_i of the neuron. In the following hidden layer (if existing), the outputs of the neurons of the preceding layer are multiplied with a variable weight w_{hi} and passed on to the next layer via input, activation and output functions net_h, F_h and f_h. This procedure is repeated until the output layer is reached. This information flow is shown in Fig. 2.12 for the fourth neuron of the first hidden layer in Fig. 2.11.

The neuron shown in Fig. 2.12 receives its input from all neurons of the preceding layer, multiplied with variable weight factors w_{ij}. This weighted input is then summed and passed on to the following layer via activation and output function. Most commonly used input functions are:

Sum $\quad \text{net}_i = \sum_j w_{ij} \cdot o_j$

Product $\quad \text{net}_i = \prod_j w_{ij} \cdot o_j$

Maximum $\quad \max(w_{ij} \cdot o_j)$

Minimum $\quad \min(w_{ij} \cdot o_j)$

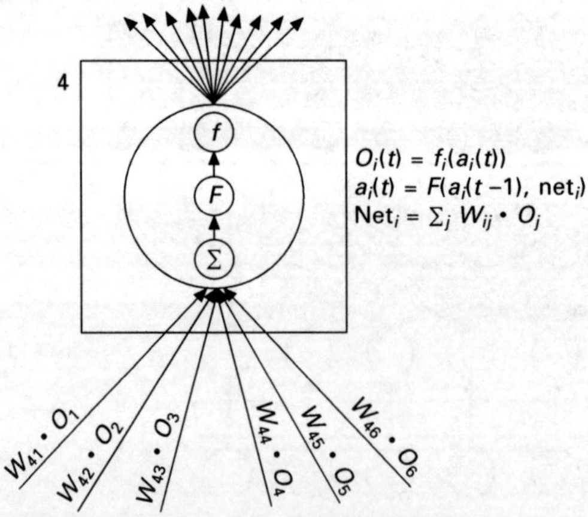

2.12 Information flow within a neuron.

Most commonly used activation functions are:

Linear $a_i = \text{net}_i$

Sigmoid $a_i = \dfrac{1}{1 + e^{-g\text{net}_i}}$

The activation function must be derivable for reasons which are explained below. Any function meeting this requirement can be used. Normally, functions are applied that spread the data sets in order to simplify their classification. Most commonly used output functions are:

Direct $o_i = a_i$

Winner takes all $\begin{cases} o_i = O_i(a_i), \text{ if } a_i = \max(a_i) \\ ok = 0 \text{ otherwise} \end{cases}$

where O_i is a user-defined function, e.g. $O_i(a_i) = 1$.

2.3.2 Learning rule

Backpropagation networks are trained with the generalized delta learning rule. The generalized form is derived here. Figure 2.13 depicts the prediction error of a simple neural network with only two weight factors depending on their respective values. It is the aim of the training to reach the minimum of the error area, in the ideal case zero. Then, the neural network would classify all fed data sets correctly. The error function is defined as

$$E(W) = E(W_1, w_2, \ldots, w_n) \qquad [2.3]$$

with W = weight vector of a layer and n = number of weights per layer.

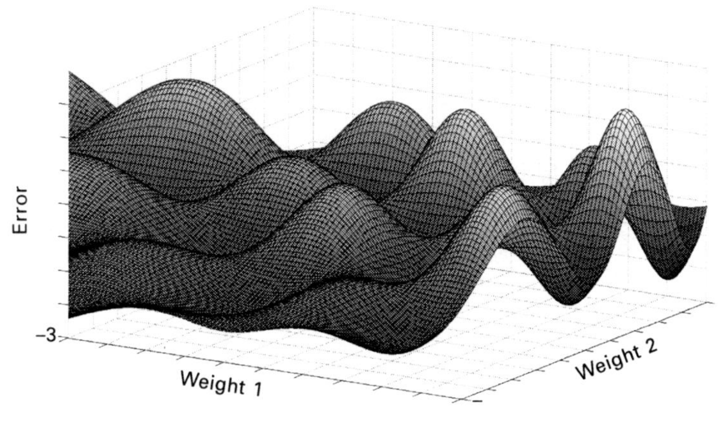

2.13 Typical error area of a simple neural network with two weight factors.

The overall error of a given pattern is normally defined as the square difference between actual and desired output of the output layer:

$$E_p = \frac{1}{2} \cdot \sum_i (z_i - o_i)^2 \qquad [2.4]$$

Here, z_i represents the desired output of neuron i and o_i stands for the actual output. The factor ½ is introduced to later simplify to derive the function. A minimization of half the error implies the minimization of the error as a whole so this does have no negative effect.

The backpropagation algorithm is a gradient descent procedure. That means that the aim is minimizing the error by systematically changing the weight factors. The weight change in each training step by a fraction of the negative gradient of the error function E is given by:

$$\Delta W = -\sigma \nabla E(W) \qquad [2.5]$$

From

$$\Delta w_{ij} = -\sigma \cdot \frac{\partial}{\partial w_{ij}} E(W) \qquad [2.6]$$

and

$$\text{net}_i = \sum_j w_{ij} \cdot o_j \qquad [2.7]$$

with

$$o_j = F(\text{net}_i) \qquad [2.8]$$

we get the second factor in equation [2.9]

$$\frac{\partial E}{\partial w_{ij}} = \frac{\partial E}{\partial \text{net}_i} \cdot \frac{\partial \text{net}_i}{\partial w_{ij}} \qquad [2.9]$$

to

$$\frac{\partial \text{net}_i}{\partial w_{ij}} = \frac{\partial}{\partial w_{ij}} \sum_j o_j \cdot w_{ij} = o_j \qquad [2.10]$$

From equation [2.6] it follows, using the chain rule:

$$\frac{\partial E}{\partial w_{ij}} = \frac{\partial E}{\partial o_i} \cdot \frac{\partial o_i}{\partial w_{ij}} \qquad [2.11]$$

Per definition we have

$$\delta_i = -\frac{\delta E}{\partial \text{net}_i} \qquad [2.12]$$

Insertion of [2.10] and [2.11] into [2.9] then leads [2.6] to:

$$\Delta w_{ij} = \text{s} \cdot o_j \cdot \delta_i \qquad [2.13]$$

From equation [2.12] it follows

$$\delta_i = -\frac{\partial E}{\partial \text{net}_i} = -\frac{\partial E}{\partial o_i} \cdot \frac{\partial o_i}{\partial \text{net}_i} \qquad [2.14]$$

For the second factor in [2.14] we get from [2.8]

$$\frac{\partial o_i}{\partial \text{net}_i} = \frac{\partial}{\delta \text{net}_i} F(\text{net}_i) = F'(\text{net}_i) \qquad [2.15]$$

For the output layer we get the first factor in [2.14], then to

$$-\left(\frac{\partial E}{\partial o_i}\right) = (z_i - o_i) \qquad [2.16]$$

Hence for the hidden layers we get:

$$-\frac{\partial E}{\partial o_i} = -\sum_h \frac{\partial E}{\partial \text{net}_h} \cdot \frac{\partial \text{net}_h}{\partial o_i} = \sum_h \left(\delta_h \cdot \frac{\partial}{\partial o_i}(o_i \cdot w_{hi}\right) = \sum_h (\delta_h \cdot w_{hi})$$

[2.17]

This shows that the error of the neurons in the hidden layer i can be derived from the error of the neurons in the subsequent layer h which are connected with this neuron by weights w_{hi}.

For the neurons in the output layer we have:

$$\delta_i = F'(\text{net}_i) \cdot (z_i - o_i) \qquad [2.18]$$

For the hidden layers we get then:

$$\delta_i = F'(\text{net}_i) \cdot \sum_h (\delta_h \cdot w_{hi}) \qquad [2.19]$$

In order to solve [2.18] and [2.19], the activation function must be derivable as mentioned above.

According to [2.6], we get the final form of the learning rule to adapt the weight factors as:

$$w_{ij}(t+1) = w_{ij}(t) + \sigma \cdot \delta_i \cdot o_j \qquad [2.20]$$

with the factor δ from [2.18] and [2.19]. This rule is also called the Hebb learning rule. The error of the output layer is therefore propagated back to the first hidden layer, which gives this algorithm its name: backpropagation.

2.3.3 Structure and capacity

Figure 2.11 shows all typical features of a neural network based on the backpropagation algorithm. The input layer is completely connected with all the neurons of the subsequent hidden layer and the neurons of the hidden layers themselves are also completely connected to the neurons of the subsequent layer. This not only increases the capacity of the neural net but also the prediction accuracy. If such a neural network is implemented into hardware, the failure of single connections between neurons does not then seriously affect the overall behaviour of the network as information is stored in a large number of connections (weights) between neurons. The last hidden layer is normally also fully connected to all neurons of the output layer.

The capacity of a neural network of this type depends mainly on the number of weights between the neurons. The more weights, the more information can be stored. For a given number of neurons, the maximum number N of weights can be calculated as follows: let n_i be the number of input parameters, n_o the number of output parameters and k the overall number of neurons. For a network with two hidden layers, we get the optimum number of neurons s_1 and s_2 in each layer for N weight factors with:

$$N = n_i \cdot s_1 + s_1 \cdot s_2 + s_2 \cdot n_0 \qquad [2.21]$$

$$s_1 + s_2 = k \qquad [2.22]$$

Insertion of [2.22] into [2.21] leads to:

$$N = n_i \cdot s_1 + s_1 \cdot (k - s_1) + (k - s_1) \cdot n_o \qquad [2.23]$$

N reaches a peak value for $\frac{\partial N^!}{\partial s_1} = 0$

After deriving [2.3], we get:

$$s_1 = \frac{1}{2} \cdot (n_i - n_o + k) \text{ and } s_2 = \frac{1}{2} \cdot (k - n_i + n_o) \qquad [2.24]$$

Example: For $n_i = 5$ and $n_o = 9$ with $k = 20$, we get $s_1 = 8$ and $s_2 = 12$ with $N = 244$.

If a neural network with three hidden layers is to be optimized, two layers are considered as one and the calculation is carried out as described above. Then the neurons for the combined layers are distributed among the two layers in a similar way. In most cases, the capacity of a neural network with one or two hidden layers with 10–20 neurons each is sufficient to represent hundreds of data sets. Based on the author's experience, more than three hidden layers are never necessary. For the prediction of such a network to be accurate, a capacity which is adjusted to the complexity of the task at hand is crucial. If the capacity is too high or too low, the results of the prediction of the network for previously unknown input can be very imprecise.

2.3.4 Learning with the backpropagation algorithm

In this section, typical strategies to train a neural network using the backpropagation algorithm are explained and solutions for typical problems are discussed.

Initialization of the weight factors

At the beginning of the training stage, all weight factors w_{ij} are initialized with random values. Here, it is important to note that these values are truly random so that no symmetrical patterns occur. A symmetrical distribution of the weights would reduce the capacity of the network as several regions of the network would have identical weights. This cannot be changed during the training due to the learning rule derived above.

The delta learning rule according to Hebb is given by:

$$w_{ij}(t + 1) = w_{ij}(t) + \sigma \cdot \delta_i \cdot o_j \quad [2.25]$$

This implies that the weight change in a single learning step is increased with a greater error δ_i. This error is directly proportional to the derivative of the activation function F. The speed of the learning process can thus be increased by high values of the derivative of F. In most cases, the activation function F is the sigmoid function. Hence, all input values are transformed into the interval [0;1]. Values closely together and near zero are spread and thus separated for easier classification. The amount of spreading can be adjusted with the factor g, the so-called gain.

From the equation of the sigmoid function given by:

$$F(\text{net}_i) = \frac{1}{1 + e^{-g\text{net}^i}} \quad [2.26]$$

we get the derivative to:

$$F'(\text{net}_i) = F(\text{net}_i) \cdot [1 - F(\text{net}_i)] \quad [2.27]$$

For $g = 1$, both functions are shown in Fig. 2.14. The value of the derivative assumes especially large values in the interval [–1;1] and reaches a maximum for net = 0. Hence, the weight factors are normally initialized with values within the interval [–1;1] or the interval [0;1].

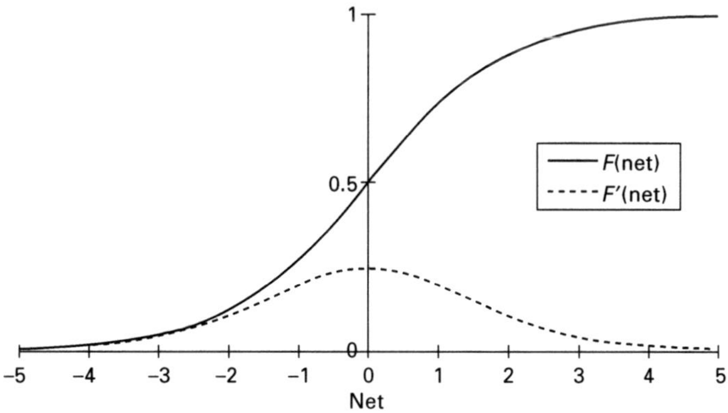

2.14 Sigmoid function F and derivative F'.

The influence of the gain g on the function is shown in Fig. 2.15. A value of $g > 1$ results in a greater spreading of the functional values in [–1;1], the absolute value of the derivative is then very large. With values of $g < 1$, the functional values are closer to each other and the absolute value of the derivative is hence smaller. The weight factors are changed during the training but will nevertheless remain close to the interval [–1;1], normally in a range of [–3;3].

Problems during training

Similar to all other gradient descent processes, one problem is that a local minimum is found which cannot be left by the algorithm. With increasing complexity of the network and hence a larger number of neurons and weight factors, the error area becomes more fissured. This in turn increases the probability of getting stuck in a local minimum. The following three cases are especially difficult to deal with.

Flat plateau

According to equation [2.18], the size of the weight change depends on the error δ_i. If this value was close to zero, the weight change becomes smaller and smaller and many learning steps are needed to leave such a minimum. In extreme cases, the local minimum cannot be left and the training process stops. Another difficulty is shown in Fig. 2.16. The gradient of the weight change for a flat plateau is close to zero as is the case for the absolute minimum. To distinguish one from the other is hence very difficult, if not impossible.

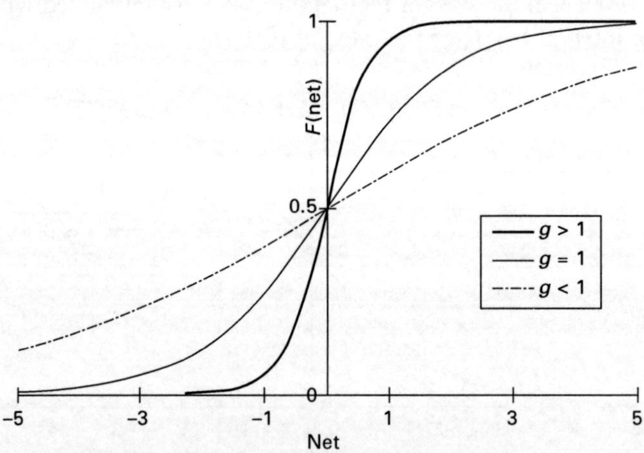

2.15 Influence of gain g on the sigmoid function.

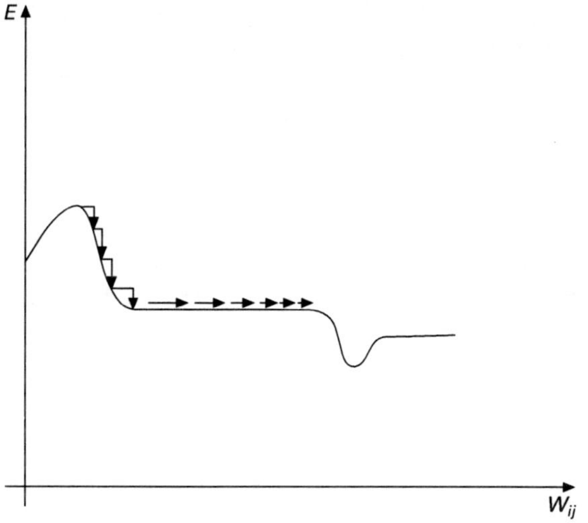

2.16 Flat plateau of the error function (adapted from Schöneburg *et al.*, 1990).

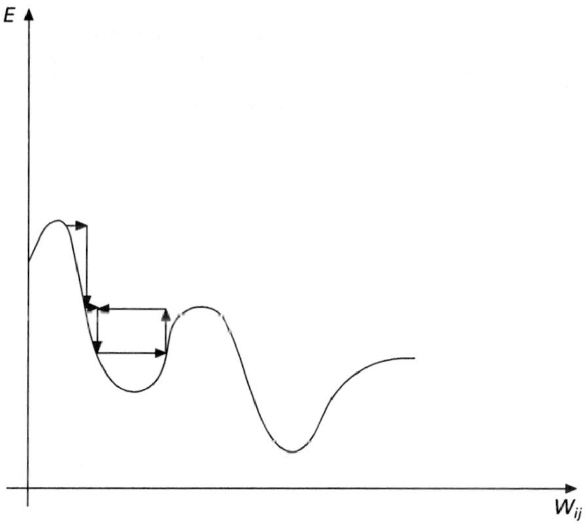

2.17 Steep crevices of the error function (adapted from Schöneburg *et al.*, 1990).

Steep crevices

As shown in Fig. 2.17, steep crevices of the error area can also be a severe problem. If the weight change is large, the algorithm results in a large step of the error function crossing the crevice. If the crevice is also steep on the other side, the error function will cause the algorithm to jump back to the side where it started from and so forth, it is very difficult to leave this vicious circle, requiring an adjustment of the learning rule during the training.

Missing a minimum

A very large weight change gradient when descending into a minimum (Fig. 2.18) may cause the algorithm to make a large step, thus crossing the sought minimum without having a chance to find it again. There are several ways to avoid or to overcome this problem. They are explained in the following section.

Optimizing the training

Learning rate

Assuming an error landscape with large plateaus, many valleys and mountains, a sensible approach is to start with a large training rate, e.g. $\sigma = 0.8$, hence crossing and investigating the landscape in large steps to quickly find the region where the optimum is located. In order not to leave that area again, in later stages of the learning process, this value is gradually reduced to smaller values, e.g. $\sigma = 0.1$. In some cases, it is better to start with smaller values for the learning rate, that is a question of trial and error.

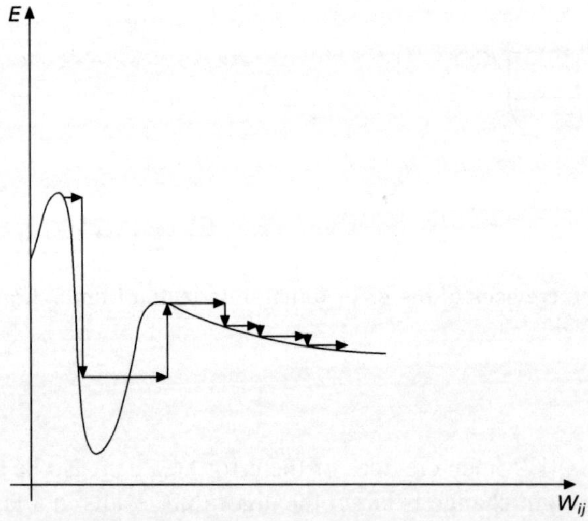

2.18 Missing a minimum of the error function (adapted from Schöneburg *et al.*, 1990).

Momentum

In order to cross large plateaus (Fig. 2.16) quickly, the so-called momentum μ is another option. It is added to the delta learning rule which is then:

$$w_{ij}(t+1) = w_{ij}(t) + \sigma \cdot \delta_i \cdot o_j + \mu \cdot \Delta w_{ij}(t) \qquad [2.28]$$

where $\Delta w_{ij}(t)$ represents the weight change of the preceding learning step.

With the momentum factor, the general direction of the weight change is more likely to remain unchanged, thus avoiding local minima. Besides, when encountering steep crevices (Fig. 2.17), the learning speed decreases, hence large minima (at the bottom of the crevice) are more likely to be recognized. Small global minima can be missed though. Normally, μ values are in the range of 0.1 to 0.2.

Constant factor added to derivative

Another problem can arise when the sigmoid function is used as activation function. As was shown in Fig. 2.14, the maximum value of the derivate of the sigmoid function is $F'(\text{net}) = 0.25$. Hence, even if the difference between actual output and desired output is very large, resulting in a large $(z_i - o_i)$ value, the actual weight change is still comparatively small. The most simple method to overcome this problem is to add a constant factor c to the derivate $F'(\text{net})$, e.g. $c = 0.1$. This should only be considered at the beginning of the training as it can cause an unwanted prolongation of the training in later stages.

Weight decay

If the weight factors become too large, this can lead to oscillations and uncontrolled jumps across the error landscape. This can be avoided when the learning rule is modified as suggested by Paul Werbos as early as 1974 (Werbos, 1974). He introduced the weight decay d which leads to the following learning rule:

$$w_{ij}(t+1) = w_{ij}(t) + \sigma \cdot \delta_i \cdot o_j + \mu \cdot \Delta w_{ij}(t) - d \cdot w_{ij}(t) \qquad [2.29]$$

Here, a certain percentage of the previous weight value is deduced from the current weight change in order to keep the weight factors themselves from becoming too large. Normally, d is in the range of 0.005 to 0.03. This variant bears the risk of slowing down the learning process at an early stage as it may prevent the weight factors from becoming large enough to cross the error landscape quickly. It should therefore be used carefully and preferably in later stages of the training.

Data submission

Another important factor influencing the behaviour of the neural network during the training stage is the data submission method. For some applications, it makes sense to form groups of related data sets which are then fed into the network subsequently in those groups. For other applications where those groups cannot be formed, it is more sensible to feed them at random and not necessarily in the order in which they were gathered in experiments. If only a small network is to be used, the data sets can be divided into subgroups (at random) which are fed and trained subsequently. In this case, the neural network is initialized at the beginning, then the first subgroup is fed and trained and this trained network is then trained with the next data sets group, etc. It is apparent that it is difficult to define the optimal approach to train a neural network. The following examples are intended to illustrate typical training methods and their results.

2.3.5 Determination of suitable training parameters

After having calculated the optimum number of input and output nodes and the optimum network size, the following factors need to be determined:

- learning rate σ;
- momentum factor μ;
- number of learning stages and cycles;
- truncation error ε.

According to the author's experience, it is sensible to start with a large learning rate (e.g. $\sigma = 0.8$). Hence, the weights are changed quickly and thus new patterns are learned and 'memorized' more easily. The learning rate is then gradually decreased until $\sigma = 0.1$. This consolidates the knowledge in later stages once it is acquired in the beginning. Momentum factors should be set in the range of 0.1 to 0.2. The truncation error determines when a training stage ends and the subsequent stage begins. Its value depends on the quality of the data submitted; typical values are in the range of 0.3 to 0.5 at the beginning (representing approximately 30–50 % overall prediction error) and 0.1 in the final training stage. The maximum number of generations is often set to 1000 at the beginning and then increased to values in the range of 3000–5000 in later stages.

Learning rate σ

The learning rate determines the weight change in each learning step. Large values lead to new knowledge being more quickly learned but also to existing knowledge being overwritten. At the beginning of the training stage, a large learning rate normally leads more quickly to the area where the optimum is located. If the learning rate is too large for too long, it is possible that the area of the optimum is completely missed. It is hence crucial to decrease it at the right time.

This is shown in Fig. 2.19. A large learning rate leads to a bumpy learning curve whereas smaller values of σ result in a smooth curve. It is noteworthy that the final result of the overall learning error (y-axis) is similar and not as dependent upon the learning rate as one would expect. A smooth course of the curve indicates that the neural network is in a stable state when the learning rate is small. An oscillating curve is a clear sign that the network alternates between meso-stable states. This implies that the predictions of such a network would not be very reliable.

Figure 2.19 also shows that although the training itself is a gradient descent process, the overall error can increase during the training. Peaks indicate that new data sets are fed into the network that do not match previously acquired knowledge thus leading to large weight changes according to Hebb's learning rule.

Momentum factor

The momentum also affects the training course as is apparent in Fig. 2.20. A large momentum factor often leads quicker to a viable result as the direction of the weight change is maintained which allows, for example, large plateaus to be crossed more quickly. If the momentum factor is too high as in Fig. 2.21, the learning progress comes to a stop and the learning error oscillates.

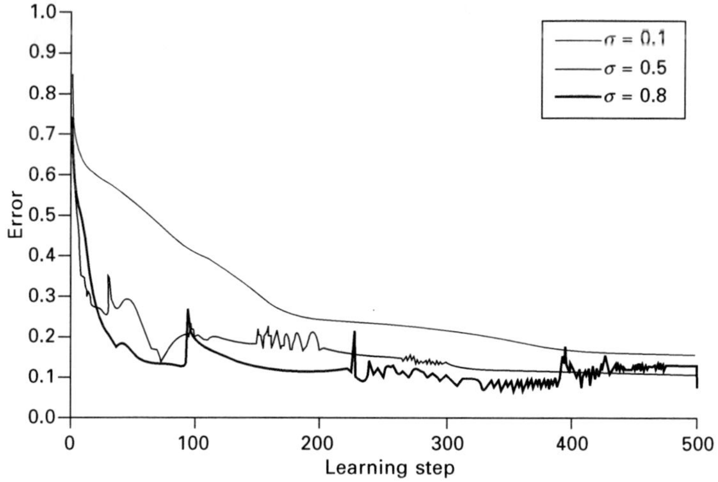

2.19 Influence of learning rate on training.

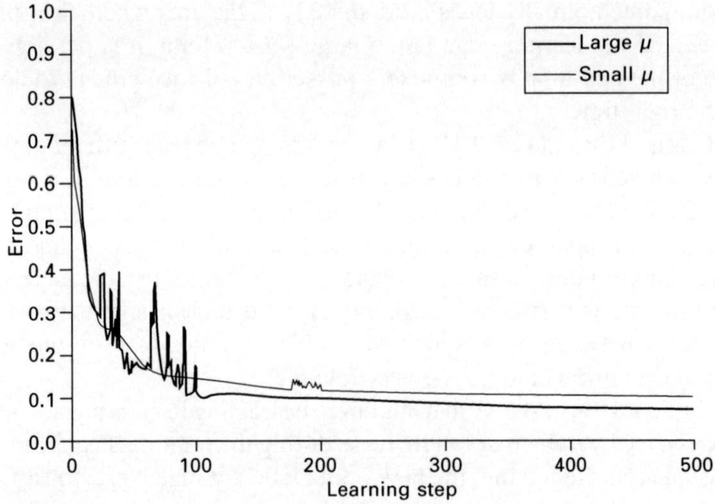

2.20 Influence of momentum factor on training.

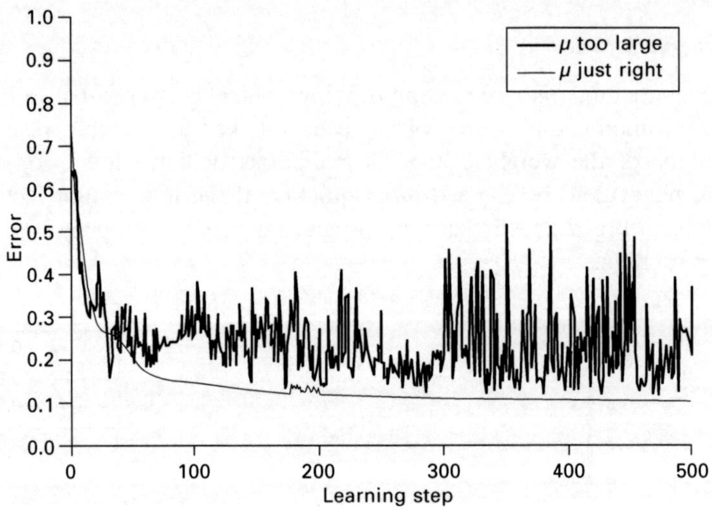

2.21 Effect of too high a momentum factor on training.

Multi-stage training

Table 2.3 gives an overview of typical values for a training scenario consisting of five stages; while Fig. 2.22 shows the course of the learning error over time. The curves oscillate several times before the course becomes smooth and levels out at the end of the training. This is a typical behaviour of the network and due to changing training factors (e.g. learning rate) or when the network overcomes a local minimum.

Table 2.3 Typical values for a training scenario with five stages

Training stage	Learning rate σ	Momentum μ	Number of cycles	Truncation error
1	0.80	0.2	250–1000	0.3
2	0.65	0.2	250–1000	0.2
3	0.50	0.1	500–1000	0.1
4	0.35	0.1	500–1000	0.05
5	0.20	0.1	1000–5000	0.01

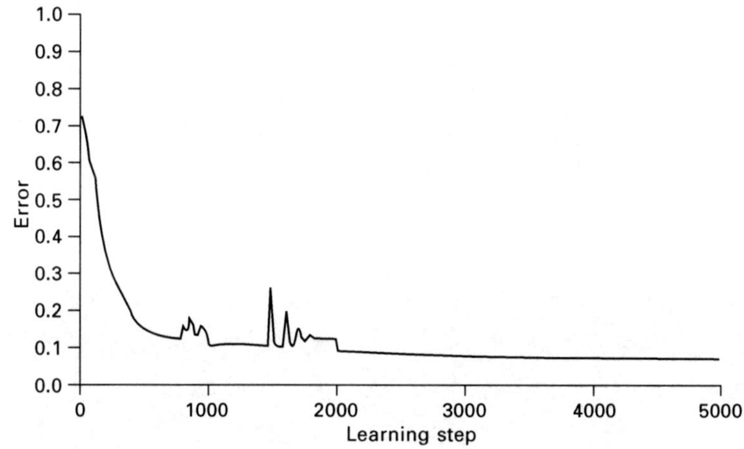

2.22 Typical training result for a scenario with five stages.

Figure 2.23 illustrates the gradual weight changes during the training. Each crossing point represents one weight factor. Different weight values are indicated by different grey values. Information flow in this example is from left (8 input nodes) to right (10 output nodes). The area in the upper left corner is therefore empty throughout the training as there are no weight factors. It is obvious that the weight changes at the beginning of the training are large and become smaller over time.

Typical example – texturing process

By texturizing synthetic yarns (POY – partially oriented yarns), they receive crimp and elasticity hence mechanical properties similar to staple fibre yarns (Fig. 2.24). The texturizing process comprises heating and cooling of the yarn, applying twist and winding up of the final product (DTY – draw textured yarn). The design of a texturizing machine is shown in Fig. 2.25.

Finding the optimum machine setting of a texturing machine is very difficult due to the large number of parameters that influence the process

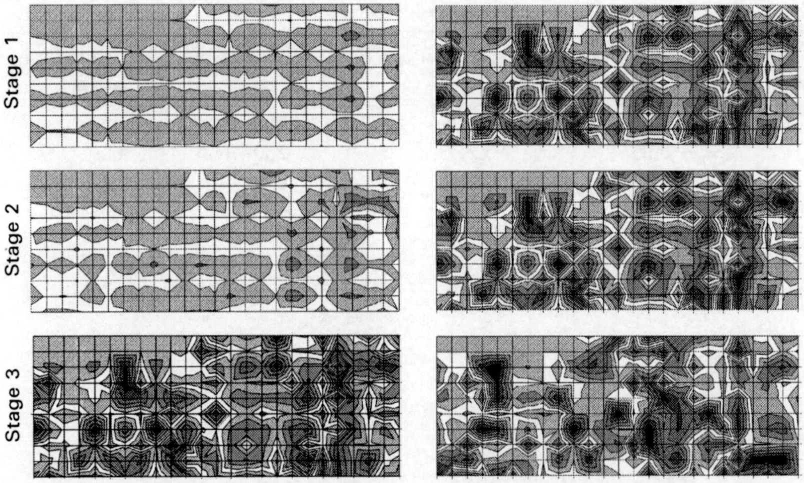

2.23 Gradual weight change over time during the training.

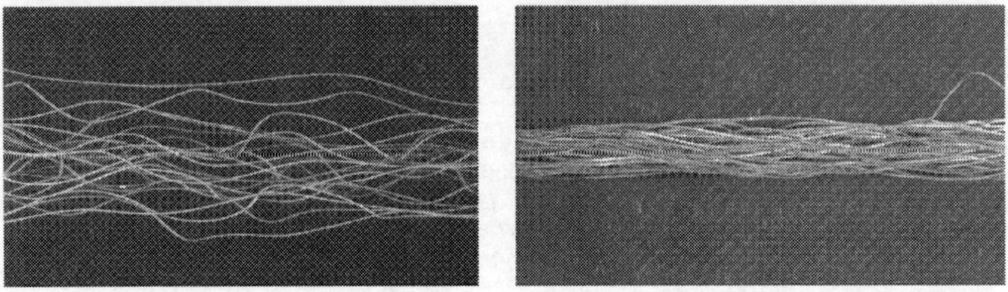

2.24 Texturized yarns (left: set yarn, right: highly elastic yarn).

(e.g. yarn speed, heater temperatures, draft, POY, *d*/*y*-ratio) and hence the final yarn characteristics (e.g. tenacity, elongation, crimp). It has proven to be impossible to create a mathematical model that analytically describes the relations between machine settings and resulting yarn characteristics. Hence, when changing the yarn type or the yarn characteristics, a pilot test has to be carried out in order to determine the new machine settings.

In order to simplify this process, a neural network was employed to determine the relation between yarn characteristics and machine settings. Hence, it was possible to calculate the optimum machine setting for given yarn characteristics and also to predict the expected yarn characteristics depending on the machine settings (Veit, 1998; Veit *et al.*, 1998).

Structure of the neural network

The neural network consisted of two hidden layers with 12 and 8 neurons each, similar to Fig. 2.11. The input function net_i was the sum function, the

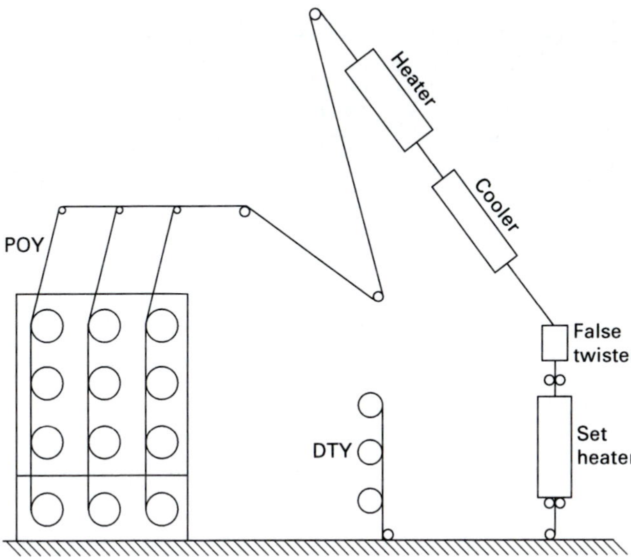

2.25 Design of a texturizing machine and major components.

activation function was the sigmoid function and the output function was linear.

Predicting the yarn characteristics depending on the machine setting

The network was trained with 36 data sets representing 36 different machine settings. Each set consisted of 15 machine parameters representing such factors as yarn speed, heater temperature, *d*/*y*-ratio and draft, and of 10 yarn parameters representing, for example, tenacity, elongation and crimp. From these sets, 35 data sets were used to actually train the network, and the remaining data set was used for the recall stage. This was repeated for all data sets. Hence, 36 neural networks were trained successively. A typical result is shown in Fig. 2.26. It is apparent that it was possible to train the neural network so that it could accurately predict the yarn characteristics from the respective machine settings.

Determining the optimum machine settings to achieve given yarn characteristics is much more difficult, as shown in Table 2.4. The prediction differs considerably from the expected result with which the yarn was actually produced. In Table 2.4, machine settings 12 and 18 and the resulting yarn profiles are shown which are similar to the desired yarn profile. It is apparent that these machine settings lead to a yarn profile similar to the one in question. The predicted machine setting resembles these settings and was therefore correct. The reason for this is typical for this kind of application: it is possible to achieve similar yarn profiles while using very different machine

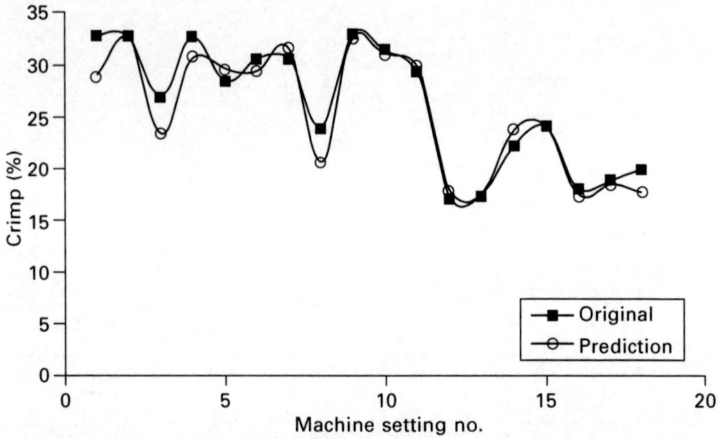

2.26 Prediction of crimp of the yarn depending on machine setting.

settings. During the training stage, the neural network normally learns to connect one of these possible settings with a certain yarn profile and hence makes a seemingly wrong prediction when fed this yarn profile. In fact, the predicted machine setting is one of many possibilities to achieve the same yarn characteristics.

In contrast, a certain machine setting always leads to the same specific yarn profile. Hence, when predicting the yarn characteristics from the machine settings, this problem never occurs. The system does not only work in laboratory scale but has also been successfully employed in industry. Here, even the machine type and other parameters (e.g. describing the mechanical behaviour of the POY) not taken into account in the laboratory model, could be predicted correctly.

2.4 Counterpropagation

Despite their similar names, this algorithm is not related to the backpropagation algorithm. It was developed in the mid-1980s by Robert Hecht-Nielsen (1987, 1990) combining the earlier models devised by Teuvo Kohonen (1981) and Stephen Grossberg (1988). This algorithm is of the feed-forward type. The advantage of this neural network type is its learning speed, which is about 10–100 times higher than that of the backpropagation algorithm. Its drawback is the much lower capacity to represent patterns.

2.4.1 Network structure

The design of a counterpropagation network is shown in Fig. 2.27. The input data are normalized and the length of the input vector is set to 1. The

Table 2.4 Comparison of predicted and actual machine settings to achieve certain yarn characteristics

	Machine settings				
	d/*y*-ratio	Draft	Yarn speed (m/min)	Yarn temperature (°C)	Yarn tension (cN)
Original values	1.6	1.53	700	174	68
Predicted values	2.1	1.53	752	171	43

	Machine settings					Yarn parameters			
	d/*y*-ratio	Draft	Yarn speed (m/min)	Yarn temperature (°C)	Yarn tension (cN)	Yarn count	Elongation (%)	Tenacity (cN/tex)	Crimp (%)
Actual	1.6	1.53	700	174	68	172	15	22	34
Prediction	2.1	1.53	752	171	43				
Data set #12	2.0	1.53	700	174	41	171	16	22	35
Data set #18	2.0	1.53	800	167	41	171	15	22	31

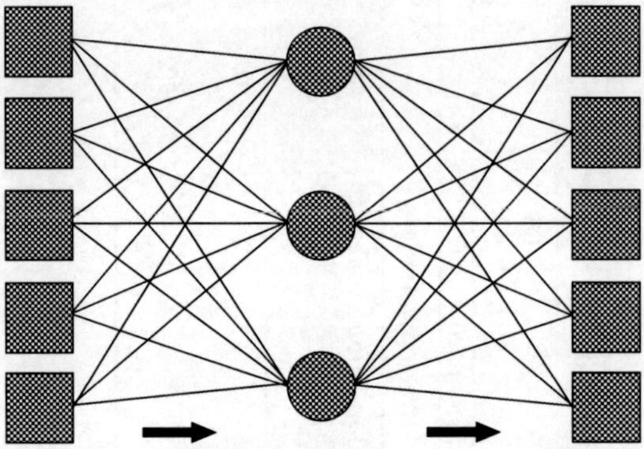

2.27 Network structure of the counterpropagation algorithm.

Kohonen layer receives weighted input, and its weight factors are variable and can be trained. The length of the weight vector of each Kohonen neuron is set to 1. Each Kohonen element is connected to all elements of the input layer. The Kohonen neurons propagate their data to the Grossberg layer via variable weight factors where the length of the weight factor of each Grossberg element is again set to 1. All Grossberg neurons are connected to all Kohonen neurons.

2.4.2 Kohonen layer

The elements of the Kohonen layer are of the 'winner takes all' type: only that element with the greatest input becomes active whereas all other elements remain inactive. We have

$$\text{net}_i = \sum_{j=1}^{n} w_{ij} \cdot o_j \qquad [2.30]$$

which is equivalent to

$$\text{net}_i = \vec{w}_i \cdot \vec{x} \qquad [2.31]$$

with $\vec{w}_i$ = weight vector of the Kohonen element. $i(|\vec{w}_i| = 1)$ and $\vec{x}$ = input vector of the Kohonen layer. Hence:

$$o_i = 1, \text{ if } a_i = \max(a_i), \text{ with } a_i = \text{net}$$

$$o_i = 0, \text{ else}$$

Since $|\vec{w}_i| = 1$ and $|\vec{x}_i| = 1$, we get:

$$\text{net}_i = \cos \beta \qquad [2.32]$$

Figure 2.28 shows a graphic representation of weight vectors of Kohonen elements in two-dimensional space. The smaller the angle between weight vector and input vector the greater net_i and thus a_i. This implies that the Kohonen neuron which best resembles the input vector has the greatest activity. The input is thus classified according to its resemblance with the weight vectors of the Kohonen neurons.

2.4.3 Grossberg layer

For the Grossberg neurons, we have

$$\text{net}_i = \sum_{j=1}^{n} w_{ij} \cdot o_j \text{ and } o_i = \text{net}_i$$

Since there is only one Kohonen element active at any given time, we get:

$$o_i = w_{ik}$$

The output of the Grossberg element i is equivalent to the weight factor between this element and the Kohonen element k.

2.4.4 Learning in the Kohonen layer

Learning in a Kohonen layer is non-supervised – also called 'discovering learning'. The output vector must therefore not be known to the Kohonen elements. This is one of the major differences to the backpropagation algorithm. According to the 'winner takes all' rule, we have:

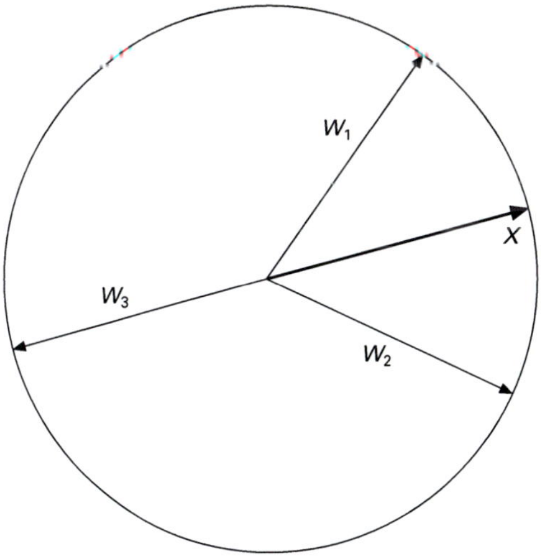

2.28 Weight vectors of a Kohonen element in two-dimensional space.

$$w_{ij}(t+1) = w_{ij}(t) + \sigma \cdot [o_j - w_{ij}(t)] \quad [2.33]$$

and

$$\vec{w}_i(t+1) = \vec{\omega}_i(t) + \sigma \cdot [\vec{x} - \vec{w}_i(t)] \quad [2.34]$$

Figure 2.29 shows a two-dimensional graphic representation of the learning process of the Kohonen neurons. The learning rate σ is normally smaller than 1 in the beginning and is subsequently decreased until it reaches the value $\sigma = 0$. The length of the weight vector can decrease during the training. The smaller the distance between weight vector and respective input vector, the smaller their difference in length. Hence, at the end of the training, all weight vectors have again a length of 1.

The weight vectors of the Kohonen elements move towards those input vectors which they resemble best. Hence, all weight vectors of the Kohonen neurons change. The change increases with the activity of the Kohonen elements which in turn increases the more the respective weight vector resembles the particular input vector. At the end of the training stage, the weight vectors of the Kohonen elements are located roughly in the middle of those input vectors which they represent best. This implies that the Kohonen layer mirrors all learned input vectors not only in terms of location but also statistically. Hence, in areas with many input vectors, the number of Kohonen vectors is also large. With n = number of Kohonen elements, the probability that a certain Kohonen neuron becomes active during a learning step becomes $1/n$. If n equals the number of input vectors, the training is completed after only one step. In this case, each input vector would exactly be represented by one weight vector.

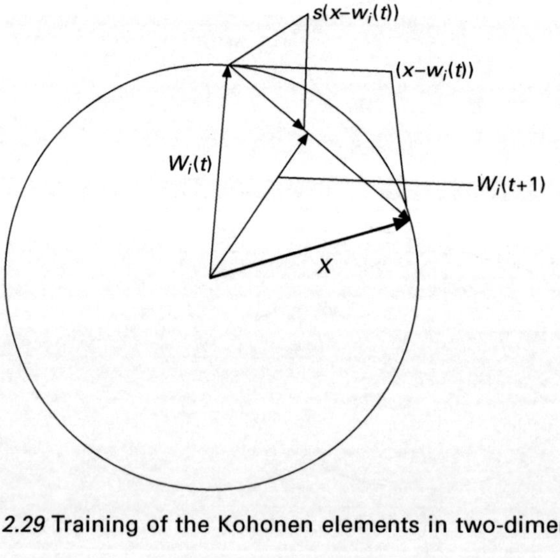

2.29 Training of the Kohonen elements in two-dimensional space.

2.4.5 Initializing the weight factors of the Kohonen layer

Normally, the weights of the Kohonen elements are initialized with small, randomly chosen values with the norm 1. A typical problem arising then is the uneven distribution of weight and input vectors that is ideally identical. As shown in the example in Fig. 2.30, only a few weight vectors represent input vectors; several weight vectors represent nothing. It is therefore an absolute necessity to distribute weight vectors and input vectors evenly. Normally, though, the distribution of the input vectors is unknown or difficult to determine. Hence, several methods were developed to overcome this problem. They are described in the following sections.

Convex combination

All weight vectors are initialized with the unit vector:

$$\vec{w}_i = \begin{pmatrix} \overline{1/n} \\ 1/n \\ 1/n \\ \cdots \\ 1/n \end{pmatrix}; |\vec{w}_i| = 1 \qquad [2.35]$$

The input vectors are then fed into the vector according to

$$\vec{x} = \alpha \cdot \vec{x} + (1 - \alpha) \cdot \vec{i} \qquad [2.36]$$

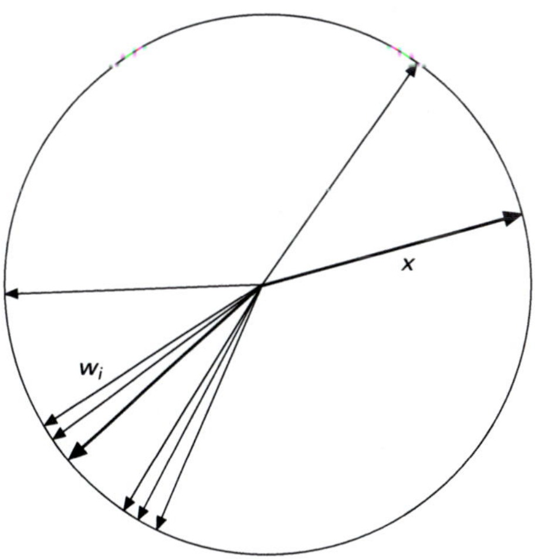

2.30 Unfavourable distribution of weight vectors in two-dimensional space.

In the beginning, the value α is close to zero and increased gradually during the training stage until $\alpha = 1$. This leads to weight vectors being dragged along with the input vectors, thus resulting in even weight and input vector distributions. The advantage of this method is the good representation of the input vectors by the weight vectors. The major drawback is the time needed to train all input patterns as they change during the course of the training.

Random noise

A random noise can be added to the input vectors. Hence we get

$$\vec{x}_{\text{new}} = \vec{x}_{\text{old}} + \vec{R} \qquad [2.37]$$

Thus in principle, all Kohonen elements can win, even those which resemble the input vector less than others. The input vectors themselves do not change during the training. This method is even more time-consuming than the convex combination. But all weight vectors partake in the learning process which can be advantageous and normally avoids weight vectors left in the void without representing anything.

Radius

All weight vectors within a given radius around the input vector partake in the learning process. If the radius is sufficiently large, all Kohonen elements can learn. The radius value decreases during the training until it reaches zero, so that finally the only Kohonen element that learns is the one which resembles the input vector best. This leads to a good representation of the input vectors but is very time-consuming.

Conscience

Kohonen elements that win more often than $1/n$ (n = number of Kohonen elements) are excluded from the training process for a certain time. Thus, all Kohonen elements can adjust their weights equally frequently.

2.4.6 Interpolative mode in the Kohonen layer

The output of the Kohonen layer normally consists of the output of the winning element. If not only one but several Kohonen elements actually win, then we have

$$o_k = b \cdot \vec{w}_k \cdot \vec{x} \qquad [2.38]$$

The norm of the output of the Kohonen layer is still equal to one. This value is distributed among all winning Kohonen elements according to their activity, hence

$$b = \frac{1}{\sum |\vec{w}_k \cdot \vec{x}|} \text{ or } b = \frac{1}{\sqrt{(\vec{w}_k \cdot \vec{x})}} \qquad [2.39]$$

The learning rule of the Kohonen elements can then be written as

$$\vec{w}_i(t + 1) = \vec{w}_i(t) + \sigma \cdot (\vec{x} - \vec{w}_i(t)) \cdot o_i \qquad [2.40]$$

The smaller the distance between $\vec{w}_i$ and $\vec{x}$ at point in time t, the greater o_i and hence the more the weights are adjusted at point in time $t + 1$.

2.4.7 Training of the Grossberg layer

The elements of the Grosberg layer learn supervised as the desired output is known. The learning rule is then:

$$w_{ij}(t + 1) = w_{ij}(t) + \sigma \cdot [z_i - w_{ij}(t)] \cdot o_j \qquad [2.41]$$

with z_i = desired output of Grossberg elements i and o_j = output of Kohonen element j. It is assumed that only one Kohonen element is active at any given time, hence there is only one $o_j \neq 0$.

For $\sigma = 1$, the weight factors w_{ij} of the Grossberg layer to this Kohonen element j become equal to the desired output z_i. Normally, σ assumes very small values (e.g. $\sigma = 0.1$) which are gradually increased until $\sigma = 1$.

At the end of the training stage, a certain group of input vectors is represented by one Kohonen element. The weight vector of this Kohonen element connecting it to the Grossberg layer represents those output vectors, to which the input vectors correspond.

2.4.8 Network structure

The typical structure of a counterpropagation network is shown in Fig. 2.31. Input and output layers are divided into two parts. Hence, there are virtually two input and two output layers that are identical. The input is fed into layers x and y; the resulting output is then X' and Y'. The name counterpropagation relates to another graphical representation of the network as shown in Fig. 2.32. If the vectors x and y are doubled (X', Y') and drawn one on each side of the Kohonen layer, the information flow is inward bound, thus giving this network type its name.

The training of this network is auto-associative. This means that the vectors X and Y are associated with themselves. Hence, if X and Y are fed into the network during the recall, the network will produce X' and Y' as output. If only the vector X is fed into the network and all values of vector Y are set

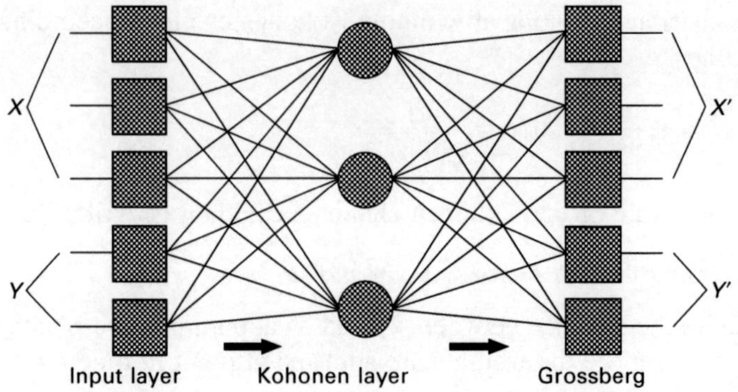

2.31 Structure of counterpropagation network.

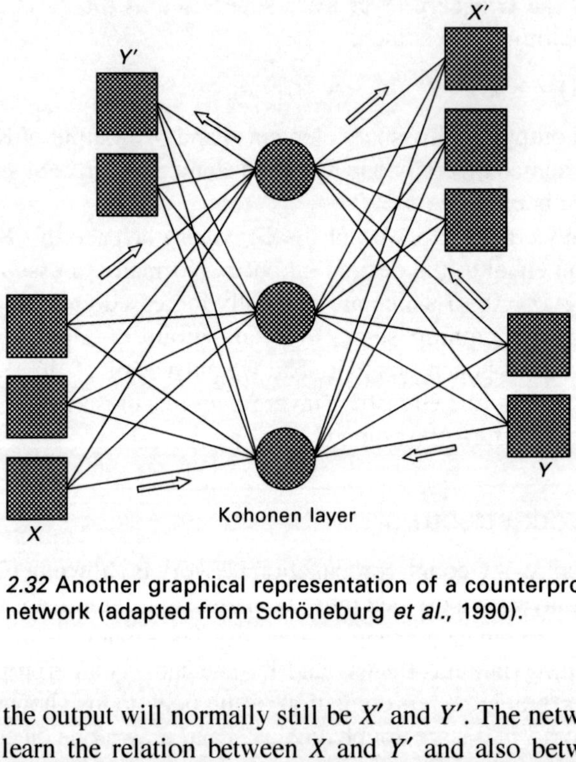

2.32 Another graphical representation of a counterpropagation network (adapted from Schöneburg *et al.*, 1990).

to zero, the output will normally still be X' and Y'. The network is therefore able to learn the relation between X and Y' and also between its inverse function, Y and X'. This is very useful for pattern recognition applications as it allows an easy reversal of the contrast of a picture. In general, this type of neural network is especially suitable for all classification problems where groups of data sets need to be clustered and only the number of clusters is known but not their exact boundaries. Typical applications in textiles are pattern recognition or classification of faults in fabrics.

2.5 Other types of neural networks

Apart from the previously mentioned network types, there are also a range of other kinds of networks suitable for special applications.

2.5.1 Hopfield networks

This network is of the feedback type. In the first step, input is fed into the network which generates output. This output is then fed into the network as new input. This process is repeated until all patterns reproduce themselves. This is a stable state which can be defined using an energy function. This kind of network can be used to recognize patterns with a random noise. Another interesting feature is the ability of this network type to eradicate learned patterns from its memory. This can be helpful when training patterns over a large period of time when the first patterns should be deleted and replaced by later ones.

2.5.2 Simulated annealing

This is another interesting type of network to determine global minima. It works in a totally different way from backpropagation and counterpropagation algorithms, but is nevertheless considered an ANN. During the training stage, it is possible to leave local minima, thus deteriorating the quality of the output. Hence, it is possible to find the global minimum which is more difficult with Hopfield networks. There are a number of other types of neural networks, but they are currently not used to solve technical problems in textile production.

2.6 Applications of neural networks to textile technology

In recent years, neural networks have been used successfully to solve numerical problems for a wide range of applications. These include pattern recognition (Wei *et al.*, 2010), language recognition (Ververidis and Kotropoulos, 2006), material inspection (Su *et al.*, 2008), machine optimization (Veit, 1999), the prediction of product performance (Liu and Qu, 2008) and many more. In textile technology, there have also been a wide range of successful applications as the number of published papers shows (Fig. 2.33). In the following sections, typical examples for such applications are briefly described. They all use the backpropagation algorithm unless otherwise stated. Many more papers have been published in recent years, but often with questionable results as verifiable data were not given or unscientific evaluation methods were used. These are hence not cited in this section.

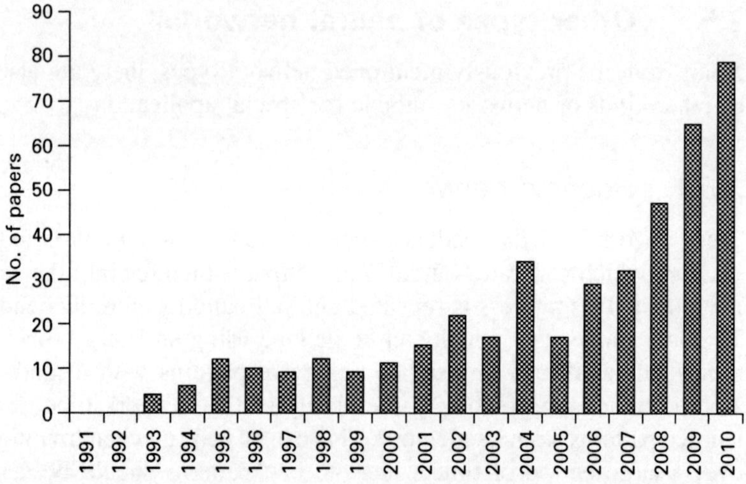

2.33 Number of published papers on neural network applications in textiles (Anon., 2011).

2.6.1 Pattern recognition

A typical example for this kind of application is the detection of trash particles in fibre webs. The influence of the machine settings in the cleaning room on the cleaning efficiency was analysed by Veit *et al.* (1996a, 1996b). Fibre samples were taken before and after each cleaner and analysed applying digital image processing in connection with a neural network. The neural net was used to classify the trash particles (e.g. neps, seed coat fragments, leaf and stem fragments). Figure 2.34 shows the digital image of a fibre web.

The backpropagation neural network consisted of five neurons in the output layer, each representing one kind of trash particle. Three hidden layers were used with 10, 15 and 20 nodes, the input layer consisted of 10 nodes each representing parameters that describe the trash particles. In the initial stage, the neural net was trained with 1500 data sets, each describing a typical trash particle with regard to size, circumference, shape, etc. Prediction accuracy after the training stage was close to 95%. Then, particle data sets unknown to the network were fed into the system and the particles were classified with an accuracy of approximately 85%. The comparison of manually classified particles and the results of the neural network analyser showed a high degree of consistency, as can be seen in Fig. 2.35.

It was very difficult to distinguish between seed coat fragments, leaf particles and stem particles as these kinds of particle appeared very much alike to the system. In order to further improve the accuracy, a fuzzy algorithm was implemented in the output layer of the neural network. This led to a significantly higher accuracy as even when two output neurons showed a

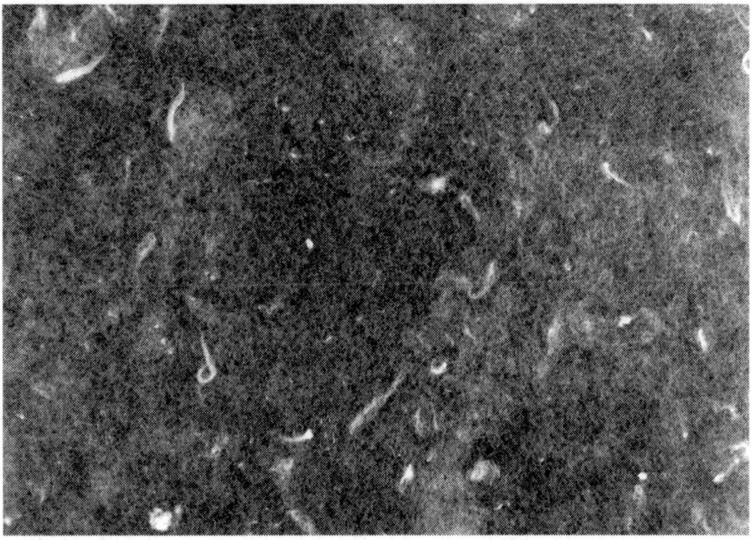

2.34 Digital image of a fibre web (Veit *et al.*, 1996b).

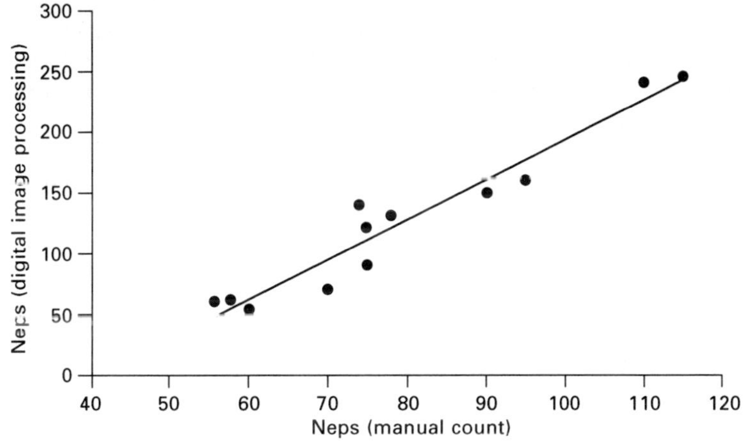

2.35 Comparison of classification result for neps: manual vs. digital image processing with neural network analyser (Veit *et al.*, 1996b).

similar activity, a fuzzy rule could help to determine the actual kind of particle.

The major advantage of this neuro-fuzzy system compared with conventional methods was the classification time which was only a matter of seconds for several hundred data sets gathered when screening one digital image of a fibre web. Conventional systems (e.g. nearest neighbour classification) took several minutes with a considerably lower accuracy (e.g. approx. 75% for nearest-neighbour algorithm).

2.6.2 Draw frame

Farooq and Cherif (2008) developed a system to determine the levelling action point (LAP) of an auto-levelling draw frame. Input parameters comprised, for example, feeding and delivery speed, break and main draft gauge. Several different network structures, all based on the backpropagation algorithm, were used. As Fig. 2.36 shows, the prediction accuracy was very high and reached 98% on average in this example. Here, two hidden layers with 12 and 10 neurons each were used. The maximum deviation between predicted and actual optimum LAP values does never exceed 3 mm which falls within the range of negligible differences when determining the LAP, as practice trials showed.

2.6.3 Hairiness of worsted yarns

Wang *et al.* (2009) developed a prediction model for the hairiness of worsted yarns. They used one hidden layer with only one hidden node, a learning rate of $\sigma = 0.3$ and a momentum factor of $\mu = 0.9$. A sensitivity analysis (see Section 2.7.1) showed that several of the originally used input parameters did not have a decisive effect on the hairiness (Fig. 2.37). Hence, only yarn twist, ring size, hauteur, mean fibre diameter, yarn count, short fibre proportion and spindle speed were needed to predict the expected yarn hairiness very accurately as can be seen in Fig. 2.38.

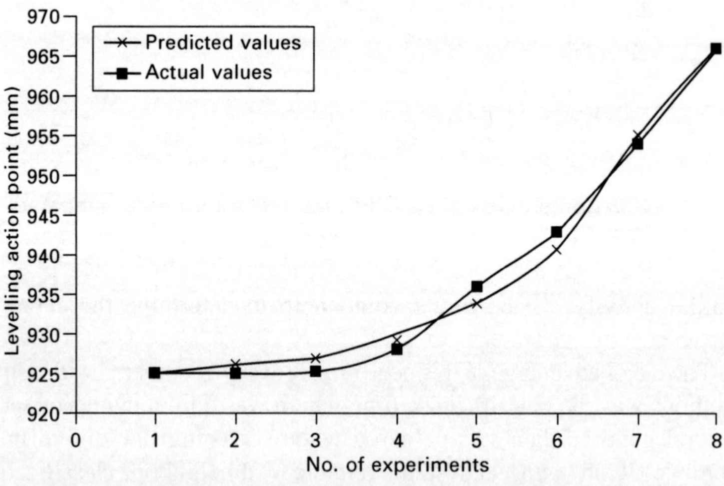

2.36 Comparison of predicted and actual LAP (adapted from Farooq and Cherif, 2008).

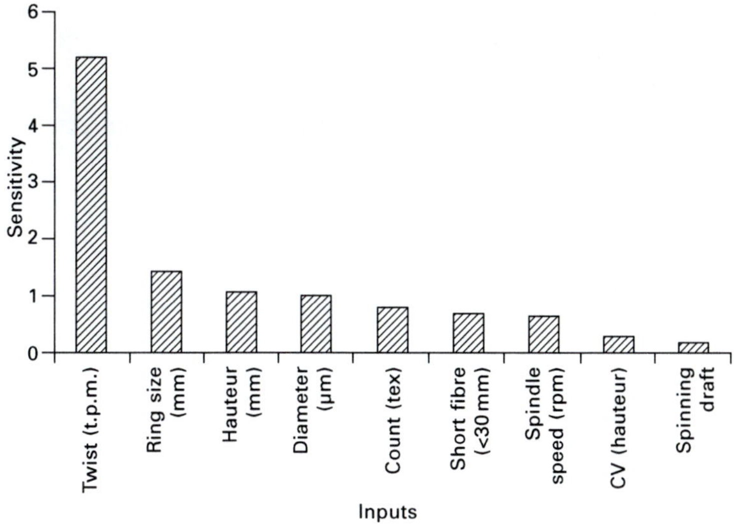

2.37 Typical sensitivity values (adapted from Wang *et al.*, 2009).

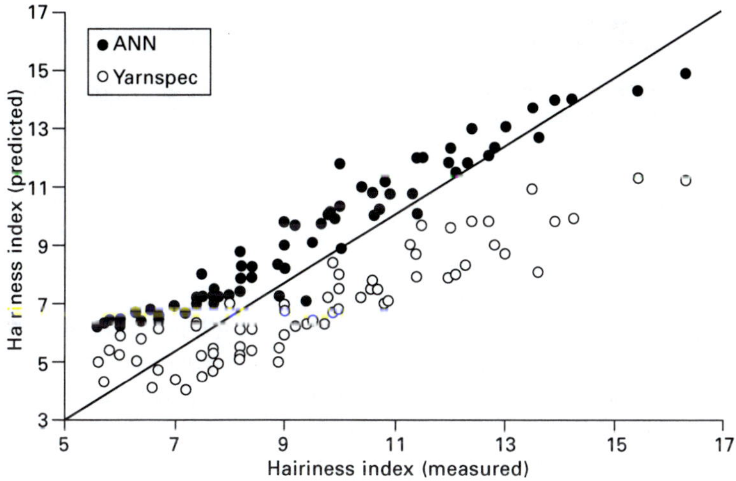

2.38 Comparison of predicted and actual hairiness values (adapted from Wang *et al.*, 2009).

2.6.4 Draw-winding process

Draw-winding is an important process to produce high tenacity yarns, mainly from polyester and polyamide. The principle is shown in Fig. 2.39. Small- and medium-sized enterprises often apply this process but rely heavily on the expert knowledge of mostly senior engineers. When they retire, their knowledge is often lost. This led to the development of a system based

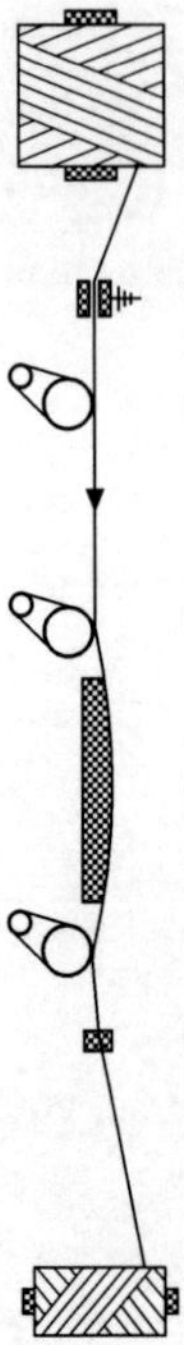

2.39 Basic design of the draw-winding process (adapted from Ramakers, 2006).

on a neural network that was able to determine the machine settings to achieve certain yarn profiles and vice versa (Ramakers, 2006). Each of the 50 data sets used consisted of parameters describing the machine set-up (e.g. temperature of godets, draft ratios) and others representing the yarn profile (e.g. yarn fineness, tenacity, shrinkage). Figure 2.40 shows the basic approach to training a neural network with these data sets.

In this case, the first results showed that the neural network was not able to learn the relation between yarn profiles and machine settings. Thus, a design of experiments (DoE) approach was applied where the machine settings were varied in a very structured way hence covering a much wider range of the parameter values. This greatly improved the accuracy of the prediction, as is shown in Fig. 2.41. The average prediction error was below 3%.

2.6.5 Weaving process

In recent years, the machine speed in the weaving process has been increased considerably. This has led to ever-greater difficulties to ensure a high fabric quality. Especially after a machine stop, the so-called stop marks in the fabric (caused by an uneven spacing between weft threads, Fig. 2.42) are a

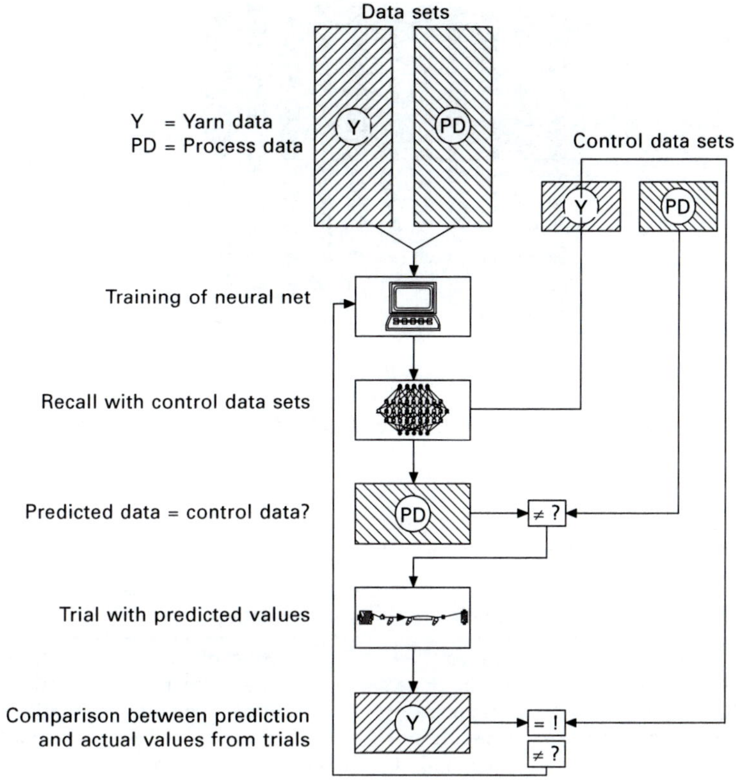

2.40 Basic approach to training the neural network (adapted from Ramakers, 2006).

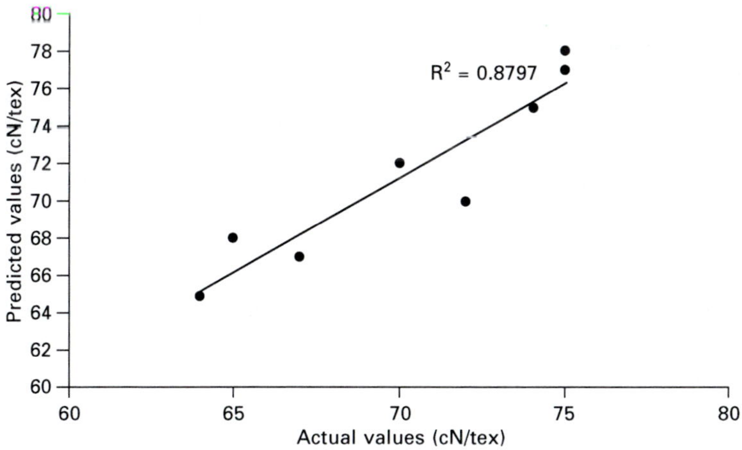

2.41 Comparison between predicted and actual yarn tenacity in the draw-winding process (adapted from Ramakers, 2006).

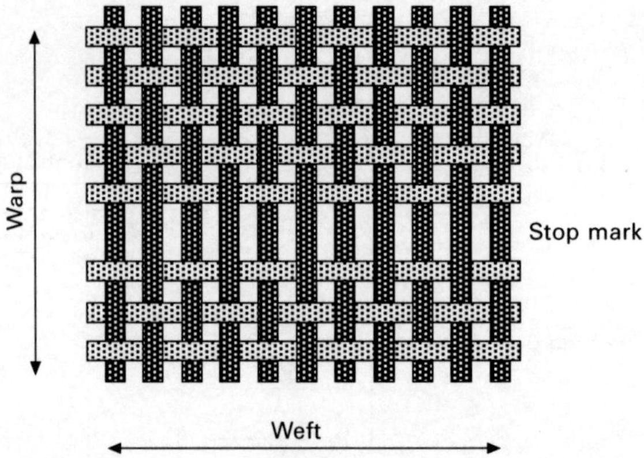

2.42 Stop mark in a fabric (adapted from Strauf Amabile and Gries, 2005).

Table 2.5 Parameters influencing the characteristics of a stop mark

Fabric		Machine	
Material	Article	Set-up	Elements
Yarn quality	Weave	Start rpm	Gear box
Yarn structure	Fabric density	Start offset	Direct drive
Stress–strain behaviour of yarns	Forces at crossing point	Start position	Clutch–break combination
Yarn bulk	Relaxation behaviour of fabric	Stop position	
Yarn hairiness	Cloth fell	Break deceleration	
Relaxation behaviour of warp		Miss pick	
Cop building of weft yarn		Basic level of warp tension	
Warp beam density		Backrest position	
Sizing of warp		Backrest movement	

fault which is no longer acceptable owing to increased market requirements. Hence, a mathematical model was developed by Strauf, Amabile and Gries (2005) describing the relations between machine settings, fabric properties and stop mark characteristics.

Table 2.5 gives an overview of typical parameters having an influence on the characteristics of a stop mark. It is apparent that there are many different parameters to be taken into account which makes an analytical description of the relations almost impossible. Hence, a neural network was used and trained with data sets representing typical machine settings, fabric properties and resulting stop mark characteristics. This was used to determine an optimum machine setting in order to minimize the stop mark hence minimizing the effect of a machine stop on the fabric quality. Table 2.6 shows a comparison of predicted and actual machine settings during the training. Figure 2.43 shows the predicted characteristics of a stop mark (here: spacing between weft threads after a machine stop) and the measured data. Since the system apparently works, it is now used in industry by weaving machine manufacturers.

2.6.6 Yarn breakage rate during weaving

Yao *et al.* (2005) published a paper on the prediction of warp breakage rate using a backpropagation network. Input values included size add-on, abrasion resistance, breaking strength and hairiness. The authors claim to have reached a prediction accuracy for the number of yarn breaks close to 90%, but no verifiable data base is provided. Bo (2010) used a similar approach, but again no verifiable data were presented and the used input values were not specified. In general, according to the author's experience,

Table 2.6 Comparison of predicted and actual machine settings to avoid stop marks

	Start rpm (%)	Synchronizing rpm (%)	Start offset warp (no. of picks)	Start position fabric (no. of picks)
Prediction	129	101	–0.02	0
Actual values	130	100	0	0

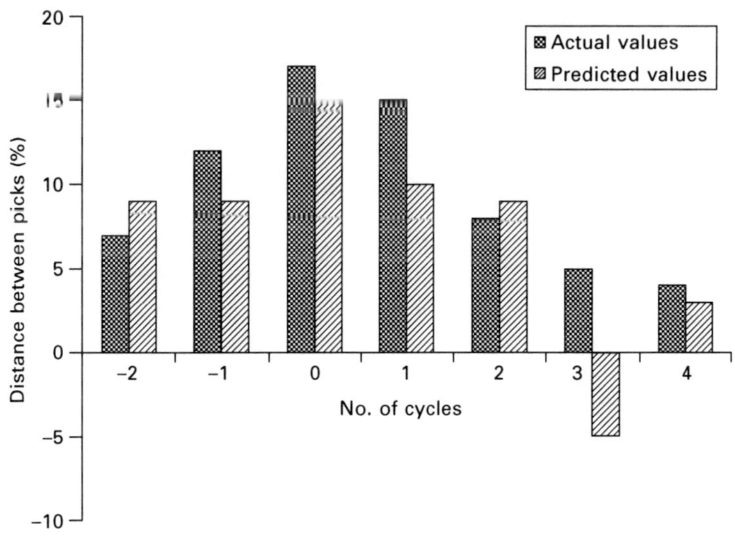

2.43 Predicted and actual stop mark characteristics (adapted from Strauf Amabile and Gries, 2005).

it is very difficult to predict breakage rates as there is a strong influence of non-measurable factors which normally cannot be neglected.

2.6.7 Yarn shrinkage

Lin (2007) developed a system to predict the shrinkage of weft and warp yarns depending on the cover factors of warp, weft and fabric as input. This is a very difficult problem as shrinkage strongly depends on the deformation of the yarns and various other factors that are very hard to keep constant. He used one hidden layer with 16 neurons, a learning rate of $\sigma = 0.9$ and a momentum factor $\mu = 0.9$ which are relatively high values. Lin trained the network with 13 data sets and used 13 others for the recall stage. The results are shown in Fig. 2.44. The value for $R^2 = 0.8723$ in this example is high and indicates that applying a neural network is a sensible approach to tackle this difficult problem.

2.6.8 Spirality of cotton fabrics

Murrells *et al.* (2009) used a neural network to predict the spirality of fully relaxed single jersey fabrics made of 100% cotton. As input factors, they used twist liveliness, yarn type, yarn linear density, fabric tightness factor, number of feeders, rotational direction, gauge of the knitting machine and dyeing method. After determining the most important factors to be used as input values, they trained a feed-forward network (probably backpropagation) with one hidden layer and a different number of neurons with 40 data sets, 26 were later used for the recall. The results were very good, as can be seen

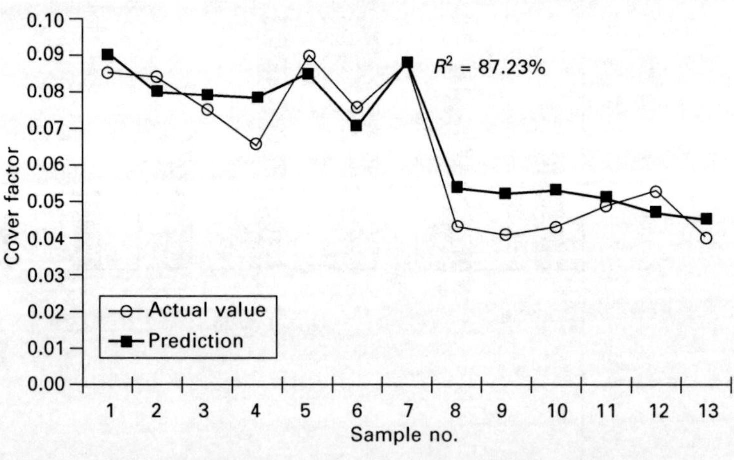

2.44 Comparison between measured and predicted warp shrinkage (adapted from Lin, 2007).

in Fig. 2.45. The correlation coefficient was $R = 0.976$ compared to $R = 0.970$ for a multiple regression model.

2.6.9 Textile fabric appearance

Cherkassky and Weinberg (2010a) used a probabilistic neural network approach (related to Kohonen nets) to determine the grades of knitted fabric appearance. This is a typical classification problem for which this type of neural network is perfectly suited. Photographs were taken from the samples and analysed. The input values of the network were the calculated fabric brightness, the average number of pills and fuzzy and the average and standard deviation for all captured profiles. The output layer consisted of five nodes with each representing a different grade. As Fig. 2.46 shows, the predicted grades coincide very well with the grades given by experts (Cherkassky and Weinberg, 2010b). Taking into account that the expert grades may also not all be correct, the achieved results are excellent.

Another paper dealing with this subject focusing on polar fleece fabrics was published by Xin *et al.* (2010). Chen and Zhang (1999) applied a backpropagation network for fibre content analysis. They developed a system that could distinguish between yak and cashmere hair based on microscopic images that reached a classification accuracy of 92%. A similar system was published by Shi and Yu (2008). It was used to identify animal fibres, namely wool and cashmere, where it reached a prediction accuracy exceeding 90%. The network used was very small with only one hidden layer containing five nodes. This shows that ANNs do not need to be complicated to solve complex problems.

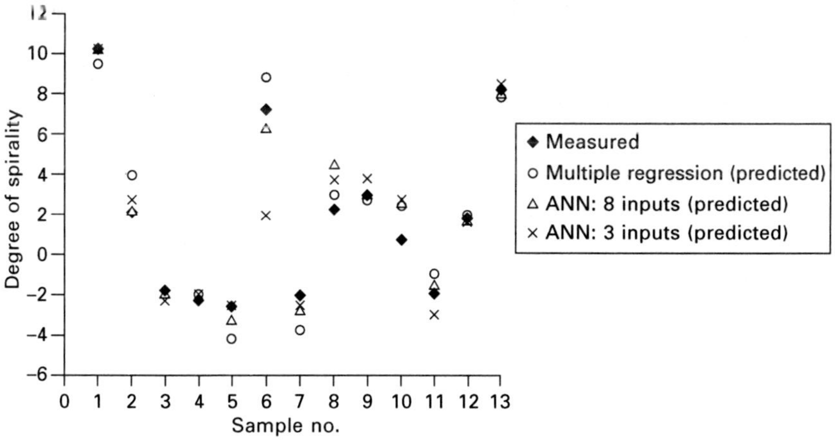

2.45 Predicted and measured degree of spirality for eight input factors (adapted from Murrells *et al.*, 2009).

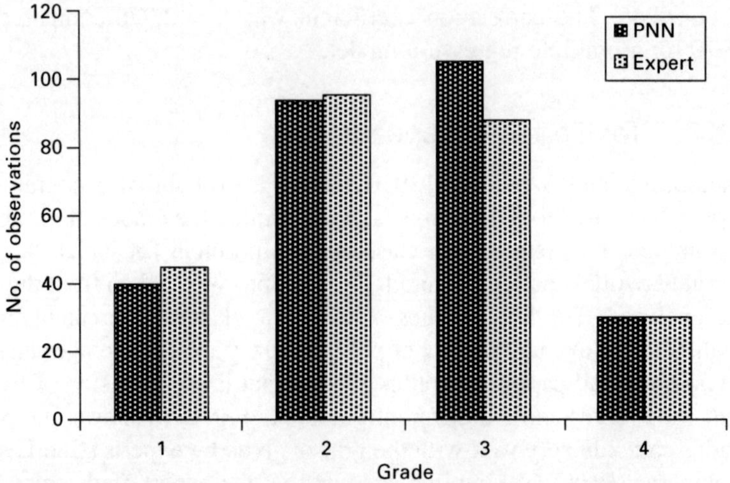

2.46 Comparison of predicted and actual grades for the knitted fabrics (adapted from Cherkassky, 2010b).

2.6.10 Fabric inspection

Yuen *et al.* (2009) used the backpropagation algorithm for the automatic defect detection, namely for evaluating fabric stitches and seams. Pictures of the fabrics were taken and nine characteristic parameters were used as input values. Since the value ranges differ, a data preprocessing method was required which transforms all input values into a similar range. The output layer consisted of only two neurons. They could assume only binary values leading to four different output patterns, 00, 10, 01, 11, of which three were used to associate the image with one of the three classes: 'seams without defects', 'seams with pleated defects' and 'seams with pucking defects'. One hidden layer was used with 10 neurons. The training set consisted of 22 samples for each class and an additional 10 samples were used in the recall stage. Although the output nodes did not always lead to values of 0 or 1 respectively (Table 2.7), the rounded results were always 100% correct. This example shows that the backpropagation algorithm is also very suitable to solve classification problems.

A similar approach to detect local textile defects using a neural network was published by Kumar (2003). It gives a comprehensive overview of feature detection from photographs aimed at training a neural network and is therefore recommended reading for anyone interested in fabric inspection. Other papers on this subject worth reading include those of Liu and Qu (2008), Yin *et al.* (2009) and Chandra *et al.* (2010). Papers by Pan *et al.* (2011) and by Kuo *et al.* (2010) deal with woven fabric pattern recognition in a similar way. Both papers provide extensive information on the methods applied.

Table 2.7 Comparison of calculated outputs and actual values

Defect classes	Index	Ideal output	Actual outputs
Seams without defects	1	(0, 0)	(0.0073, 0.0016)
	2	(0, 0)	(0.0072, 0.0024)
	3	(0, 0)	(0.0062, 0.0034)
	4	(0, 0)	(0.0061, 0.0042)
	5	(0, 0)	(0.0061, 0.0038)
Seams with pleated defects	1	(0, 1)	(0.0048, 0.9985)
	2	(0, 1)	(0.0072, 0.9984)
	3	(0, 1)	(0.0038, 0.9996)
	4	(0, 1)	(0.0039, 0.9988)
	5	(0, 1)	(0.0036, 0.9996)
Seams with puckering defects	1	(1, 0)	(0.9983, 0.0000)
	2	(1, 0)	(0.9971, 0.0000)
	3	(1, 0)	(0.9924, 0.0236)
	4	(1, 0)	(0.9903, 0.0233)
	5	(1, 0)	(0.9988, 0.0001)

Source: adapted and shortened from Wang *et al.*, 2009.

Table 2.8 Average error for different output parameters

Sr. no.	Performance parameters	Error (%)
1	Air permeability	36.44
2	Bursting strength	11.49
3	Edgecomb resistance warp	6.29
4	Edgecomb resistance weft	17.15
5	Tearing strength warp	21.95
6	Tearing strength weft	16.26
7	Tensile strength warp	5.73
8	Tensile strength weft	3.31
9	Elongation warp	5.62
10	Elongation weft	4
11	Packability	3.85
	Mean error (%)	12.01

Source: adapted from Behera and Goyal, 2009.

2.6.11 Design of airbag fabrics

Behera and Goyal (2009) applied a backpropagation network to predict the performance parameters for airbag fabrics. The input parameters include fabric areal density, ends and picks per inch, warp and weft yarn count, strength and elongation. The results for the different output parameters are shown in Table 2.8. The values were calculated as the mean of the results of five samples that were used in the recall stage.

It is apparent that the average error strongly differs for different parameters: air permeability is very hard to predict whereas the tensile strength in the warp and weft directions is calculated very accurately. Apart from physical explanations for this problem, another reason could be the application of a single neural network to predict all parameters at once. In cases such as this, the use of separate neural networks where each predicts only one parameter normally improves the overall result. This is because, firstly, each neural network structure could then be tailored for each parameter and, secondly, that the relations between input and output values do not affect each other during training.

2.6.12 Thermal resistance and thermal conductivity of textile fabrics

Kothari and Bhattacharjee (2007) applied two different neural networks to predict the steady-state thermal resistance and the maximum instantaneous heat transfer Q_{max} of woven fabrics. The input parameters included weave type, weft and warp count, weft and warp density. The neural network had two hidden layers. At first, for each output parameter, a separate network containing a small number of neurons in each hidden layer was trained and the prediction compared with the measured values. Then, a network consisting of two hidden layers with a higher number of nodes was used to predict both output values simultaneously. This increased the mean error from 8.6% to 10.4% (see also Section 2.6.1). Figures 2.47 and 2.48 show the results for both output values for both cases. Both thermal resistance and heat transfer could be predicted with sufficient accuracy with thermal resistance showing better results.

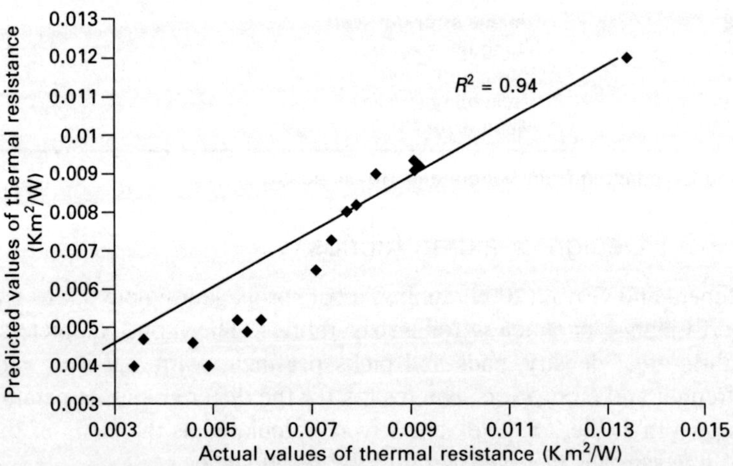

2.47 Comparison of calculated thermal resistance and actual values (one neural net) (adapted from Kothari and Bhattacharjee, 2007).

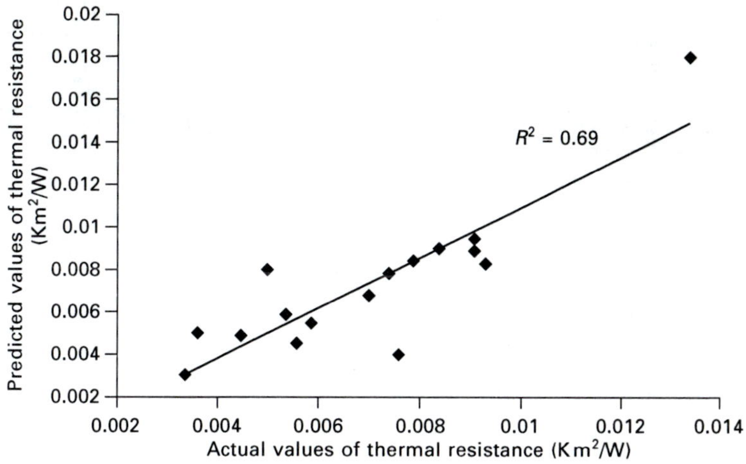

2.48 Comparison of calculated heat transfer and actual values (two neural nets) (adapted from Kothari and Bhattacharjee, 2007).

Majumdar (2011) describes a model based on the backpropagation algorithm to predict the thermal conductivity of knitted fabrics made of cotton–bamboo yarns. As input parameters, he used the structure, the thickness and the area weight of the knitted fabric, the yarn count and the bamboo fibre content. The fabric structures were encoded using –1 , 0 and +1 to represent the three different kinds. One hidden layer with three nodes was sufficient to accurately predict the thermal conductivity of the fabrics. A trend analysis showed the influence of each input parameter on the conductivity value.

2.6.13 Protective textiles

Ramaiah *et al.* (2010) investigated the possibility to predict the performance of ballistic fabrics made of Kevlar 29 using their material properties as inputs. In order to avoid overfitting, great care was put on the selection of suitable data sets representing the whole range of parameter values. The input parameters included the specific modulus of the fibre and its tenacity, the fibre density, warp and weft yarn count and impact velocity of the projectile. The output parameters were penetration depth and dissipated energy by the fabric. One hidden layer was sufficient to give excellent results as can be seen in Figs 2.49 and 2.50. Only one network was used to predict both values simultaneously. The achieved results can be used to design and develop new ballistic protection fabrics without having to carry out time-consuming and costly destructive tests.

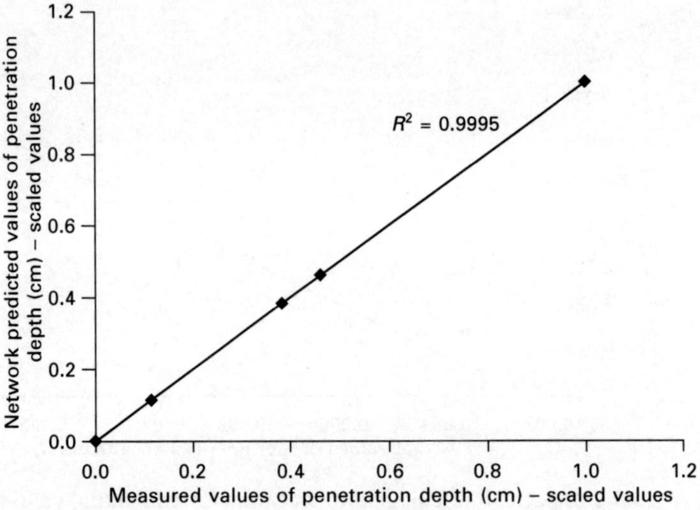

2.49 Comparison of predicted and measured values for penetration depth (adapted from Ramaiah *et al.*, 2010).

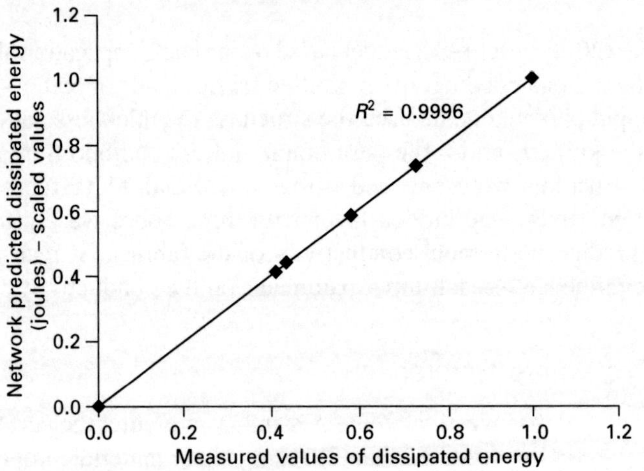

2.50 Comparison of predicted and measured values for dissipated energy (adapted from Ramaiah *et al.*, 2010).

2.6.14 Emotion-based textile indexing

A truly innovative application of neural networks in textiles was published by Kim *et al.* (2007). They used a neural network to predict the emotional response of humans on specific textile patterns. At first, they developed a system using wavelet transformation techniques to extract special features out of textile patterns. Then, they trained a network with this information to relate it to descriptive, emotional terms such as dynamic, modern and casual. A comparison of the calculated emotional response on a wide range of patterns with the actual answers of test persons showed an accuracy of about 90%. This example shows the wide range of applications of neural networks.

2.6.15 Smart carpet

Another application was developed by Infineon Technologies, Germany. A self-organizing, fault-tolerant microcontroller network based on RFID technology (radio frequency identification) was integrated into a carpet. Through connected sensor elements, it is possible, for example, to detect the position of a person on the carpet or to measure the temperature on a certain spot. Each sensor node or neuron is connected to its four nearest neighbours to which it can pass on information. Upon initializing the system, each neuron determines its position relative to the others. In case of a sensor failure, the system still works as this neuron is then disregarded by its neighbours. Hence, the carpet can be cut into pieces without impairing its functionality (Anon., 2003).

2.6.16 Other applications

Other interesting papers on the application of neural networks in textiles in a broader sense that are not described in detail here include one on the stationary solution of the geometrical shape of the ring-spinning balloon in zero air drag by Tran *et al.* (2010) and another on the modelling of an industrial plant for the biological treatment of textile wastewaters by Molga and Cherbański (2006).

2.7 Practical advice in applying neural networks

In this section, advice is given on how to tackle typical problems that are often encountered when trying to apply a neural network.

2.7.1 Significance of input parameters

For many applications, the number of possible input parameters is too huge for practical use. Therefore, it is essential to determine the most important input parameters before actually starting the training stage. In addition, connected parameters should be determined and reduced to the one parameter that can be measured or set easiest by eliminating all the others. Then, all remaining parameters should be analysed with regard to their significance on the output values. A simple method to achieve this was published by Wang *et al.* (2009). They calculate a so-called sensitivity value S_k for each input parameter by varying its value around its mean value and observing the change of the calculated output for all patterns with all weight factors unchanged according to:

$$S_k = \frac{\sum_{p=1}^{P} \sum_{i=1}^{o} (y_{ip} - \overline{y}_{ip})}{\sigma_k^2} \qquad [2.42]$$

where p is the pattern number, o is the number of network outputs, k denotes the input parameter, y_{ip} stands for the trained output and σ_k^2 represents the variance of the input perturbation. This simple method quickly shows which parameters may have a significant influence on the output parameters and should therefore considered when training the network. A typical example, taken from Wang *et al.* (2009), is shown in Fig. 2.37. It is apparent that the input values CV_H and CV_D as well as spinning draft can be neglected without affecting the prediction too much.

2.7.2 Network structure and size

The performance of a neural network during the training stage can be judged by the size of the summed difference between calculated and desired output. Since the desired output is known to the network, this overall training error is normally less than 10%. Whether a neural network has actually recognized the relations between input and output parameters or just learned 'by heart' to relate these parameters can be determined only in the so-called recall stage (see Fig. 2.3). If the network has actually learned the relations, it is called well generalized. If the network is not sufficiently complex (e.g. too low number of neurons in a too low number of hidden layers), despite a good training result, the recall accuracy will be poor (below 50%). The network is then called underfitted. A more common problem is an overfitted network. In this case, the network structure is too complex, thus detecting relations between input and output parameters that are caused by noise in the data rather than by actual and sensible differences in the respective values. A sensitivity test as described above can help to reveal an overfitting network. However, only a test with real data shows whether a neural network is well generalized, hence leading to reliable and accurate output or not. The optimum network size for one or two hidden layers can be calculated as described in Section 2.3.3.

2.7.3 Database

In many cases, the available database is taken from production data sheets and hence has often not been collected systematically. This can lead to training data sets that overrepresent certain regions of the parameter space and completely lack data sets in others. This can be avoided with data sets that are gathered using experimental design (factorial design) thus ensuring that boundary values are taken into account during the training stage (Fig. 2.51). As it is much easier for a neural network to interpolate than to extrapolate, this greatly improves the performance of a neural network since extreme values for each parameter are part of the training set.

2.7.4 Recall stage

An important question which has to be answered before beginning training is the proportion of data sets which has to be set aside during the training to be used later in the recall stage. In general, a minimum of 10% of the available data sets should not be used during training. If these data sets are then fed into the network during the recall, a vast majority of the predictions should be sufficiently accurate in order to have a degree of certainty that the network is well generalized. If the number of data sets is very small, a different approach is to use all but one of the data sets for the training and to feed the remaining in the recall stage. This has to be repeated with different nets (same structure) for all patterns (see Fig. 2.52). Hence it will be apparent whether the neural network has learned the relations between input and output and can thus predict each output correctly.

A commonly made error which can be found in many papers not cited here is to compare the calculated with the actual output and determine a

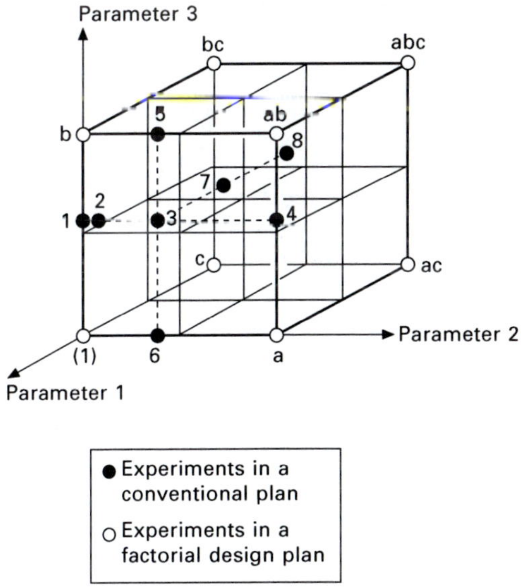

2.51 Experimental design vs. conventional experimental plans.

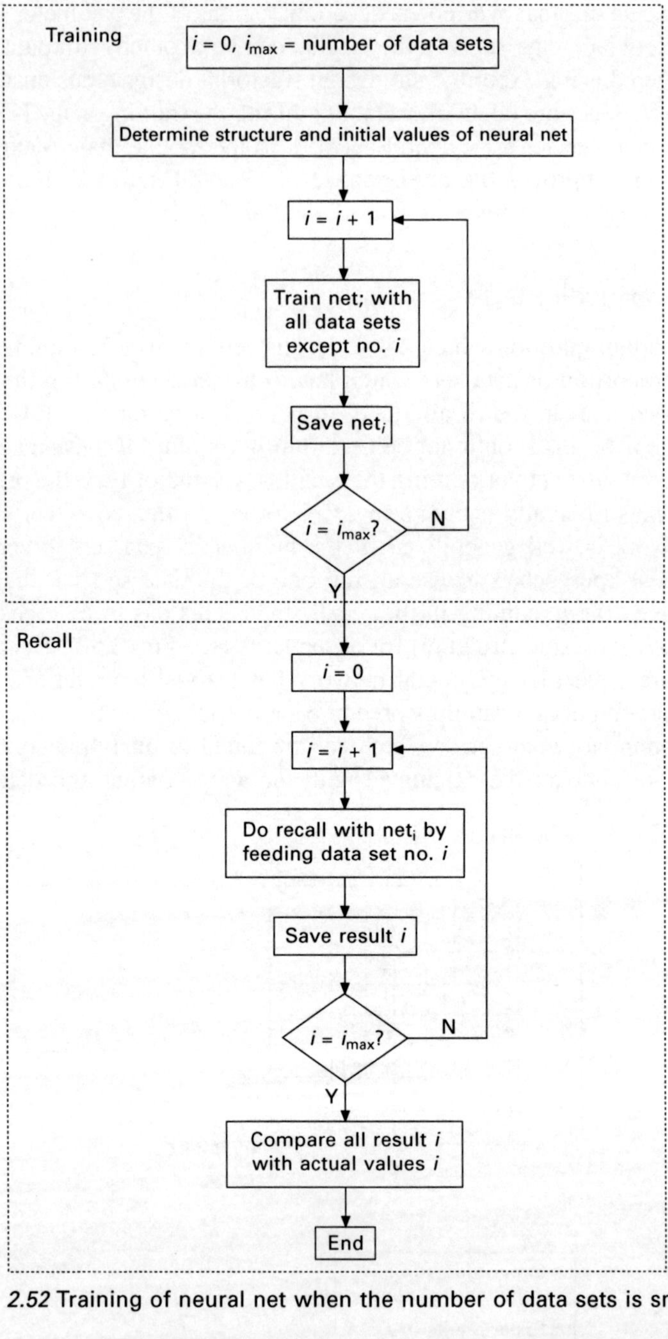

2.52 Training of neural net when the number of data sets is small.

relative error as shown in Table 2.9. At first glance one would think that the predictions of the neural network are very precise. In reality, it is obvious that the neural network has learned to relate input value 2 with the output

Table 2.9 Example for commonly made errors when assessing the quality of a neural network

Input		Output		
No. 1	No. 2	Actual	Prediction	Difference (%)
38.62	22.89	45.48	46.02	−1.19
39.71	21.76	44.32	44.21	0.25
34.56	24.87	46.43	45.99	0.95
35.80	23.84	44.40	43.84	1.26
35.71	22.46	43.76	45.56	−4.11
37.58	9.34	18.43	18.65	−1.19
38.64	10.23	18.46	18.55	−0.49
36.91	10.49	19.91	19.23	3.42
37.34	10.41	18.72	18.52	1.07
35.95	9.50	18.62	18.88	−1.40

value. The deviation of the predicted output values is in fact the same as the deviation of the actual output values. This result does not prove that the network has actually recognized any relation. This problem is related to the one described in Section 2.7.3 and could have been solved using factorial design to have a representative database.

2.8 Summary

As the examples described show, neural networks can be a powerful tool to solve a wide range of problems. Their application offers a wide range of advantages compared to other models when solving numerical problems. The main advantage of these systems is there commercial availability which allows the user to concentrate on providing data sets rather than having to design a new computer model.

But there are also a number of risks of which the user should be aware. A major drawback is the fact that it is not possible to understand in a mathematical sense the relations between input and output data which the network has found. A test run with patterns that are unknown to the network can give an indication but no certainty whether the predictions of the trained network will be accurate. A famous example of how not to train a neural network was carried out in the USA in the 1960s. A system was devised to recognize armoured vehicles even when camouflaged. The neural network was trained with photographs of landscape with and without vehicles. In the lab-recall stage, the prediction accuracy was 100%. A practice trial ended in disaster as the system failed completely. All photographs with vehicles had been taken in sunshine, all the others with overcast sky. The system had learned to perfectly distinguish between different weather conditions but not to detect camouflaged vehicles. Hence, all prospective users are strongly advised to always test their systems in practice and not only in laboratory scale before relying on the predictions and making any promises to customers.

2.9 Web references for software tools

There is a wide range of software tools and general information on the world wide web. The following links are a selection of useful websites.

Neural network software

http://www.kdnuggets.com/software/classification-neural.html
http://www.simtel.net/category/Neural-Network-Programs/207.html
http://www.dmoz.org/Computers/Artificial_Intelligence/Neural_Networks/Software/
http://www.neuronalesnetz.de/links3.html

Free neural network software

http://www.membrain-nn.de/main_en.htm
http://sourceforge.net/projects/joone/index.html
http://www.ra.cs.uni-tuebingen.de/software/JavaNNS/welcome_e.html
http://axel-arnold.de/index.php?id=68&L=2
http://www.justnn.com/
http://www.bcp.psych.ualberta.ca/~mike/Software/
http://cormactech.com/neunet/download.html

Commercial neural network software

http://www.crgraph.de/index.html
http://www.mathworks.com/products/neuralnet/
http://www.neurosolutions.com/
http://www.easynn.com/
http://www.neuroxl.com/
http://www.alyuda.com/neural-networks-software.htm
http://www.npsl1.com/

Other sources

A list of publications dealing with the application of neural networks is given here:
http://synapse.cs.byu.edu/publications.php
One of the early standard books about neural networks covering a wide range of net types and theoretical aspects is available online here:

http://page.mi.fu-berlin.de/rojas/neural/

A forum for forecasting applications using neural networks can be found here:

http://www.neural-forecasting.com/

2.10 References and further reading

Anon. (2003), 'Magic carpet ride', *Expert Systems*, Vol. 20, No. 5, 312.

Anon (2011), Literature survey. http://www.engineeringvillage2.com [accessed 17 May 2011].

Behera, B. K., Goyal, Y. (2009), 'Artificial neural network system for the design of airbag fabrics', *Textile Research Journal,* vol. 39, no. 1, 45–55.

Bo, Z. (2010) 'The prediction of warp breakage rate of weaving by considering sized yarn quality using artificial neural network theory', *2010 International Conference on Computer Design and Applications, ICCDA 2010*, vol. 2, V2526–V2529.

Chandra, J.K., Banerjee, P.K., Datta, A.K. (2010), 'Model modification and prediction of mechanical behavior of fabrics in uniaxial tension', *Journal of the Textile Institute*, vol. 101, no. 8, 699–706.

Chen, C.H., Zhang, X. (1999) 'Using neural networks for fiber content analysis', *Proceedings of the International Joint Conference on Neural Networks*, vol. 6, 3913–3916.

Cherkassky, A., Weinberg, A. (2010a), 'Objective evaluation of textile fabric appearance part 1: Basic principles, protrusion detection, and parameterization', *Textile Research Journal*, vol. 80, no. 3, 226–235.

Cherkassky, A., Weinberg, A. (2010b), 'Objective evaluation of textile fabric appearance part 2: SET Opti-grade tester, grading algorithms, and testing', *Textile Research Journal,* vol. 80, no. 2, 135–144.

Farooq, A., Cherif, C. (2008), 'Use of artificial neural networks for determining the leveling action point at the auto-leveling draw frame', *Textile Research Journal,* vol. 78, no. 6, 502–509.

Grossberg, S. (1988), 'Nonlinear neural networks: principles, mechanisms, and architectures', *Neural Networks*, vol. 1, 17–61.

Halleb, N. (2010), 'Model modification and prediction of mechanical behaviour of fabrics in uniaxial tension', *Journal of the Textile Institute*, vol. 101, no. 8, 707–715.

Hebb, D. (2002), *The Organization of Behavior: A Neuropsychological Theory*, Lawrence Erlbaum Assoc Inc. (new edition of book published in 1949).

Hecht-Nielsen, R. (1987). Counterprogagation networks. *Applied Optics*, vol. 26, no. 23, 4979–4984.

Hecht-Nielsen, R. (1990), *Neurocomputing*. Addison-Wesley, Reading, MA.

Kim N.Y., Shin Y., Kim E.Y. (2007) 'Emotion-based textile indexing using neural networks' *Proceedings of the International Symposium on Consumer Electronics, ISCE.*

Kohonen, T. (1981), Automatic formation of topological maps of patterns in a self-organizing system. *Proceedings of 2SCIA, Scand. Conf. on Image Analysis*, 214–220, Helsinki, Finland.

Kothari, V.K., Bhattacharjee, D. (2007), 'A neural network system for prediction of thermal resistance of textile fabrics', *Textile Research Journal,* vol. 77, no. 1, 4–12.

Kumar, A. (2003) 'Neural network based detection of local textile defects', *Pattern Recognition*, vol. 36, no. 7, 1645–1659.

Kuo, C.-F. J., Shih, C-Y., Ho, C-E., Peng, K-C. (2010), 'Application of computer vision in the automatic identification and classification of woven fabric weave patterns', *Textile Research Journal*, vol. 80, no. 20, 2144–2157.

Lin, J.-J. (2007), 'Prediction of yarn shrinkage using neural nets', *Textile Research Journal*, vol. 77, no. 5, 336–342.

Liu, J., Zuo, B., Zeng, X., Vroman, P., Rabenasolo, B. (2010), 'Nonwoven uniformity identification using wavelet texture analysis and LVQ neural network. *Expert Systems with Applications*, vol. 37, no. 3, 2241–2246.

Liu, S.-G., Qu, P.-G. (2008), 'Inspection of fabric defects based on wavelet Analysis and BP neural network', *Proceedings of the 2008 International Conference on Wavelet Analysis and Pattern Recognition*, ICWAPR, vol. 1, 232–236.

Majumdar, A. (2011), 'Modelling of thermal conductivity of knitted fabrics made of cotton-bamboo yarns using artificial neural network', *Journal of the Textile Institute,* vol. 102, no. 9, 752–762.

McCulloch, W., Pitts, W. (1945), 'A heterarchy of values determined by the topology of nervous nets', *Bulletin of Mathematical Biophysics,* vol. 7, 89–93.

Minsky, M., Papert, S. (1969), *Perceptrons*, MIT Press, Cambridge.

Molga, E., Cherbański, R. (2006), 'Modelling of an industrial full-scale plant for biological treatment of textile wastewaters: Application of neural networks', *American Chemical Society,* vol. 45, no. 3, 1039–1046.

Murrells, C.M., Tao, X.M., Xu, B.G., Cheng, K.P.S. (2009), 'An artificial neural network model for the prediction of spirality of fully relaxed single jersey fabrics', *Textile Research Journal,* vol. 79, no. 3, p. 227–234.

Pan, R., Gao, W., Liu, J., Wang, H. (2011), 'Automatic recognition of woven fabric based in image processing and BP neural network', *Journal of the Textile Institute*, vol. 102, no. 1, 19–30.

Ramaiah, G.B., Chennaiah, R.Y., Satyanarayanarao, G.K. (2010), 'Investigation and modeling on protective textiles using artificial neural networks for defense applications', *Materials Science and Engineering B,* vol. 168, no. 1, 100–105.

Ramakers, R. (2006), 'Systematical method for development of sensor-based quality monitoring of man-made fiber processing – drawing, winding, texturing', *Proceedings of the International Man-Made Fibres Congress*, 45, Dornbirn.

Rosenblatt, F. (1958), 'The perceptron: a probabilistic model for information storage and organization in the brain', *Psychological Review*, vol. 65, 386–408.

Rumelhart, D.E., Hinton, G.E., Williams, R.J. (1986), 'Learning representations by back-propagating errors', *Nature*, vol. 323, 533–536.

Shi, X.-J., Yu, W-D. (2008) 'Identification for animal fibers with artificial neural network', *Proceedings of the 2008 International Conference on Wavelet Analysis and Pattern Recognition, ICWAPR*, vol. 1, 227–231.

Schöneburg, E., Hansen, N., Gawelczyk, A. (1990), *Neuronale Netzwerke*, Verlag Markt & Technik, München.

Strauf Amabile, M., Gries, T. (2005), 'Stand marks in warp knit fabrics – Analysis of influencing parameters', *Melliand Textilberichte*, vol. 86, no. 4, E49–E50.

Su, T., Kung, F., Kuo, Y. (2008), 'Application of back-propagation neural network fuzzy clustering in textile texture automatic recognition system', *Proceedings of the 2008 International Conference on Wavelet Analysis and Pattern Recognition*, ICWAPR, vol. 1, 46–49.

Tran, C-D., Phillips, D. G., Fraser, W.B. (2010), 'Stationary solution of the ring-spinning balloon in zero air drag using a RBFN based mesh-free method', *Journal of the Textile Institute*, vol. 101, no. 2, 101–110.

Veit, D. (1998), 'Optimization of a false-twist texturing-machine with neural networks and evolutionary algorithms', *Proceedings of the International Man-Made Fibres Congress*, 37, 1–13.

Veit, D. (1999), *Einstellung von Falschdraht-Texturiermaschinen mit Hilfe der Evolutionsstrategie und neuronaler Netze*, Shaker Verlag, Aachen, Germany.

Veit, D., Hormes, I., Bergmann, J., Wulfhorst, B. (1996a), Image processing as a tool to improve machine performance and process control', *International Journal of Clothing Science and Technology*, vol. 8, no. 1/2, 66–72.

Veit, D., Maetschke, O., Wulfhorst, B. (1996b), *Einsatz der digitalen Bildverarbeitung zur Optimierung des Öffnungs- und Reinigungsprozesses von Baumwolle*, Institut für Textiltechnik der RWTH Aachen, Aachen, Germany.

Veit, D., Batista de Sousa, P., Wulfhorst, B. (1998), 'Application of a neural network in the false-twist texturing process', *Chemical Fibers International*, vol. 48, no. 2, 155.

Ververidis, D., Kotropoulos, C. (2006), 'Emotional speech recognition: Resources, features, and methods', *Speech Communication*, vol. 48, no. 9, 1162–1181.

Wang, X., Khan, Z., Lim, A., Wang, L., Beltran, R. (2009), 'An artificial neural network-based hairiness prediction model for worsted wool yarns', *Textile Research Journal*, vol. 79, no. 8, 714–720.

Wei, G., Piyan, L., Hai, Zh., Zhan, M. (2010), 'The research and application of image recognition based on improved BP algorithm', *Proceedings – 3rd International Conference on Intelligent Networks and Intelligent Systems*, ICINIS 2010, 80–83.

Werbos, P.J. (1974), 'Beyond regression: new tools for prediction and analysis in the behavioral sciences', PhD thesis, Harvard University, Cambridge.

Yao, G., Guo, J., Zhou, Y. (2005) 'Predicting the warp breakage rate in weaving by neural network techniques', *Textile Research Journal*, vol. 75, no. 3, 274–278.

Xin, B., Hu, J., Liu, X. (2010), 'Neural network modeling of polar fleece fabric appearance evaluation', *Proceedings – 2010 6th International Conference on Natural Computation, ICNC 2010*, vol. 1, 417–421.

Yuen, C.W.M., Wong, W.K., Qian, S.Q., Fan, D.D., Chan, L.K., Fung, E.H.K. (2009), 'Fabric stitching inspection using segmented window technique and BP neural network', *Textile Research Journal*, vol. 79, no. 1, 24–35.

Yin, Y., Zhang, K., Lu, W B. (2009), 'Textile flaw classification by wavelet reconstruction and BP neural network', *Lecture Notes in Computer Science*, vol. 5552 LNCS, no. 2, 694–701.

Zell, A. (1997), *Simulation neuronaler Netze*, Oldenbourg Wissenschaftsverlag.

3
Evolutionary methods and their application to textile technology

D. VEIT, RWTH Aachen University, Germany

Abstract: Evolutionary algorithms are a powerful tool for solving optimization problems. The method is robust and can solve problems for a wide range of applications. It is especially useful for tackling problems that are impossible or hard to solve using other techniques. Evolutionary algorithms try to mimic the biological principles of reproduction and selection. This chapter gives an overview of the various methods and concludes with selected examples of successful applications of evolutionary methods to solve textile-related problems.

Key words: evolutionary algorithms, evolution strategies, genetic algorithms, textile technology.

3.1 Introduction

Evolutionary algorithms, evolution strategies and genetic algorithms are often used as synonyms in different contexts although there are distinctive differences between them. Here, we use the terms evolutionary algorithm and evolution strategy to describe the basic principles, whereas the term genetic algorithm (GA) is used to describe a particular method applying these principles which was developed by J. Holland in the mid-1970s (Holland, 1992).

A problem which is often encountered in engineering is the determination of an optimum for a set of complex equations. The experimenter in Fig. 3.1 has defined a proving ground and now searches for the optimum, in this example represented by the highest point of the surface. There are several different approaches to solve this problem. They work on a global or a local scale and they can be deterministic or stochastic.

A global deterministic search will eventually in all cases be successful but may be extremely time-consuming (Fig. 3.2). A global stochastic search will lead in almost all cases to a viable solution, normally near the optimum (Fig. 3.3) and is much quicker than the former approach. A local deterministic search with a gradient process can also produce a useful solution but can also end in a local optimum far away from the global optimum (Fig. 3.4). A local stochastic search has a greater chance to succeed with a slightly higher computing time. This strategy is based on the so-called evolutionary algorithms which are the main subject of this chapter (Fig. 3.5).

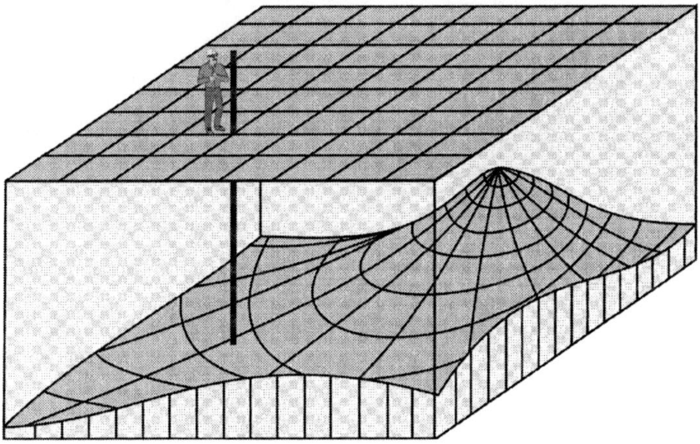

3.1 Experimenter searching for the optimum (adapted from Rechenberg, 2003).

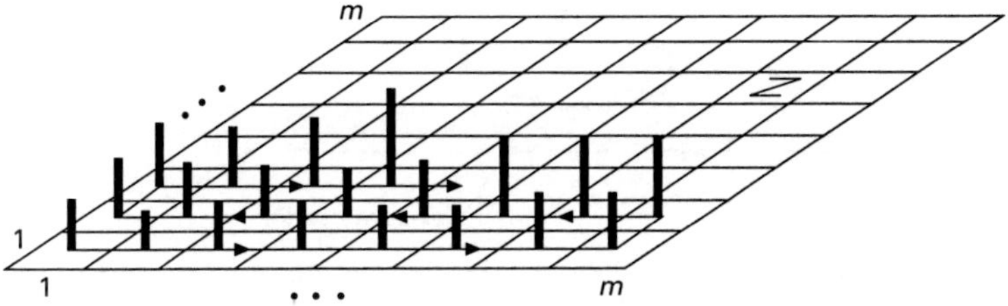

3.2 Deterministic search (adapted from Rechenberg, 2003).

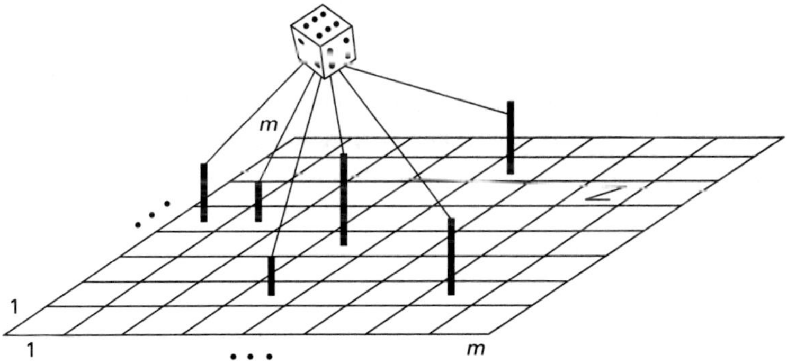

3.3 Global stochastic search (adapted from Rechenberg, 2003).

3.2 Biological background

Evolutionary algorithms follow the principles laid down by Darwin (1859) and Wallace (1871). They described the biological processes omnipresent

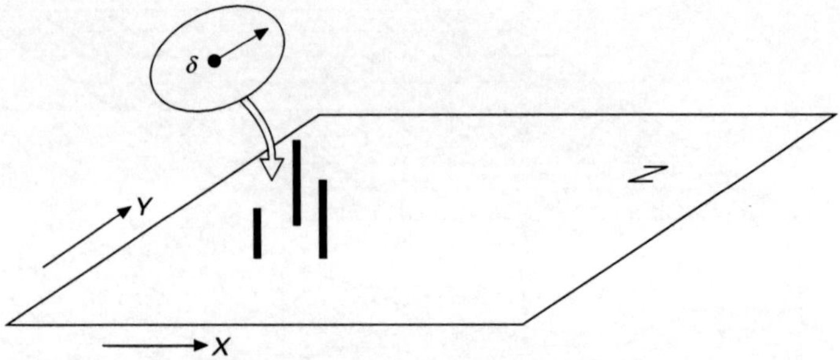

3.4 Local deterministic search (adapted from Rechenberg, 2003).

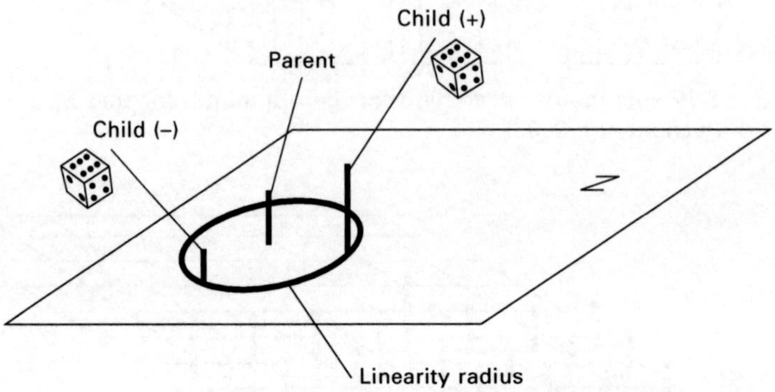

3.5 Local stochastic search with evolutionary algorithms (adapted from Rechenberg, 2003).

in nature where individuals within a population compete for resources (e.g. food). Individuals better adapted to their environment than others have a higher chance of reproducing and thus passing on the genes that made them fitter to their offspring. Within a few generations, the genetic trait favouring a certain behaviour or any other evolutionary advantage spreads within the population. This also works even if several traits are required to actually produce an advantage. This process was called by Darwin 'the survival of the fittest' (Darwin, 1859).

3.2.1 Basics of genetics

The hereditary traits of a living being are encoded in genes that are copied during reproduction and passed on to the individuals of the next generation. Owing to mutation, different variants (alleles) of these genes are generated. This leads either to a slight change of a certain trait (e.g. red instead of yellow petals of a flower) or to completely new traits (e.g. resistance against vermin). This range of variations together with a recombination of the genes gradually leads to differences between the individuals within a population despite their being the offspring of the same parents. The process of a changing relative frequency of the alleles, which leads to traits becoming more common or scarce within a population, is called evolution. This change occurs due to natural selection or by accident. Natural selection is based on different survival and reproduction ratios of the individuals owing to their different traits. The so-called genetic drift is caused by random mutations of the individuals, leading to a higher percentage of a certain trait within the population.

The principles of inheritance of traits (genetics) were discovered by Mendel (1866). He assumed that 'the parental factors, combined in the zygote, do not lose their identity, but are recombined in the next generation'. The respective genetic information is binary encoded and can be recombined during reproduction. Normally, the binary encoded genes are fully passed on to the next generation. A true mixture therefore cannot occur in nature. The offspring always resembles the parents in a specific trait. For example, combining seeds of plants with red and yellow petals does not result in orange petals but again in red or yellow petals. Figure 3.6 shows this principle for three subsequent generations. Parents with diploid genes AA and aa generate offspring with genes Aa. When this offspring is recombined, the second and subsequent generations consist of individuals with diploid genes AA, Aa and aa with differing frequency. In many cases, certain traits are generated by different genes. The combination of seeds of plants with red and yellow petals can therefore nevertheless result in plants with orange petals.

Mendel's research results were not recognized by the scientific community until the early 1900s when they were rediscovered by von Tschermak (1900), de Vries, (1906) and Correns, (1912). This later led to the establishment of genetics as a science. De Vries enunciated the theory of mutation stating that

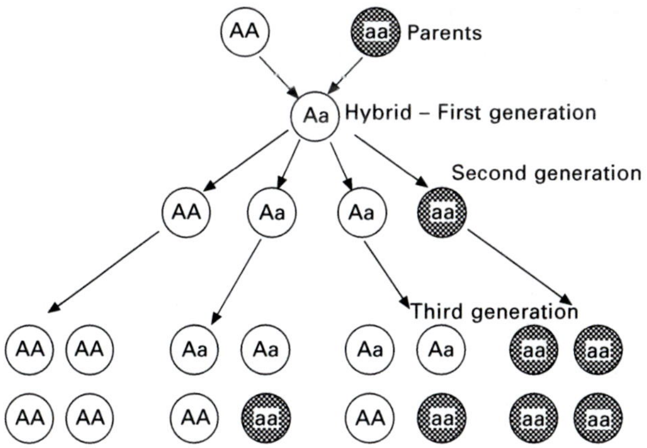

3.6 Principle of inheritance.

evolution is driven by randomly occurring variations of the genetic code of organisms. This, together with the theory of natural selection, today forms the basis of modern evolution theory.

3.2.2 Evolutionary theory

The development of the classical evolutionary theory is often attributed to Charles Darwin and his major publication '*On the origin of species by means of natural selection*' (Darwin, 1859). However, there were also other scientists, e.g. Alfred R. Wallace and Jean-Baptiste de Lamarck, who made important contributions to the theory of evolution (Wallace, 1871; de Lamarck, 1909). Wallace's approach was similar to Darwin's whereas Lamarck assumed that in the course of many generations, all animals had a tendency to develop to higher levels which implies that evolution is a constantly progressing process.

Darwin postulated:

> As many more individuals of each species are born than can possibly survive, and as, consequently, there is a frequently recurring struggle for existence, it follows that any being, if it vary however slightly in any manner profitable to itself, under the complex and sometimes varying conditions of life, will have a better chance of surviving, and thus be naturally selected. From the strong principle of inheritance, any selected variety will tend to propagate its new and modified form.

These principles are called 'survival of the fittest' and 'natural selection'. The computer model described below which is commonly used today is based upon them.

In natural evolution, the binary encoded information is changed from generation to generation by mutation and recombination. During recombination, gene sequences consisting of bases are exchanged between adjacent DNA strands (crossing-over). This leads to a change in the order of the bases and thereby can alter traits. Recombining gene sequences by crossing-over can hence be advantageous (repair of damaged gene parts) but it can also lead to unwanted effects.

3.2.3 Chemical basics of inheritance

The discovery of the structure of deoxyribonucleic acid (DNA) as the carrier of genetic information by James Watson, Francis Crick, Maurice Wilkins and Rosalind Franklin led to a deeper understanding of the chemical principles behind inheritance (Watson and Crick, 1953). They found that genetic information is stored in the so-called bases (adenine, thymine, guanine, cytosine) that always come in pairs as shown in Fig. 3.7.

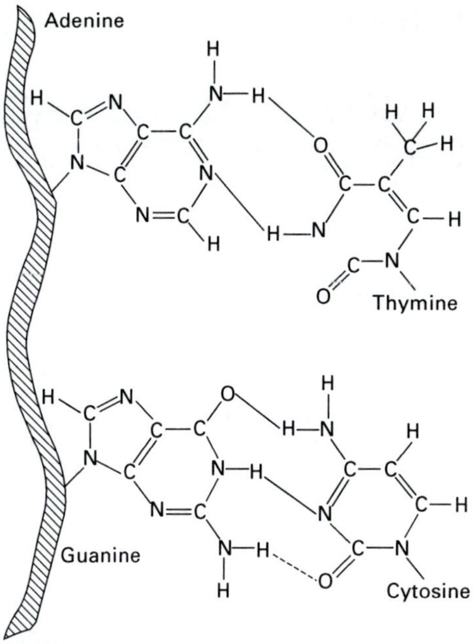

3.7 Section of DNA with base pairs.

The genetic information is stored in the sequence of the bases and can therefore be read and copied very easily. If an error occurs during reading or copying, the coded information is changed. This is called a mutation. Figure 3.8 shows two common ways to create a mutation: exchanging a base for another (top) and deleting a base thus changing also the order of all subsequent base triplets (bottom). This can cause havoc in a gene as all information behind the deleted base is completely altered.

3.2.4 Genotype and phenotype

Friedrich L. A. Weismann, a German evolutionary biologist, introduced two new terms into evolutionary science: genotype and phenotype (Weismann, 1885). The genotype comprises the genetic makeup of an individual responsible for its traits. The phenotype represents the observable characteristics of an individual, also called morphology. Changing the genotype leads to a different phenotype which in turn may lead to an individual which is better adapted to its environment. This increases its changes to successfully reproduce which in turn leads to a propagation of this genotype. The benefit of a genotypic mutation of a single individual or a group of individuals does not become apparent before the interaction with the environment.

Exchange of bases:

Before: AGT-TCG-ACT-ATT-CGA-AGT
After: AGG-TCG-ACT-ATT-CGA-AGT

Deletion of one base:

Before: AGT-TCG-ACT-ATT-CGA-AGT
After: AGT-CGA-CTA-TTC-GAA-GT...

3.8 Two methods to generate a mutation.

In contrast to Lamarck's theory that phenotypes change themselves through aiming to achieve an optimum (e.g. lengthening of a giraffe's neck to reach leaves higher up), changes in nature normally occur without purpose. By natural selection, non-purposeful alterations of genotypes can lead to phenotypes that are better adapted to a changed environment, thus indirectly optimizing the phenotype. An alteration of the genotype can of course also lead to non-viable solutions (individuals). If they survive, they decrease the quality of the genetic pool in the short run. But these individuals can turn out to be much better adapted to a changing environment than their relatives which previously had a competitive edge thus giving the whole population a better chance of survival. Figure 3.9 illustrates that: a small population (small genetic variation between individuals) can lead to a dead end if the environment changes (top); a larger population (greater genetic variation between individuals) may survive even a dramatic change of the environmental conditions because it contains viable solutions that were previously not relevant (bottom).

A typical example from nature is the peppered moth *Biston betularia*. Its light-coloured kind was perfectly adapted to birch trees, while its melanic kind was easy prey to birds. With the beginning of industrialization and huge amounts of coal being burned, birch trees in England turned black, which gave the black kind suddenly the competitive edge. So, evolution in nature is neither purely accidental (genotypic changes) nor purely goal-oriented (phenotype alteration). Only the combination of both principles leads to evolutionary progress.

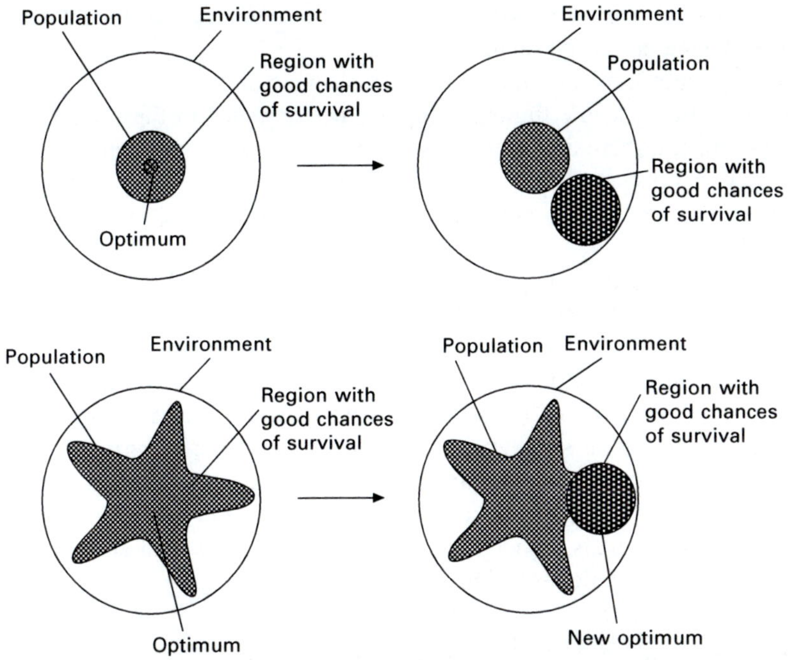

3.9 Effect of a changing environment on the fitness of individuals.

3.3 Mathematical models of evolution strategy

At first sight, it seems to be a contradiction to use the term 'strategy' in the context of 'evolution'. As has been pointed out, the power of natural evolution lies in the fact that by accidental, non-strategic changes of parameters (genes), a system (population of individuals) can be constantly improved. However, the term 'evolution strategy' is commonly used in the scientific world of optimization problems and therefore also used here. When designing a mathematical model to emulate nature, there normally is a strategy behind it, so the term is justified in this context. Applying these principles, adaptive, self-optimizing systems can be designed to determine a solution for a given system of equations.

When imitating biological evolution in a computer-based model, it is not the main goal to truly copy the processes occurring in nature. The main focus is on applying the same principles that are the basis of natural evolution or at least to come close. Depending on the problem to be solved, the model resembles more or less nature's example. Each individual in a mathematical model represents a possible solution to a given problem. The parameters to-be-optimized are normally pooled in a vector consisting of binary or real numbers. Depending on how accurately it solves the problem, the vector is assigned a so-called fitness score. This is equivalent to real organisms in nature competing for resources where some are better adapted thus having a higher chance of getting them, e.g. food.

The first mathematical models of this kind were developed in the 1960s and 1970s. Owing to the limited computer performance in these days and other difficulties, there are only a few examples of successful applications, which were not used on a broad scale until recently. This is obvious when looking at the number of publications in that field over the last 20 years (Fig. 3.10).

In general, two different approaches can be distinguished: John H. Holland of University of Michigan, Ann Arbor, USA, developed the so-called genetic algorithms (GA) that describe the heredity transmission in nature in a mathematical model using binary vectors (Holland, 1992). Ingo Rechenberg and Hans-Paul Schwefel of Technical University Berlin, Germany, invented the so-called evolution strategy which is based on the same principles of 'survival of the fittest' and 'natural selection' (Rechenberg, 1974, 1994; Schwefel, 1995). They do not use binary encoding but real numbers to describe the individuals, hence leaving the path of nature but simplifying the approach as a whole, especially for engineers.

3.3.1 Basics

Before applying the principles of natural evolution using a mathematical model, certain terms need to be introduced:

- Environment E: E stands for the environment with its boundary conditions in which the system in question is located, or, biologically spoken, in which the individuals live.

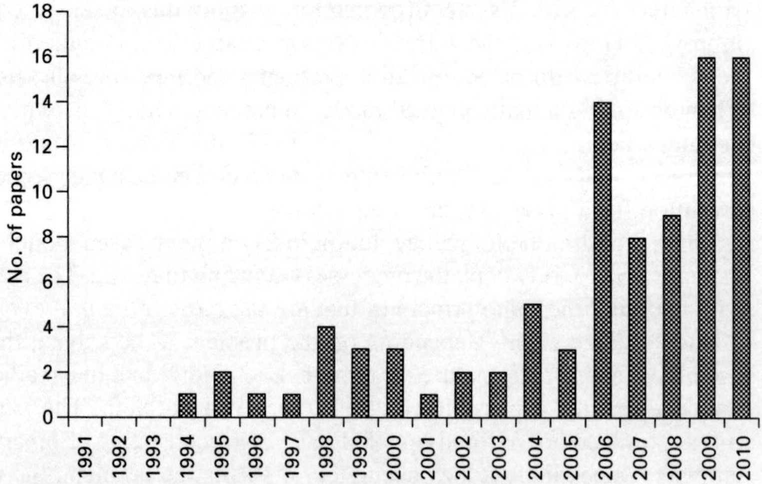

3.10 Published papers in evolutionary methods.

- Individuals k: A limited number of genotypes whose phenotypical interpretation $\pi(k)$ interacts with the environment.
- Structure-altering operators o: They represent a limited number of operators by which the genotypes of the individuals can be altered.
- Fitness score and benchmark function B: The degree of adaptation of an individual to its environment or to certain boundary conditions can mathematically be described by a fitness score f. The better adapted an individual is, the higher its fitness score. For each phenotype π, the benchmark function B calculates its fitness score f according to $f = B(\pi)$. The assessment of the fitness is based on the response R of the environment E on the phenotype π of the individual.
- Adaptation algorithm A: This algorithm modifies the genotype k of an individual through operators o as a reaction to the success of the phenotype π in the environment.

Starting with the genotype $k(t)$ of the individual at point t, the phenotype $\pi(k(t))$ is derived. The interaction of the phenotype π with the environment E results in a reaction $R(t)$ of the environment. In the next step, based on this reaction, the fitness value f of the phenotype is calculated applying the benchmark function $B(\pi)$. After the selection of the fittest individuals according to f, the adaptation algorithm A changes the genotype $k(t)$ and creates a new individual $k(t + 1) = o(k(t))$.

Figure 3.11 shows three subsequent steps of this mathematical model. It is important to note that the increment of the evolutionary changes in each step need not necessarily be constant. Moreover, the step size can be

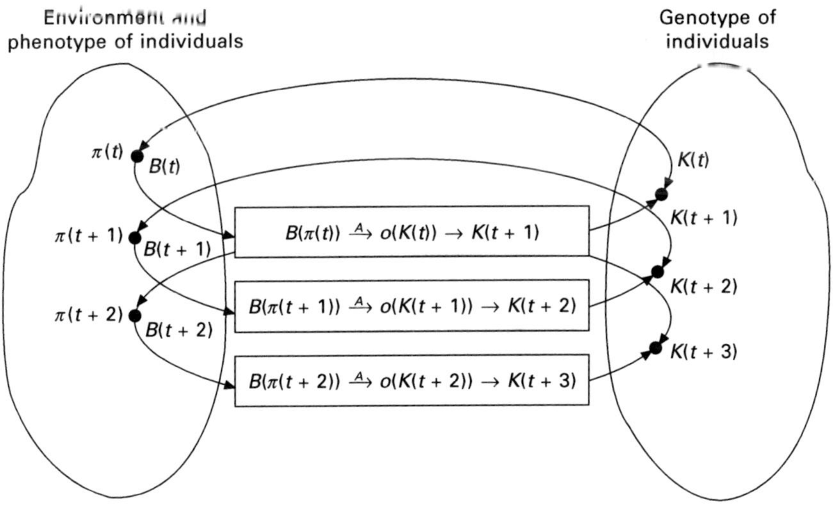

3.11 Principle of three evolution steps.

influenced on purpose by adjusting the operators *o*. Out of the genetic pool $G(t)$ at point *t*, the adaptation algorithm *A* creates a new (modified) genetic pool $G(t + 1)$. This in turn leads to a new population of individuals $p(t + 1)$ where the population $p(t + 1)$ is better adapted to the environment (or at least not worse) than the population $p(t)$. This adaptation process does not stop until a previously set termination criterion $T(p)$ is achieved. This is the case when a certain number of evolutionary cycles is completed or at least one individual has a fitness score *f* which reaches a pre-set value. This leads to the scheme of the evolutionary algorithm, as shown in Fig. 3.12. For the initialization of the genotypes, either a random combination of numbers or an already viable solution can be used to create the first generation of the individuals (vectors).

3.3.2 Evolution strategies

The principles as described above form the basis for all successfully applied evolution algorithms or strategies. In order to clearly characterize the different strategies, Rechenberg and others devised a commonly used notation which is described in this section (Rechenberg, 1974) and illustrated in Fig. 3.13.

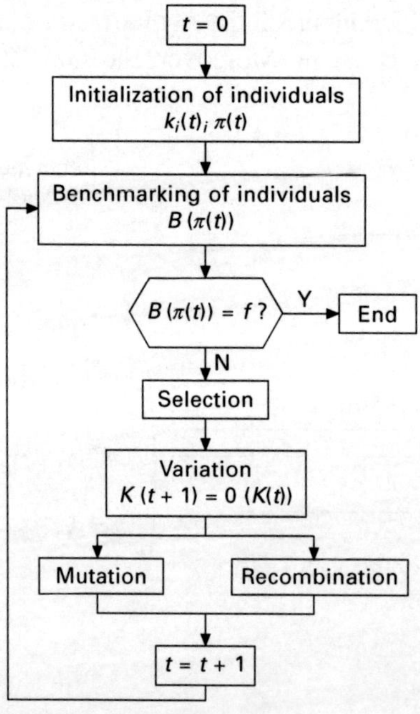

3.12 Principle of the evolutionary algorithm.

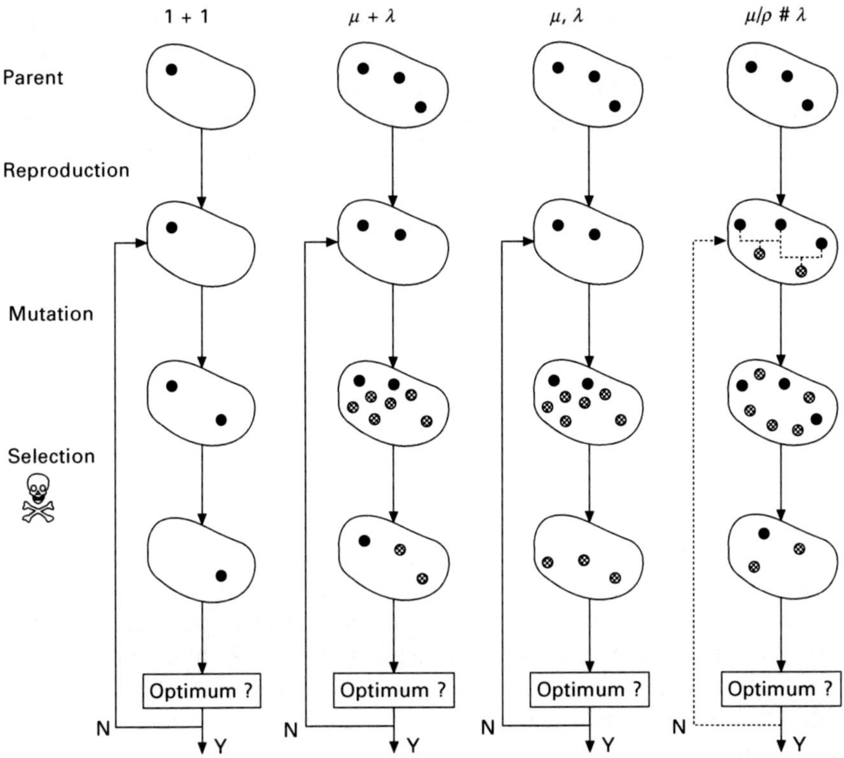

3.13 Evolution strategies according to Rechenberg (1974) and Schwefel (1995).

(1 + 1) evolution strategy

This is the simplest version of an evolution strategy. In the first step, a single individual is created and a copy of it is reproduced. Both are vectors consisting of real numbers (see above). In the next step, the copied individual is slightly changed (mutation) by adding small positive or negative values to the vector's components if it is encoded in real numbers. If it is binary encoded, a small number of 0 are replaced by 1 and vice versa. Both individuals are assessed using the benchmark function B. The better an individual meets the set criteria for a viable solution, the higher its fitness value f. The principle of survival of the fittest determines which individual is promoted to the next round (selection), where an exact copy of it is produced again so that the next evolutionary cycle can start. If both individuals, original and copy, have the same fitness value f, one is chosen at random and the other one is eliminated from the gene pool.

These steps are repeated until an acceptable solution is found. This is a commonly used approach if it is unclear whether the best individual at the end of the evolutionary cycles, the best solution, is a global or only a local optimum. Then, in most cases, it is absolutely acceptable if the solution is sufficiently near the optimum where 'sufficiently near' is a matter of perspective. Other common abortion criteria are the completion of a preset number of cycles (generations) or the maximum computing time for the simulation as a whole. This kind of evolution strategy does not have a population of individuals in the true sense but gradually changes the genotype of a single individual by mutation.

(μ + λ) evolution strategy

This approach is a generalization version of the (1 + 1) evolution strategy where not only a single individual but a whole population is created. μ denotes the number of parental individuals, λ stands for the number of children created in each cycle or generation, with $\lambda \geq \mu \geq 1$. The single steps are similar in principle to the (1 + 1) evolution strategy. In the first step, λ individuals from the parent individuals are selected with equal probability and hence λ children individuals are created which are copies of their respective parents. A repeated selection of the same parent individual is possible and even necessary in the case of $\lambda > \mu$. In the second step, the λ children are altered by mutation and assessed together with all parent individuals. The best μ individuals are then selected as parents for the next generation.

This strategy ensures that the quality of the best individual of the population after each cycle cannot decrease since the best individual always survives even if it has not been selected to create a child individual (Fig. 3.13). Therefore, the (μ + λ) strategy often ends in a local optimum with most individuals close to it.

(μ, λ) evolution strategy

The (μ, λ) strategy overcomes this problem. After λ individuals have been selected for reproduction in the first step, all parent individuals are deleted. Then, the λ children are assessed according to their fitness value f and μ individuals are selected as parents for the subsequent generation. This approach is closer to nature as there are no immortal individuals thus making it easier for the algorithm to leave a good, but still local, optimum. The progress of the simulation over time can be assessed using a benchmark function which calculates the fitness value f of the best individual after each cycle (generation).

Figure 3.14 shows a typical example for the ($\mu + \lambda$) and the (μ, λ) strategies over time. It is apparent that the quality of the best individual oscillates for the (μ, λ) strategy since parent individuals that have a better fitness compared to their children are nevertheless deleted. In contrast, with the ($\mu + \lambda$) strategy,

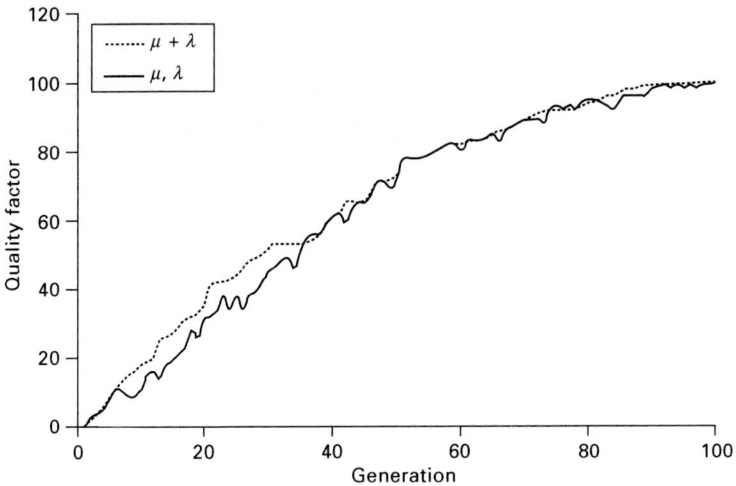

3.14 Development of best fitness score for (μ + λ) and (μ, λ) strategies.

the fitness f of the best individual of each generation can never decrease. In the end, the final result is very similar for both approaches, with the (μ, λ) strategy probably leading to a more stable result as it normally avoids local optima. The size of the oscillation of the quality function depends on the mutation increment. If this value is small, normally the deviation of the quality function is also small.

In many cases, an evolution strategy is carried out using several independent populations of individuals. After each step, only the 'best' population survives whereas the others are deleted. Each surviving population creates daughter populations which are then subjected to mutation and benchmarked after the next cycle where only the best population survives and so forth, as shown in Fig. 3.15. Both the ($\mu + \lambda$) and the (μ, λ) strategies can be combined to the (μ # λ) strategy if the selection process is not important. The symbol # then denotes either '+' or ';'.

($\mu/\rho\#\lambda$) evolution strategy

In contrast to the approaches as described above, this strategy applies the concept of sexual reproduction. The principle is very similar to those above, the main difference being that children are not produced by copying single parent individuals but by combining the genotypes of ρ parent individuals with $\rho >$ 1 (Fig. 3.13). Thus, a whole group of parents takes part in the reproduction process of a single child. The parents are selected either at random (roulette wheel) or depending on their fitness value f where fitter parents have a greater chance to reproduce than the others (see next page).

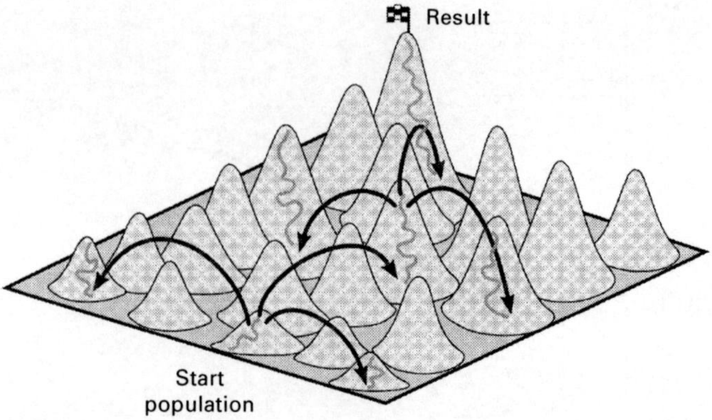

3.15 Evolution strategy with independent populations (adapted from Rechenberg, 2003).

3.3.3 Recombination

Since each genotype vector consists of numbers, there is a wide range of possibilities to generate children out of the selected ρ parent individuals. A commonly used approach is the roulette wheel where the size of the section that each parent individual occupies is directly related to its relative fitness value. Hence, the chance of fitter individuals to be selected as parents when the ball rolls is higher than for less adapted individuals. Nevertheless, even badly suited individuals have a chance to reproduce and to pass on their genes, which in the end may be crucial to find the optimal solution. This especially applies to multi-scale problems where not only one single gene but many genes determine the overall fitness of a solution. As shown in Fig. 3.16, we have:

$$\varphi_{\text{rel1}} = \frac{f(P_1)}{\sum_i f(P_i)} \cdot 2\pi \qquad [3.1]$$

with $f(P_i)$ = fitness value of phenotype of individual i.

Also, an element of chance can be introduced by selecting some of the parent individuals at random from the whole population while disregarding their respective fitness values. This can increase the convergence speed of the whole population, when, for example, only one or two individuals of a population of 30 are generated this way.

In Fig. 3.17a, the single genes of both parents are combined completely thus generating a child consisting of genes of both parents. In this example, the genes (in order from the top) 1, 4 and 7 are taken from parent individual 1 whereas the others come from parent individual 2. In Fig. 3.17b, the genes of the child consist in equal parts of the parental genes according to:

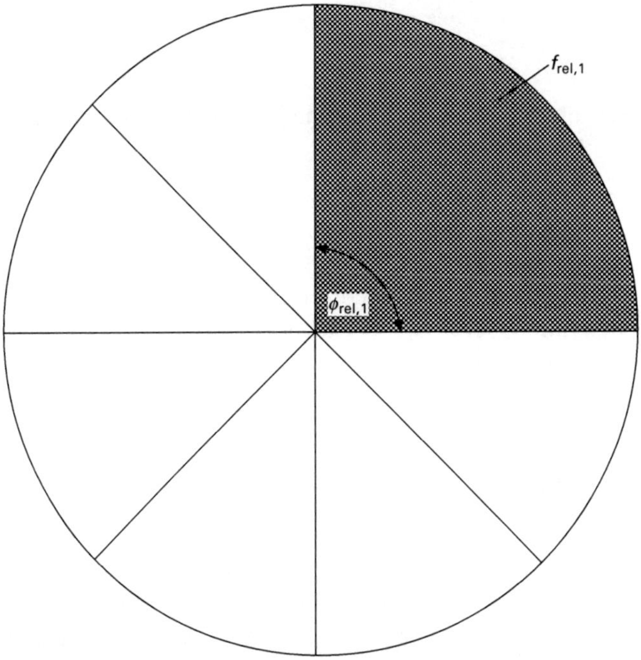

3.16 Example for roulette wheel method to select parent individuals.

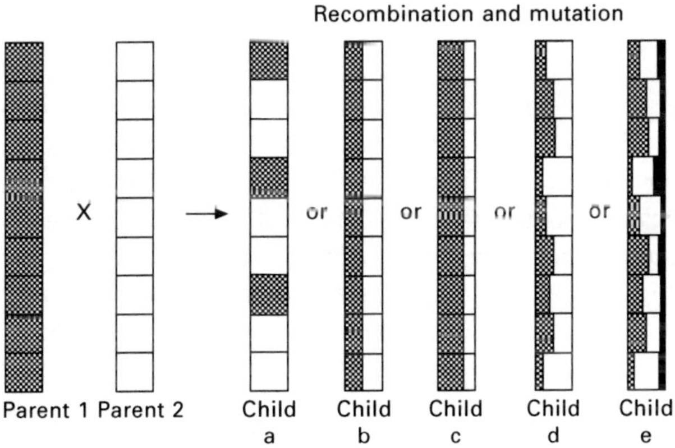

3.17 Principles of generating children by recombination.

$$\text{gene}(i)_{\text{child}} = \frac{\text{gene}(i)_{\text{parent1}} + \text{gene}(i)_{\text{parent2}}}{2} \qquad [3.2]$$

with i = position of gene in vector.

Figure 3.17c shows the creation of child genes with different percentages for each parent:

$$\text{genes}_{\text{child}} = \rho_1 \cdot \text{genes}_{\text{parent1}} + \rho_2 \cdot \text{genes}_{\text{parent2}} \tag{3.3}$$

with $\rho_1 + \rho_2 = 1.$ [3.4]

If the percentage of each parental gene passed on to the child differs from gene to gene, then we have

$$\text{gene}(i)_{\text{child}} = \rho_{i,1} \cdot \text{gene}(i)_{\text{parent1}} + \rho_{i,2} \cdot \text{gene}(i)_{\text{parent2}} \tag{3.5}$$

with $\rho_{i,1} + \rho_{i,2} = 1$ [3.6]

and ρ_i = recombination factors, according to Fig. 3.17d.

The recombination factors ρ_i themselves are normally calculated by taking into account the fitness scores f_1 and f_2 of the parent individuals according to

$$\rho_i = \frac{f_i}{f_1 + f_2} \tag{3.7}$$

An alternative is to set the recombination factors at random though this often increases the number of generations needed to reach the optimum but may reduce the number of mutations required (see below). In any case, this approach where information of single genes (numbers) is actually mixed leaves the path of nature. Nevertheless, in many cases it works better than the classic approach where genes are only exchanged.

3.3.4 Mutation

Mutation plays an important part in evolutionary processes. Without mutation, in a mathematical model, children could never be created outside the space spanned by the parent individuals. When adding a small percentage of mutation to a child, this is possible. The mutation of single genes in nature is a process that normally takes place in small steps. In other words, children usually resemble their parents. Large, erratic alterations are very rare. Thus, in a mathematical model of evolution, the so-called mutation increment is kept to small values. Often, normally or uniformly distributed values are used.

Figure 3.18 depicts a reproduction process similar to that as shown in Fig. 3.17d, with the addition of a mutation component. Hence

$$\text{genes}_{\text{child}} = \text{genes}_{\text{parent},j} \pm \vartheta \cdot \text{genes}_{\text{parent},j} \tag{3.8}$$

and

$$\text{genes}_{\text{child}} = \text{genes}_{\text{parent},j} \pm [\vartheta_1 \cdot \text{gene}(1), \ldots, \vartheta_{i_{\max}} \cdot \text{gene}(i_{\max})]_{\text{parent},j} \tag{3.9}$$

The parameter ϑ is called mutation increment. Its value is normally in the range of $\vartheta = [0; 0.2]$, but greater values are not uncommon. Especially at

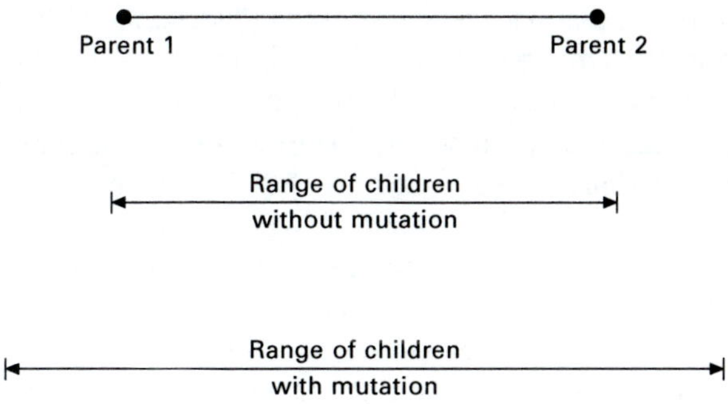

3.18 Generating children by mutation.

the beginning of the reproduction process, this can lead to quick progress towards the global optimum. Theoretical considerations by Rechenberg (2003) and Schwefel (1995) show that on average, about one-fifth of all mutations should lead to an improvement of the population fitness. If this value is exceeded, the mutation increment should be increased; if it is below that value, the mutation increment needs to be decreased.

In self-regulating mutative increment control, the mutation increment needs to be adjusted depending on the specific problem at hand. Since this is very time-consuming, a better approach is to automatically optimize it by applying evolution strategy. Hence, the mutation increment becomes itself part of either each vector of the individuals or of the population vector as a whole and is thus optimized together with the population consisting of all individuals.

3.3.5 Selection pressure and population waves

These are two crucial factors influencing the outcome of the evolution process. Selection pressure is defined as

$$S = (\mu/\lambda) \quad [3.10]$$

This allows the value S to be varied by adjusting μ and λ between the extreme values 0 (highest pressure) and 1 (lowest pressure). If the number of children λ exceeds the number of parents μ, then there is a surplus of children and hence the selection pressure is high. If the number of children equals the number of parents, then, for example in the $(\mu + \lambda)$ strategy, all children survive and the selection pressure is nil.

With population waves, a changing number of individuals can be simulated. This is achieved by either systematically or randomly changing the number μ of surviving individuals after each cycle which become the parents of the next generation. If the selection pressure is to be kept constant, λ must be adjusted to keep S constant.

In the context of population size and lifespan, depending on the problem to be solved, normally a number of 10 to 50 individuals is fully sufficient to find a viable solution within a couple of hundred generations. If it is unclear what the solution looks like, it can be sensible to start with a larger number of individuals in order to quickly reach the area of the optimum solution. When that is reached, which can be seen by the slowing down of the progress of the benchmark function, the number of individuals is reduced. The lifespan of individuals can be set to just one generation or be limited to a certain value in the range of 2 to 10 generations. Hence, very fit individuals are kept for a certain time to include them into the reproduction process thus generating fit children. On the other hand, this can lead to a complete standstill of the process as a local optimum may attract the majority of individuals. This local optimum is then very difficult to leave. This behaviour is also called unwanted genetic drift.

3.3.6 Summary

In general, according to Rechenberg (1994), Schwefel (1995) and the author's own experience, the number of children λ should exceed the number of parents by a factor of between 5 and 10. Selection pressure should be set to values in the range of 0.2 to 0.3. For easy-to-solve problems, the (1 # 5) evolution strategy normally works well. For more difficult problems, the (1 # 10) evolution strategy together with a self-regulative mutative increment control is more advisable (see Section 3.3.4). Figure 3.19 summarizes the

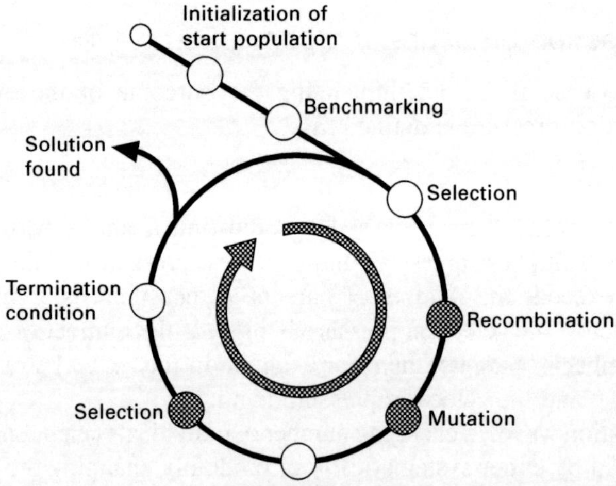

3.19 Steps during evolution (recombination, mutation, selection) (adapted from Rechenberg, 2003).

evolution strategy and shows the different steps as described earlier that are carried out subsequently.

3.4 Genetic algorithms

John H. Holland developed the so-called genetic algorithms (GA) parallel to Rechenberg's evolution strategy. There, the individuals are not encoded with vectors consisting of real numbers but binary as sequences of 0 and 1. The individuals are often called chromosomes. They are created from parents using the crossing-over method where whole sequences of 0 and 1 are exchanged to produce new vectors. Mutations are also possible, when 0 is converted to 1 and vice versa (Fig. 3.20). The selection process follows the same principles as described earlier.

A fundamental difference is the selection of the parent individuals following the principle of 'survival of the fittest'. This means that parents with a good fitness value automatically have a better statistical chance to be selected for reproduction than those with bad fitness values. This aspect can also be taken into account in the evolution strategy. A self-regulating mutative increment control is difficult to implement, which is also different from Rechenberg's approach.

There are a large number of publications describing applications in engineering and textiles in particular that use GAs and only a very few that apply Rechenberg's evolution strategy (see Section 3.6). The reason for this may be due to the fact that Holland's method was easier to implement in a computer program at the time it was developed and thus spread quicker than Rechenberg's approach. With today's computing power however, the higher complexity of the latter method no longer poses an obstacle. This, together with its great advantages, which lie especially in a much better and easier fit of a 'real numbers problem', should give Rechenberg's method a competitive edge in many applications.

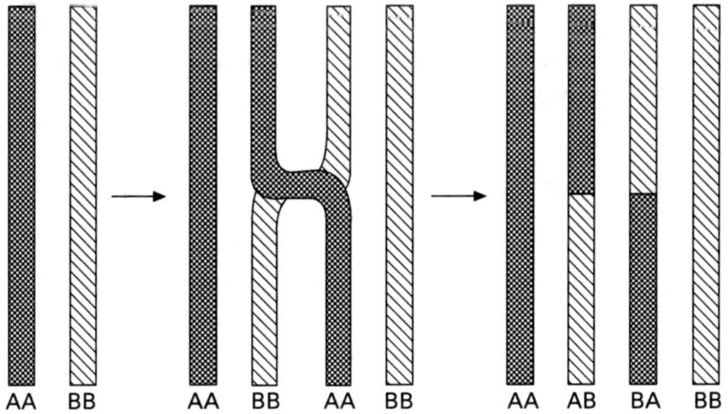

3.20 Generating offspring by crossing-over and mutation.

3.5 Comparison of evolutionary algorithms and iteration processes

There is a range of aspects in which classical iteration processes and evolutionary algorithms compete. Since the optimum solution is normally not known, usually all equations of the system at hand are set to zero for all these methods. All solutions found can then be compared with the nil-vector and hence easily be benchmarked.

3.5.1 Speed

Iteration processes, the most common being the so-called hill-climbing method, are very slow compared with evolutionary algorithms. This is due to the fact that in iterations, the solution is improved within a large number of normally very small steps. This implies that a large number of solutions needs to be calculated until a viable solution is found. Even if the distance between the current solution and the desired solution is large, the steps are often still small and all solutions in the vicinity of the current solution are also calculated.

Here, evolutionary algorithms have a competitive edge: if the current solution greatly differs from the prospective solution, then solutions near the current individual are no longer considered and solution areas a certain distance away from the current position are investigated. This behaviour can be controlled by setting the recombination and the mutation factors to certain values. Then, only those regions are further analysed, in which the difference between an individual and the optimum solution is below a given value. This often leads to either the global optimum or a local optimum which is often a sufficiently good solution.

3.5.2 Accuracy

Applying evolutionary algorithms always leads to a viable solution if carried out properly and if such a solution exists at all. But this solution is almost never the global optimum for most engineering problems.

3.5.3 Reproducibility

A variation of the starting individuals (vectors) of the evolutionary algorithm normally leads to different results. Even if the starting individuals were the same, since mutation changes the genes at random, the final results normally differ. With iteration processes, this problem is smaller. On the other hand, this phenomenon can be used to calculate different solutions to the problem which then serve as the starting points of another evolutionary cycle. In most cases, the deviation between the solutions is very small which shows that reproducibility normally is within acceptable limits.

3.5.4 Boundary conditions

In order to increase the processing speed of an iterative calculation or to even start the calculation, often boundary conditions are required. This in turn leads to a limitation of the parameters investigated and of the area in which the solution is searched, hence possible solutions are excluded. When using evolutionary algorithms, there are no boundary conditions needed as the number of parameters that can be optimized simultaneously is much larger. In addition, the solution area is not limited so the chance of excluding a good solution from the start is nil. This is a major advantage of evolutionary algorithms.

3.5.5 Summary

In summary, the main advantages of evolutionary algorithms over iteration processes are:

- computing speed is much greater;
- no boundary conditions are required.

The major disadvantages are:

- lower reproducibility of the solution;
- possibly a greater difference between found solution and global optimum.

3.6 Applications of evolutionary algorithms to textile technology

In recent years, evolutionary algorithms have been used successfully to solve technical problems for a wide range of applications. That includes the optimization of design of aircrafts (Hansen and Horst, 2008), the bows of cargo ships (Rechenberg, 2003), antennae (Kim *et al.*, 2011), formula 1 racing cars (Anon., 2004) and many more. In textile technology, there have also been several successful applications that are briefly described in this chapter. Most approaches employ Holland's method of binary encoding. This is surprising as Rechenberg's principles in these cases would be much easier to use and to program than to Holland's model, according to the author's experience.

3.6.1 Texturing process

Texturing is a process to give crimp to flat filament yarns in order to improve their elasticity and handle. The yarn is twisted, heated and cooled to set the twist (Fig. 3.21). In the predominantly used false-twist texturing process, the yarn is damaged during twist insertion caused by friction between yarn and texturing discs in the false twister. Several attempts have been made to create a mathematical model of the actual twist insertion. The respective equation systems were solved using iterative methods (Olbrich, 1993). Major emphasis was put on the optimization of the design of the discs (size, shape, number) to minimize friction and hence the yarn damage thereby increasing production speed (Veit *et al.*, 1998). A significant drawback of applying these iterative methods was the necessity to take boundary conditions into account, which in turn limits the possible solutions as described previously. Therefore, an evolutionary algorithm was applied.

Genotype and phenotype

Experiments show that the yarn takes the shortest route through the false twister. Hence, the yarn is divided into small sections and the coordinates (*x*, *y*, *z*) of the middle of these sections are regarded as the genotype vector. The length of thc yarn path through the false twister is then taken as the phenotype. The shorter the path of an individual the better its fitness and thus the greater are its chances of passing on its genotypical information to

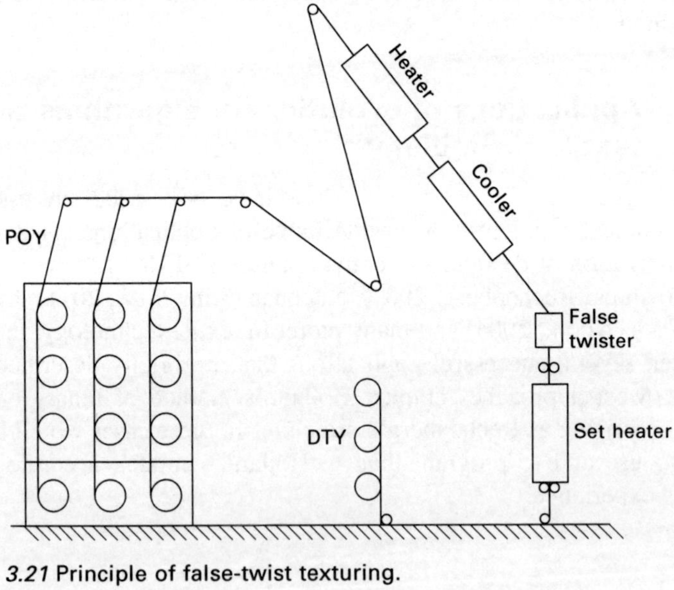

3.21 Principle of false-twist texturing.

the next generation. The genotypes are varied by changing the coordinates of the vector. The only boundary condition to be taken into account is that the yarn cannot be inside the discs. It is hence allowed for the yarn to either touch the discs or to hover slightly above them. This is not possible with iterative methods and opens up a whole new range of possible solutions. Figure 3.22 shows a disc within the false twister together with a typical yarn path which is divided into single points.

Number of individuals and lifespan

In this example, the lifespan of individuals was not limited in order not to lose a good solution. The evolution process itself was stopped after 5000 cycles (generations) as the improvement of the best solution in each step was then only marginal. Figure 3.23 shows the results of the yarn path length after 1000 generations for a different number of individuals. It is apparent that in this case 30 individuals are sufficient.

Recombination factor

Children were generated using the principles as described above with a variable percentage of the genes of each parents (see Fig. 3.17d). In Fig. 3.24, the yarn length of the best individual after 5000 generations is depicted together with the number of parents that were used to generate children. It is obvious that using only two parents results in the best children. A larger number of parent individuals leads to worse solutions, as non-optimal genes are also taken into account, thus downgrading the result in this example. A

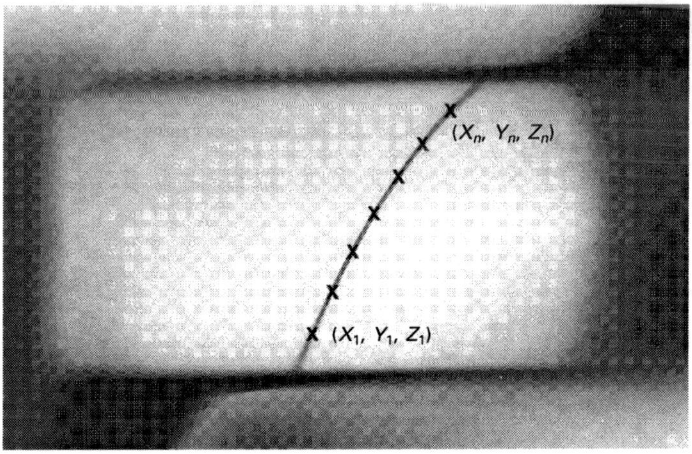

3.22 Yarn path across disc in the false twister with single points.

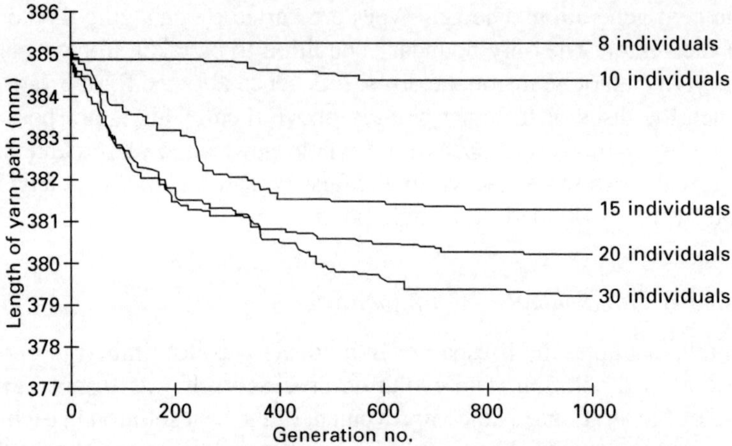

3.23 Influence of number of individuals on final result.

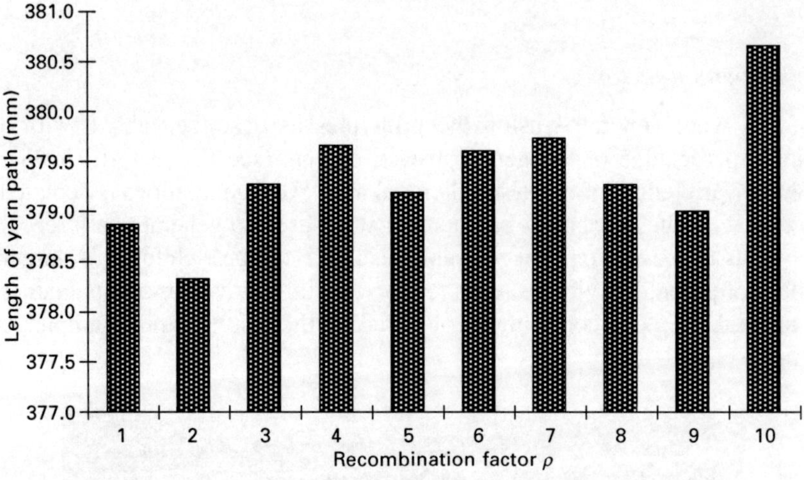

3.24 Influence of recombination factor on final result.

pure mutation is apparently also a sensible approach leading to the second best result in this example.

Size of the gene pool

Another important factor is the size of the gene pool from which the parents are taken to create a child. Figure 3.25 shows the yarn length of the best individual after 5000 generations depending on the size of the gene pool. When the two parents are selected out of the best six individuals of the previous generation following the principles as described previously, the

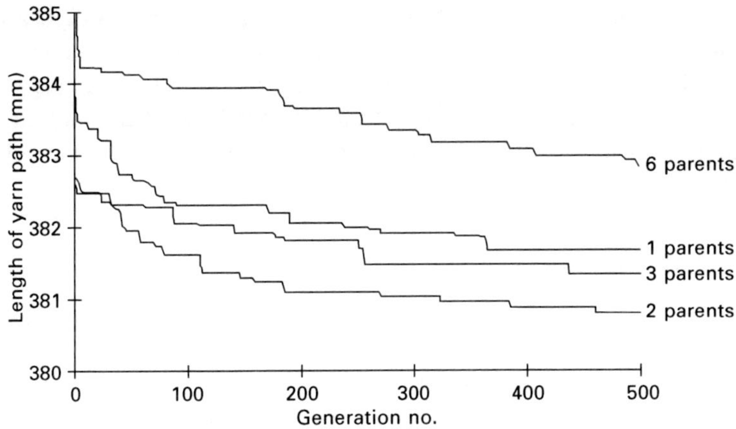

3.25 Influence of size of gene pool on the final result.

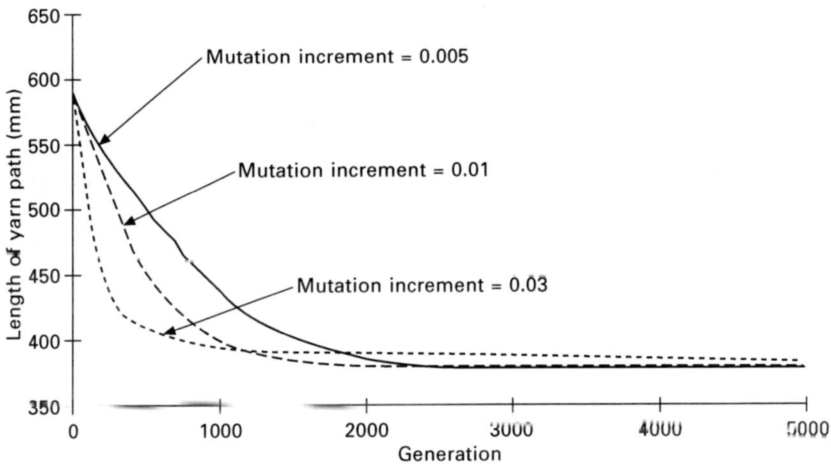

3.26 Influence of mutation increment on the final result.

result is considerably worse in comparison when only the two best parental individuals of the gene pool are selected to create a child. This is due to the fact that the variety of the parental genes is large. Hence, here the genes of 'unfit' individuals have a negative effect on the children's fitness.

Mutation increment

In order to further improve the fitness of the children, a mutation component was added to the children. The value of the mutation increment (ratio) has great influence on the course of the evolution process but not so much on the final result (see Fig. 3.26). A large mutation increment quickly leads to a good result. After 5000 generations, this plus has disappeared when comparing the final result with those achieved with a smaller mutation step. Therefore, in the beginning of the evolution process, a large mutation increment can be advantageous but should then be gradually decreased when the area where the optimum is located is reached in order to avoid missing a good solution by accident.

Number of mutants

Children can be created using the principles of recombination (see Fig. 3.17a–d) or by simply mutating a single parent individual. The number of these pure mutants relative to the number of children also determines the outcome of the evolution process as is shown in Fig. 3.27. Here, 15 children were created in each generation with the number of pure mutants varying between 1 and 10. It is apparent that a too large percentage of mutants leads to unsatisfying results whereas a small number (1 to 3) leads to equally good solutions.

Importance of mutants for the evolutionary progress

Mutants are often crucial for the progress of the evolutionary process as is shown in Fig. 3.28. Here, the length of the yarn path of the best individual of each generation if shown for 2500 generations. An upright black line indicates that the best individual of each generation is a mutant, a white line indicates that it is a child created by pure recombination. It is apparent that with rare exceptions, the best individual of each generation is a mutant. This

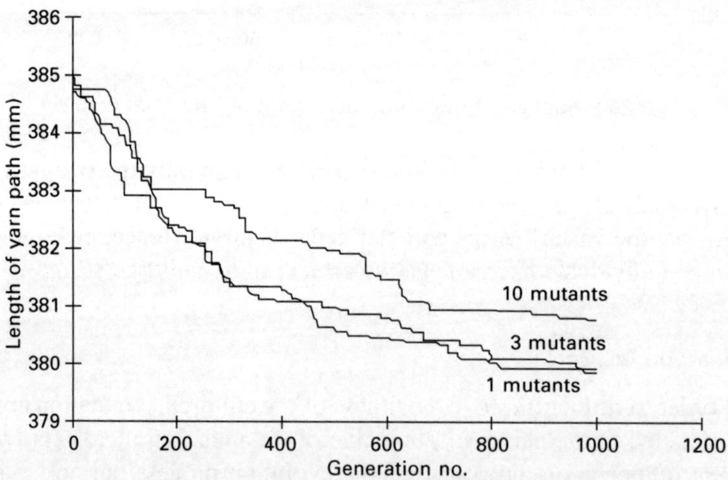

3.27 Influence of the number of true mutants on the final result.

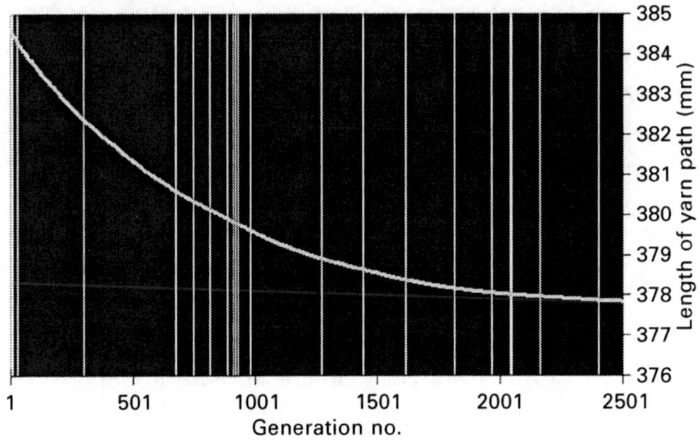

3.28 Influence of mutants on the evolution process.

Table 3.1 Settings of the different evolution stages

Stage	Population size	Gene pool size	Recombination factor	Mutation increment	Percentage of mutants	Number of yarn sections (resolution)
1	15	2	2	0.03	20	100
2	20	2	2	0.01	10	500
3	30	2	2	0.005	3	1000

does not imply, though, that children created by recombination have a low fitness value. In contrast, a thorough analysis shows that the mutants which are the best individual of each generation are mostly themselves descendants of parents that were created applying pure recombination. Hence both, mutants and recombined children, are crucial to ensure evolutionary progress.

Final algorithm and settings

Based on the findings as described above, the evolutionary process was divided into several stages as shown in Table 3.1. The last column shows the resolution of the yarn path. This stands for the number of sections into which the yarn path is divided. The coordinates of the respective middle of each section (x, y, z) are used to create the genotype of each individual yarn path.

Reproducibility of the results

The element of chance plays an important role in evolution as was previously discussed. Hence, the final result of an evolution process always differs even if the starting point was the same for two cycles. The differences are normally very small after a certain amount of generations and can therefore normally be neglected.

Robustness

Since no boundary conditions are needed when applying an evolutionary algorithm, the starting point for the calculation is irrelevant for the final result. This is illustrated in Fig. 3.29. The course of the expected yarn path is reversed and taken as the starting point. Even starting from this extremely bad solution, the algorithm is able to come up with a viable result within 5000 generations.

3.6.2 Weaving machine

An evolutionary algorithm was employed to optimize the machine settings during weaving by Wolters and Gries (2001). The principle of this approach was called 'AUTOWARP' and is shown in Fig. 3.30. Fabric data are fed into a simulation program and the expected warp tension (1) is calculated based on data from weaving mills by using a neural network. It is not possible to directly relate the course of the warp tension during weft insertion with a specific machine setting. Therefore, another simulation program is used that is able to calculate the warp tension based on the machine settings. As is shown in Fig. 3.31, that calculated warp tension (2) is compared with the desired warp tension (1). If the difference is sufficiently small, the optimum machine

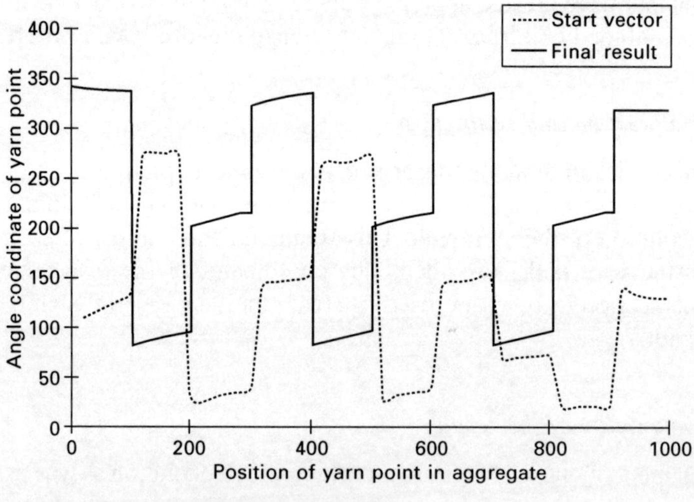

3.29 Robustness of evolution strategy.

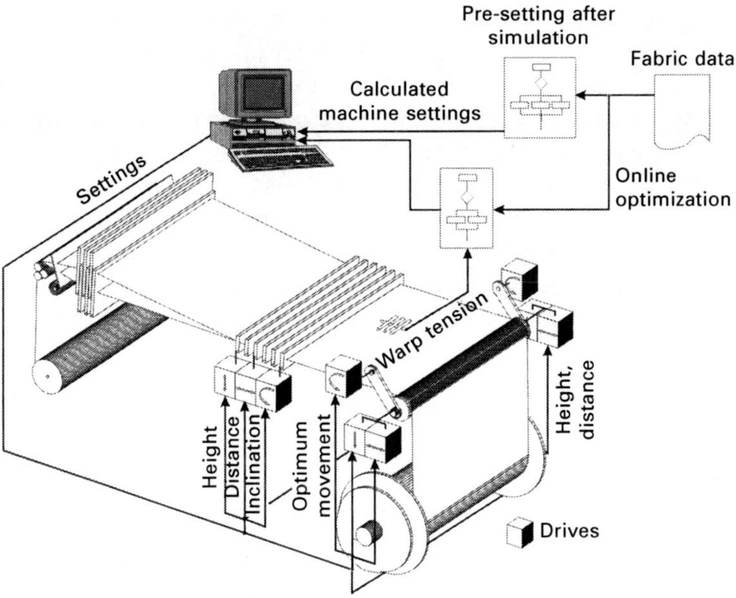

3.30 Design of AUTOWARP to optimize weaving machine settings (adapted from Wolters and Gries, 2001).

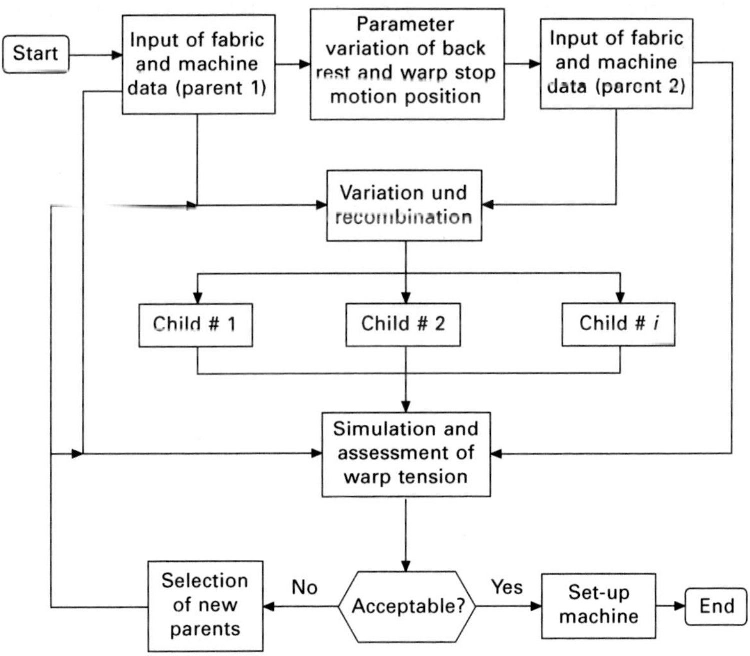

3.31 Evolutionary algorithm to determine optimized machine settings (adapted from Wolters, 2002).

setting has been found. If this is not the case, those machine settings that led to the closest correlation with the warp tension (1) are then used as parents for the next generation of machine settings. The resulting warp tension (2) is again calculated, compared with warp tension (1) and so forth until both warp tensions coincide sufficiently well. Then, the optimized machine setting to produce the desired fabric has been found. This system was successfully tested in weaving mills and is now a commercially available product. It shows the great potential of GA to help finding optimum machine settings even for highly complicated processes and machines. It is one of the rare examples of an application in textiles of Rechenberg's method.

3.6.3 Degumming of kenaf

Zheng and Du (2009) combined GAs with a neural network in order to reduce the chemical residues on kenaf fibres, thus decreasing environmental pollution. In a first step, they trained a neural network to find feasible and acceptable machine settings. Then, they defined an objective function f_{GA} given by

$$f_{GA} = (\text{residual gum} - \text{content})^2 + (\text{weight} - \text{loss} - \text{ratio})^2 \quad [3.11]$$

This they used in a second step to further optimize the machine settings. A verification of the results by experiments showed that machine set-up was indeed improved compared to the original state.

3.6.4 Staple fibre spinning

Liu and Yu (2008) also combined GAs according to Holland with a backpropagation neural network but with a different approach from the one outlined in Zheng and Du (2009). They created a so-called genetic neural network where the weight and threshold matrices were optimized using GAs. One major aim was to optimize the evenness of wool rovings (output) depending on wool top oil content, wool top moisture, fibre mean diameter and ten other parameters (input). A chromosome was defined which included the weight and threshold values of all nodes. This led to a chromosome string length of 106 as a network was used with 13 input, 7 hidden and 1 output neurons. Based on the performance of each respective neural network, gradually optimized settings for weight and threshold matrix were found. As Fig. 3.32 shows, the genetically optimized neural network led to slightly better results than a conventional network. This approach is most interesting and certainly worthy of pursuing further for a wide range of applications where the classic method to optimize the weight factors of an ANN is too time consuming.

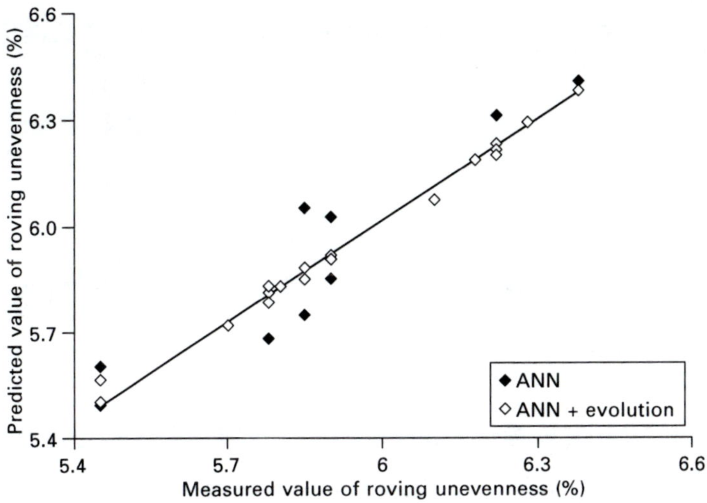

3.32 Comparison between roving evenness predicted with neural network (filled symbols) and with genetic neural network (open symbols) (adapted from Liu and Yu, 2008).

3.6.5 Carpet pattern design

Zamani *et al.* (2009) used GAs to generate new carpet patterns in a very innovative way. According to the authors, traditional methods to generate hand-made carpet patterns include artistic imitation and combination of traditional patterns, the adaptation of painting tables and the general generation of patterns through a creative process. This is time-consuming and expensive. An interesting alternative is the use of GAs. In a first step, each pattern is encoded in a chromosome consisting of information about corner, brink and medallion (centre of the carpet) motives, the respective sizes and colours. In total, 28 genes are encoded using a binary function as shown in Fig. 3.33.

In order to create a new population, different crossover and mutation operations are employed leading to new chromosomes and thus to new patterns. These patterns are then presented to the user for assessment, thus attributing a fitness value to the patterns. Depending on this value, the respective individuals of the pattern population are more likely to be selected for reproduction of the next pattern generation. The best pattern is always kept. This process leads quickly to new and innovative patterns as shown in Fig. 3.34 (initial population) and Fig. 3.25 (patterns after four generations). In order not to lose high-rated patterns, certain parts of a pattern can be set manually to a defined value by the user and thus taken out of the evolution process.

To further improve this method, a neural network is used. It is trained

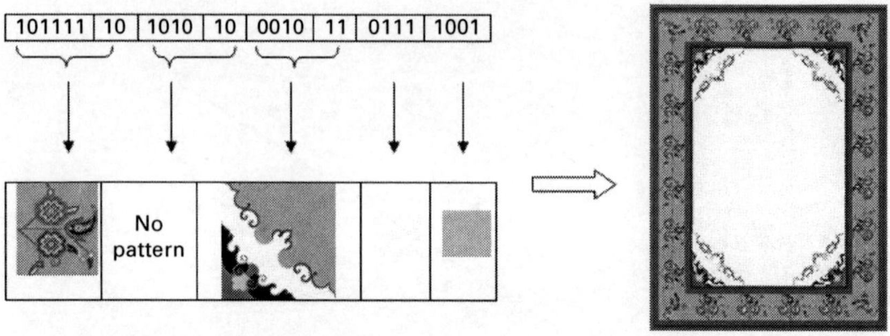

3.33 Chromosome encoding of a pattern (adapted from Zamani *et al.*, 2009).

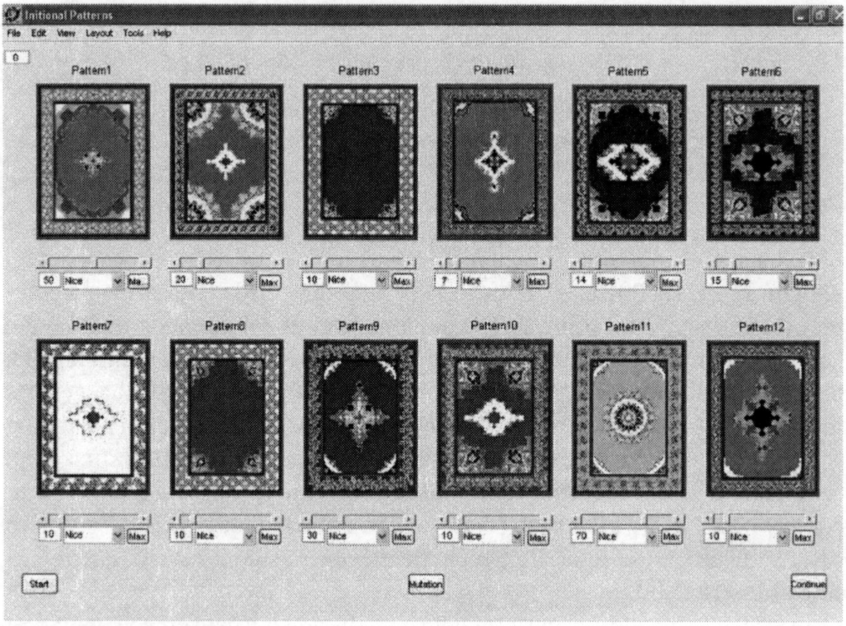

3.34 Initial population (courtesy of Professor Latifi, Amirkabir University, Iran).

with the characteristics of a pattern as input and the grading of the user as output. Hence, it is possible to actually predict the expected fitness value of a certain pattern thus avoiding unwanted patterns and directing the process towards higher graded motives.

3.6.6 Improving the structure of an ANN

In a similar approach as described above, Admuthe *et al.* (2009) used a GA to optimize the settings of an ANN which was used to link fibre properties

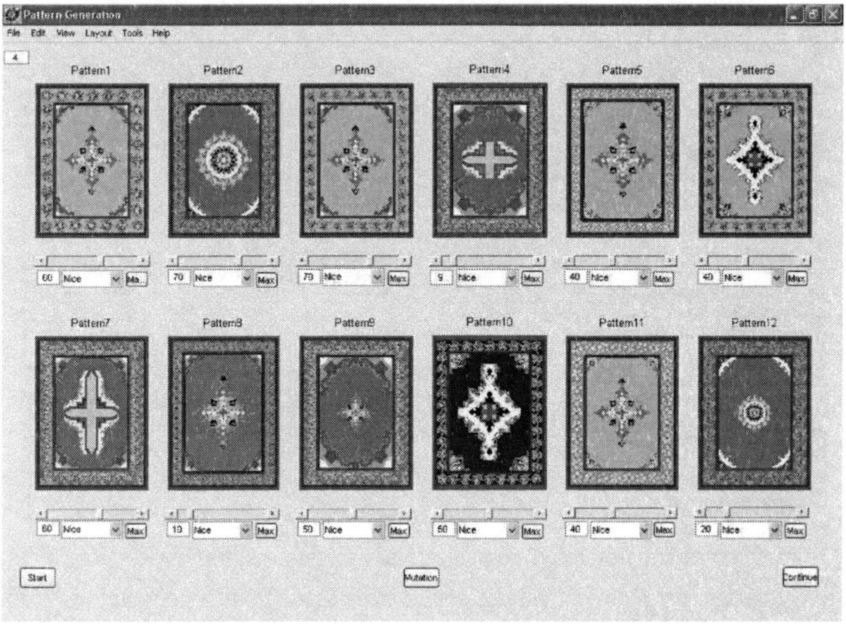

3.35 Result of evolution process after four generations (courtesy of Professor Latifi, Amirkabir University, Iran).

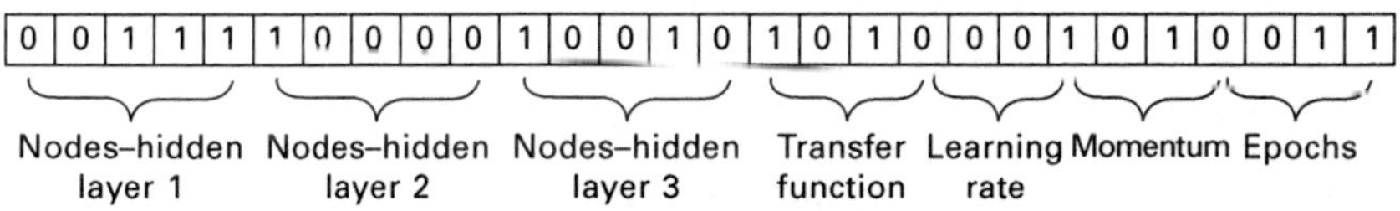

3.36 Encoding ANN structure and other parameters in a chromosome (adapted from Admuthe *et al.*, 2000).

and resulting yarn characteristics. In contrast to the method described in Section 3.6.1, they binary encoded the structural and other parameters in a chromosome as depicted in Fig. 3.36 using Holland's principle. Learning rate, momentum factor, number of generations and other parameters were assigned numbers selected from predefined values in order to limit the number of possible combinations. Then, the classic method of GA was employed as described earlier. The number of individuals was set to 100 with each representing seven parameters of the ANN setting. After only 50 generations, the achieved result with regard to mean error of the ANN for all patterns was below the set absolute threshold of 0.01 (Fig. 3.37). This method is a very useful approach for users of ANNs who lack extensive experience about adjusting the crucial factors for a successful training of an ANN properly and hence are not able to define optimal settings of an ANN directly.

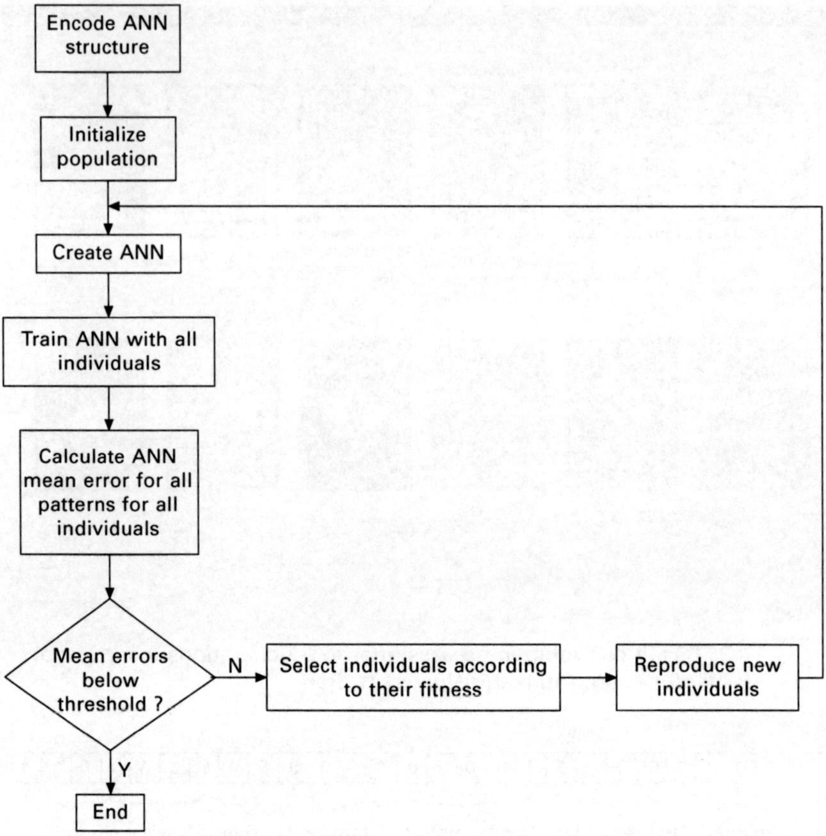

3.37 **Optimizing the training of an ANN with evolution strategy (adapted from Admuthe *et al.*, 2009).**

3.6.7 Spinning of staple fibre yarns

In another paper, Admuthe and Apte (2009) applied a genetic algorithm based on Holland's approach to further improve the results of a neural network. The basic idea is shown in Fig. 3.38. In a first step, they trained a backpropagation network with fibre properties and the resulting yarn characteristics as is common practice. They then employed a genetic algorithm to match as best as possible the predicted fibre properties with those fibres that are actually on stock. The fitness function of the GA was described by

$$\min f(x) = \sqrt{\sum_{n} (R_i - A_i)^2} \qquad [3.12]$$

where $f(x)$ is the fitness value, R_i represents the predicted property, A_i the property of available fibres and n stands for the number of parameters

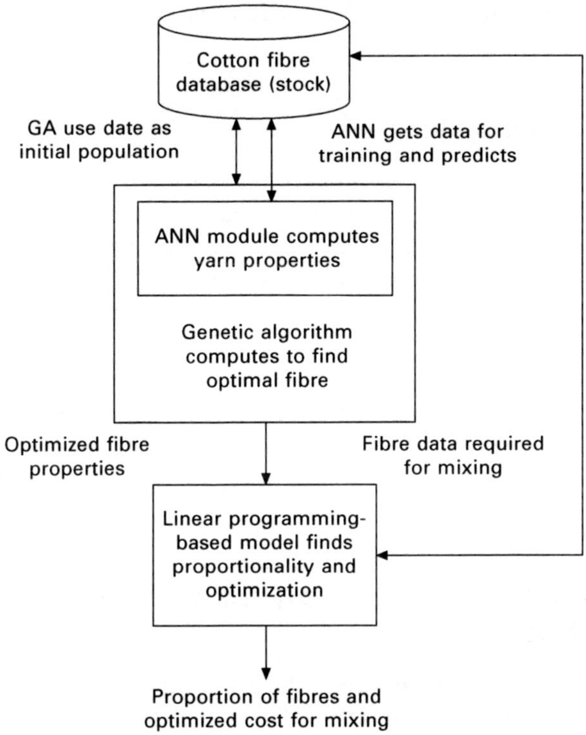

3.38 ANN-GA-LP model (adapted from Admuthe and Apte, 2009).

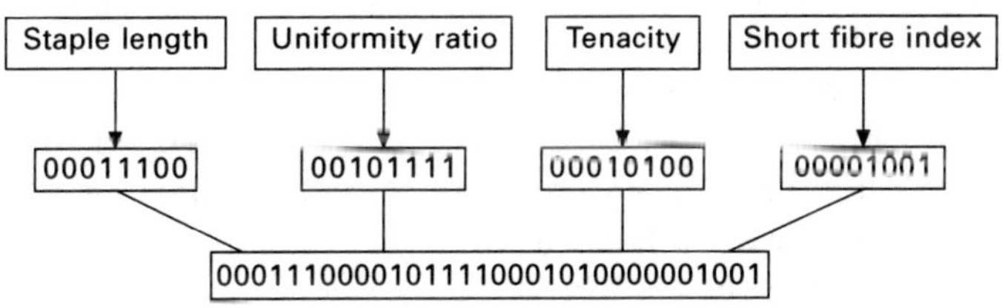

3.39 Binary encoding of fibre properties in a chromosome (adapted from Admuthe and Apte, 2009).

describing the fibre. The fibre properties themselves were encoded according to Fig. 3.39.

A linear programming (LP) method was applied to further improve the quality of the predictions by calculating an optimized blend consisting of a wide range of different fibres. This was used to reduce the cost by replacing expensive fibres with cheap fibres while still keeping the required properties of the blend. This is a novel approach that led to a distinctive improvement of the predictions with regard to costs and could, according to the authors, be especially useful for mills with a wide range of yarns from different cotton varieties.

3.6.8 Other applications

Recent papers include the optimization of the design of apparel production lines (Wang and Chen, 2010), a system that automatically separates colour printed fabrics (Kuo *et al.*, 2009), the detection of the layout of colour yarns from the colour pattern of yarn-dyed fabrics (Pan *et al.*, 2010) and a system that helps to detect the courses of breakdowns during the weaving process (Lin, 2009). The range of possible applications of GA in textiles is huge, since in many textile-related problems, there are often many parameters interacting and relations that cannot be described with mathematical equations. Most researchers use Holland's approach which is, as mentioned above, strange as Rechenberg's method appears to be easier to use but is apparently unknown to many scientists.

3.7 Practical advice in applying evolutionary methods

The settings of the evolutionary algorithm, e.g. number of individuals in the population, crossing-over factor and mutation factor, are not crucial in terms of whether an optimal solution will be found or not. The basic mechanism of evolution strategy was found to be very robust (Grefenstette, 1986) and a variation of these parameters normally only marginally affects the overall performance. Depending on the problem, either binary or real number encoding of the individuals can be advantageous. In general, one would expect real number encoding better suited to solve problems involving complex equations. If a binary problem needs to be solved, then Holland's approach may be easier to implement.

Much more critical is to find a suitable fitness function. In the ideal case, the fitness function is smooth and regular with only one optimum. In reality, this is, unfortunately, hardly ever the case. Otherwise one could just as well employ a hill-climbing algorithm. Hence, it is normally very difficult to find the global optimum when there are a large number of local optima. In order to cut short a normally very time-consuming process, the 'resolution' of the search space can be changed as was shown in Section 3.6.1. In the following steps, the resolution can be refined until it reaches the actual values. Another possibility is to define intermediate goals that will eventually lead to the overall target. This can be done by giving fitness points to solutions that reach some of the overall targets to the degree with which they fulfil them or by just counting the number of sub-targets that they achieve.

A common problem is premature convergence of the algorithm. This can be solved by introducing a sufficiently high number of mutants with a suitable mutation increment as was described earlier. This also works for slowly converging populations when all individuals of a population are located around the optimum and the gradient of the fitness scores is insufficient to actually push some individuals towards the global target.

In general, it is always advisable to run the evolution strategy more than once in order to get an idea whether the found solutions are similar or located in totally different regions of the search space which does happen at times. In the latter case, repeated runs of the algorithm will eventually lead to the global or at least an acceptable optimum. Different methods are described in detail in Beasley *et al.* (1993a, 1993b) which also give an excellent introduction into GA in general.

3.8 Web references for software tools

There is a wide range of software tools and general information on the world wide web. The following links are a selection of useful websites.

GA software

http://www.freedownloadmanager.org/downloads/genetic_algorithms_software/

Free GA software

http://jgap.sourceforge.net/
http://www.oocities.org/geneticoptimization/
http://download.cnet.com/Genetic-Algorithm-Library/3000-2070_4-10822277.html
http://www.brothersoft.com/genetic-algorithm-and-direct-search-tool-226443.html (Matlab© extension)
http://genetic-algorithm-labview.fyxm.net/ (Labview© extension)
http://www.top4download.com/free-genetic-algorithm/

Commercial GA software (selection)

http://www.wardsystems.com/
http://www.palisade.com/evolver/ (for Excel© applications)
http://www.nd.com/genetic/
http://www.rmltech.com/

Other sources

http://dces.essex.ac.uk/staff/rpoli/gp-field-guide/ ('Field Guide to Genetic Programming')

http://www.obitko.com/tutorials/genetic-algorithms/ (comprehensive tutorial)

http://samizdat.mines.edu/ga_tutorial/ga_tutorial.ps (online tutorial)

3.9 References

Admuthe, L.S., Apte, S.D. (2009), 'Optimization of spinning process using hybrid approach involving ANN, GA and linear programming', *Proceedings of the 2nd Bangalore Annual Compute Conference, COMPUTE'09*.

Admuthe, L., Apte, S.D., Admuthe, S. (2009), 'Topology and parameter optimization of ANN using genetic algorithm for application of textiles', *Proceedings of the 5th IEEE International Workshop on Intelligent Data Acquisition and Advanced Computing Systems: Technology and Applications,* IDAACS' 2009, 278–282.

Anon. (2004), 'Mit Darwin zum Formel-1-Sieg', *Der Spiegel* (*26*), 132.

Beasley, D., Bull, D.R., Martin, R.R. (1993a), 'An overview of genetic algorithms: Part 1: fundamentals', University of Wales, College of Cardiff, *University Computing, 1993*, 15(2), 58–69.

Beasley, D., Bull, D.R., Martin, R.R. (1993b), 'An overview of genetic algorithms: Part 2: Research topics', University of Wales, College of Cardiff, *University Computing, 1993*, 15(4), 170–181.

Correns, K. (1912), *Die Bestimmung und Vererbung des Geschlechtes nach Neuen Versuchen mit Höheren Pflanzen*, BiblioBazaar (new edition, 2009).

Darwin, C. (1859), *On the Origin of Species by Means of Natural Selection, or the Preservation of Favoured Races in the Struggle for Life*, Murray, London. New edition by Sterling Publishing (2008).

de Lamarck, J. B. (1909), *Philosophie Zoologique*, Verdière, Paris.

De Vries, H. (1906), *Species and Varieties, Their Origin by Mutation*, Lectures Delivered at the University of California by Hugo de Vries, Nabu Press (new edition, 2010).

Grefenstette, J.J. (1986), 'Optimization of control parameters for genetic algorithms', *IEEE Trans SMC*, *16*, 122–128.

Hansen, L., Horst, P. (2008), 'Multilevel optimization in aircraft structural design evaluation', *Computers and Structures 86(1 – 2)*, 104–118.

Holland, J.H. (1992), *Adaptation in Natural and Artificial Systems*, 2nd edition, MIT Press.

Kim, K.-T., Kwon, S.-H., Ko, J.-H., Kim, H.-S. (2011), 'An optimal design of a 19.05GHz high gain 44 array antenna using the evolution strategy', *Transactions of the Korean Institute of Electrical Engineers*, *60 (4)*, 811–816.

Kuo C.-F. J., Kao, C.-Y., Chiu C.-H. (2009), 'Integrating a genetic algorithm and a self-organizing map network for an automatically separating color printed fabric system', *Textile Research Journal*, *79(13)*, 1235–1244.

Lin, J.-J. (2009), 'A genetic algorithm-based solution search to fuzzy logical inference for breakdown causes in fabric inspection', *Textile Research Journal*, *79(5)*, 394–409.

Liu, G., Yu, W.-D. (2008), 'Integrated genetic neural network and its application in the worsted fore-spinning process', *Proceedings of the 4th International Conference on Natural Computation*, vol. 5, 554–558.

Mendel, G. (1866), 'Versuche über Pflanzenhybriden', *Verhandlungen des naturforschenden Vereines in Brünn*, Bd. IV, Abhandlungen: 3–47. English translation: Druery, C.T and William Bateson (1901). 'Experiments in plant hybridization'. *Journal of the Royal Horticultural Society*, *26*, 1–32.

Olbrich, A. (1993), 'Simulation of twist insertion in the false-twist texturing process', *Chemiefasern-Textilindustrie*, *43/95*, issue 10, 828–830.

Pan, R., Gao, W., Liu, J., Wang, H. (2010), 'Genetic algorithm-based detection of the layout of color yarns', *Journal of the Textile Institute, 102(2)*, 172–179.

Rechenberg, I. (1974), *Evolutionsstrategie*, Frommann-Holboog.

Rechenberg, I. (1994), *Evolutionsstrategie 94*, Frommann-Holboog.

Rechenberg, I. (2003), *Evolutionsstrategie*, script to lecture, TU Berlin.

Schwefel, H.-P. (1995), *Evolution and Optimum Seeking (Sixth-Generation Computer Technology)*, Wiley VCH.

Veit, D., Laourine, E., Wulfhorst, B. (1998), 'New disc-geometry in the false-twist texturing process to reduce the snow production', *Chemical Fibers International*, *48/1998*, 89–90.

Von Tschermak, E. (1900), *Versuche über Pflanzenhybriden*, Nabu Press (new edition, 2010).

Wallace, A.R. (1871), *Contributions to the Theory of Natural Selection*, Macmillan and Company.

Wang, B., Chen, S. (2010), 'The optimized design of apparel streamline based on genetic algorithm', *Proceedings of the 6th International Conference on Natural Computation.*

Watson, J.D., Crick, F. (1953), 'Molecular structure of nucleic acids: a structure for deoxyribose nucleic acid', *Nature*, 171, 737–738.

Weismann, F.A. (1885), *On Germinal Selection as a Source of Definite Variation*, BiblioBazaar (new edition, 2008).

Wolters, T. (2002), '*Verbesserte Webmaschineneinstellungen mittels Simulationsrechnungen*', Dissertation, RWTH Aachen University, Germany.

Wolters, T., Gries, T. (2001), 'Intelligent adjustment device for looms', *Proceedings of the International Textile Congress, Vol. 1*, 310 – 317.

Zamani, F., Amani-Tehran, M., Latifi, M. (2009), 'Interactive genetic algorithm-aided generation of carpet pattern', *Journal of the Textile Institute*, *100(6)*, 556–564.

Zheng, L., Du., B. (2009) 'Neural network modeling for bio-enzymatic degumming on kenaf', *5th International Conference on Natural Computation, ICNC 2009*, vol. 2, 223–226.

4 Fuzzy logic and its application to textile technology

D. VEIT, RWTH Aachen University, Germany

Abstract: Fuzzy logic uses terms similar to human language to describe values of parameters and relations. These 'linguistic rules' thus allow an easy transfer of human expert knowledge onto a computer-based control system. The term 'fuzzy' indicates the difference to Boolean logic as not only the numbers '0' and '1' but also, for example, '0.3' are possible. Fuzzy logic is used today in many technical devices and machines. This chapter gives an overview of the method and concludes with selected examples of successful applications of fuzzy logic to solve textile-related problems.

Key words: fuzzy logic, textile technology.

4.1 Introduction

The term fuzzy logic in its true sense means 'imprecise logic'. At first glance, this appears to be a contradiction in itself as in classic logic, only clearly defined states represented by either '0' or '1', true or wrong, etc. are considered. In fuzzy logic, there are also intermediate states, the so-called 'degrees of truth', represented, for example, by '0.2' or 'rather wrong' or 'rather big'. Fuzzy logic is therefore an all-encompassing logic concept in which classical logic can be regarded as a border case.

A typical example is the measurement of the temperature of ambient air. Let 'hot' be defined as $\vartheta = 40\,°C$. In classic logic, a temperature above 40 °C would be regarded as 'hot' and any temperature below would be considered as 'not hot' (or 'cold'). A temperature of $\vartheta = 39.9\,°C$ would therefore be considered 'not hot'. In fuzzy logic, there are also other states, such as 'quite hot' and 'warm'. A temperature of $\vartheta = 49.9\,°C$ would then be assigned to 'quite hot' with a degree of truth of 99% and to 'cold' by only 1%. A temperature of $\vartheta = 36\,°C$ could then be 'hot' with a degree of truth of 60% and be 'warm' with a degree of truth of 40% as shown in Fig. 4.1. Hence, fuzzy logic mirrors the way of human thinking much better than classical logic.

The curves that assign degrees of truth to the input values are the membership functions. In most cases, they are triangular (as in Fig. 4.1), but they can also follow the Gaussian curve or any other mathematical function that is suitable (see Section 4.2.1). Fuzzy logic is usually applied in order to solve

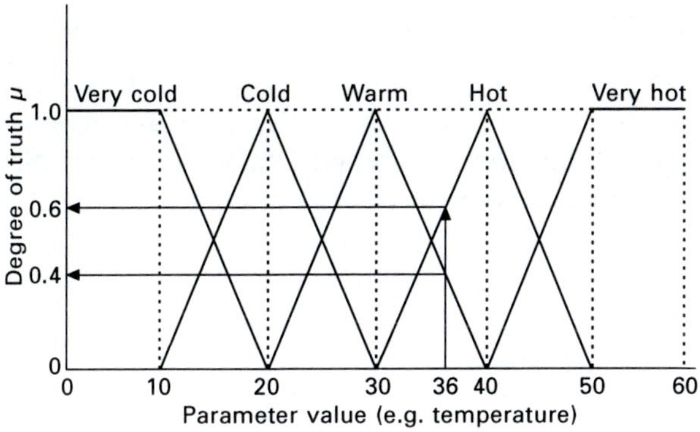

4.1 Fuzzy logic representation for air temperature.

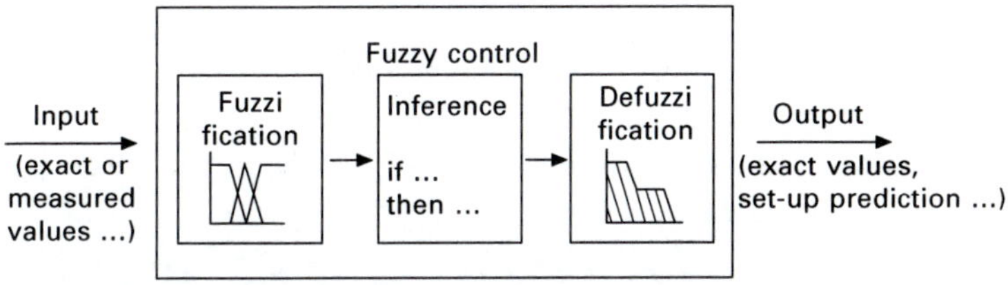

4.2 Principle of a fuzzy logic controller.

controller-related problems by formulating so-called linguistic rules. These linguistic rules are closely related to the human way of thinking, hence allowing 'imprecise' definitions and terms. It is therefore not necessary to know the mathematical background behind the relations of input and output parameters. This makes fuzzy logic controllers a very powerful tool which can easily be employed to solve a wide range of technical problems. Figure 4.2 shows the principle of a fuzzy logic controller with the three stages: fuzzification (assigning the degrees of truth to the input), inference (applying the rules) and defuzzification (calculating the output).

The concept of fuzzy logic was developed in the 1960s by Professor Lotfi A. Zadeh, University of Berkeley, USA (Zadeh, 1965). He introduced the term 'fuzzy' in the meaning of 'imprecise'. Since then, fuzzy logic has been used in a wide range of applications. Typical examples are 'intelligent' washing machines and other household appliances (Nitta, 1993). The washing machine fuzzy controller was designed to determine the required amount of washing powder and washing time depending on, for example, the mass of the clothes to be washed, the type of dirt and the hardness grade of water. Other successful applications include ABS systems in cars (Mauer, 1995; Yazicioglu, 2008) and medical diagnosis systems (Stadnik, 1986; Hayashi, 1994).

4.1.1 Development in recent years

Fuzzy logic was initially not taken seriously by many scientists and sometimes even fought against. In the 1980s and 1990s, many Japanese companies successfully employed fuzzy control systems in many applications (Nitta, 1993). In Germany, in particular Professor H.-J. Zimmermann of RWTH Aachen University, was a major contributor to the development of fuzzy controllers (Meier *et al.*, 1994; Zimmermann and Sebastian, 1994; Zimmermann, 2001, 2010). Japanese and German companies are still predominant in the industrial applications of these systems.

In contrast to neural networks, there were only a few publications about new developments of fuzzy controlled systems in textile technology up to 2007 (as can be seen in Fig. 4.3). One of the reasons was widespread disappointment caused by too high expectations with regard to the performance of these systems. Without a distinctive knowledge of the basic rules governing a certain control engineering problem, fuzzy systems cannot be applied successfully. In recent years though, the number of papers on this subject has increased, which shows that this concept has been taken up again to solve current problems.

4.1.2 Typical application

A typical example of a fuzzy logic application is the control of the water influx into a colouring bath depending on the amount of water already in

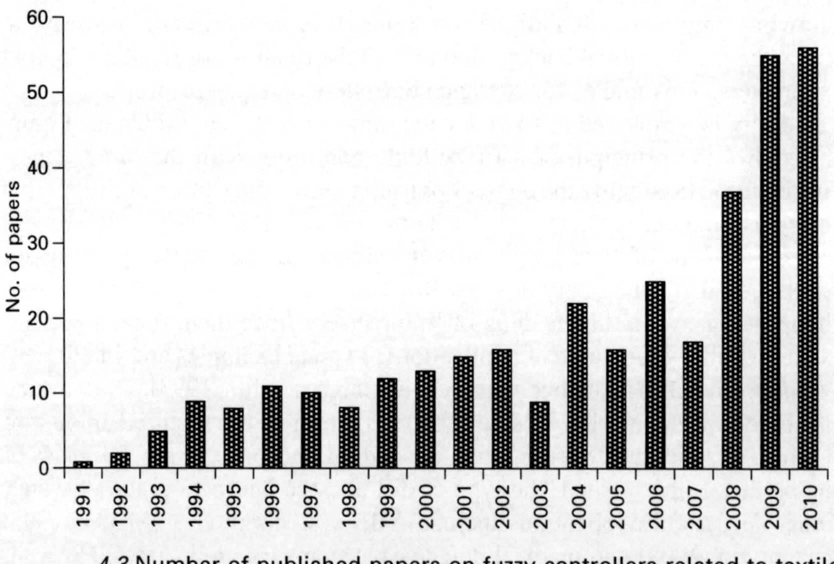

4.3 Number of published papers on fuzzy controllers related to textile technology.

the vessel and the temperature of both the water in the vessel and the water gushing out of the tap. If the vessel is 'quite full', the temperature of the water in the vessel is 'rather warm' and the water coming out of the tap is 'very hot', then the amount of water added into the vessel should be decreased 'rather quickly' and its temperature should be set to 'warm'. To achieve this with a conventional proportional-integral-derivative (PID) controller is very difficult as complex differential equations need to be devised and solved. Fuzzy logic allows this problem to be solved in a very easy and efficient way (see Section 4.6).

4.2 Imprecise mathematics

Imprecise mathematics is an extension of classical mathematics. The rules that apply are as precisely phrased as in classical mathematics. The fundamental difference is the ability to carry out calculations with imprecise values and imprecise sets.

4.2.1 Terms and definitions

This section introduces the basic definitions used in fuzzy logic and gives typical examples.

- Imprecise value: a value which has a certain degree of uncertainty due to, for example, measurement inaccuracy: 40 cN ± 0.3 cN.
- Imprecise set: consists of elements that cannot be assigned solely to a certain set but only to a certain degree (0–100%). A typical example is the set of thick books. A lexicon normally belongs to the set of 'thick' books with a degree of truth of 90% and to the set of 'rather thick' books with a degree of truth of 10%. A tourist leaflet normally belongs to the set of 'thin' books with a degree of truth of 99% and to the set of 'rather thick' books with a degree of truth of 1%.
- Linguistic variable: term for an imprecise set, e.g. 'thick', 'rather thick', 'not thick'.
- Degree of truth: a quantitative measure for the degree of membership of an element to an imprecise set. Normally, values between 0 and 1 are used. A degree of truth of $\mu = 0.7$ means that an element belongs to a certain set by 70%.
- Membership function: this function is used to calculate the degree of truth of a variable to a certain set depending on its value. Usually, linear functions (e.g. triangular) are used, sometimes also Gaussian functions, as shown in Fig. 4.4.
- Normalized set: when using a normalized set, all values for the truth values are in the interval [0; 1]. A normalized set can be obtained from

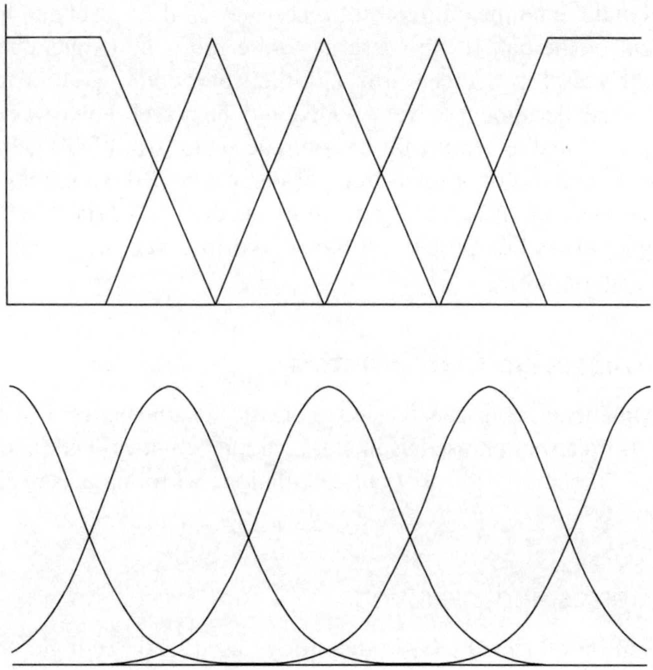

4.4 Membership functions (top: triangular; bottom: Gaussiean).

an unnormalized set by dividing all degree of truth values by the highest value.

4.2.2 Graphic representation in diagrams

Figure 4.1 shows the graphic representation of five imprecise sets and their membership functions. Normally, the y-axis is the degree of truth and the x-axis represents the absolute value of the variable in question. A single element can belong to different sets, e.g. 'warm' and 'hot'. The sum of its degrees of truth is normally $\sum_i \mu_i = 1$. It is also possible that an element belongs to more than two sets and that its sum of degrees of truth μ_i exceeds the value 1.

4.2.3 Numerical representation

The elements of a set and their degree of truth values can also be described numerically. A common definition is

$$A = \{(x_1, \mu_1), (x_2, \mu_2), \ldots\}, \forall x \in P \qquad [4.1]$$

with

$\forall$ = allquantor ('for all').

In equation [4.1], x_1, x_2 stand for the elements and μ_1, μ_2 for the respective degrees of truth to the set A. The set A in turn is a subset of the population P. Variables x_i with values for the degree of truth with $\mu_i = 0$ are normally not listed. There is a wide range of other numerical representations of fuzzy sets which are not mentioned here but can be found, e.g. in Ross (2010), Nguyen and Walker (2005), and Zimmermann (2001).

4.3 Set operations

In order to formulate rules, the sets have to be linked. This can be done using terms such as 'and', 'or', 'and not', 'or not'.

4.3.1 Terms

x	element
A, B	imprecise, normalized sets
$\mu_A(x)$, $\mu_B(x)$	degree of truth of element x to the imprecise sets A and B
P	set of all elements x (population)
min {...}	minimum operator; the lowest degree of truth μ from the variables within the brackets
max {...}	maximum operator; the highest degree of truth μ from the variables within the brackets

4.3.2 Or operator

The 'or' operator is defined as the unification operator of A and B: $A \cup B$.

Maximum function

In fuzzy logic, the 'or' operator is normally represented by the maximum function. For example:

$$P = \{a,b,c,d,e,f\}$$

$$A = \{(a;0{,}8), (c;0{,}3), (d;0{,}3), (f;0{,}5)\}$$

$$B = \{(b;1), (c;0{,}4), (d;0{,}5), (e;0{,}2)\}$$

Elements with a degree of truth $\mu = 0$, are not listed. When applying the maximum operator, we get

$$A \cup B = \{(a;0{,}8), (b;1), (c;0{,}4); (d;0{,}5), (e;0{,}2), (f;0{,}5)\}$$

Sum function

Another representation of the 'or' operator is the sum function. Then we have

$$A + B = \{x; \mu_{A+B}(x)\} \ \forall \ x \in P$$

Hence

$$\mu_{A+B} = (x) = \mu_A(x) + \mu_B(x) - \mu_A(x) \cdot \mu_B(x) \ \forall \ x \in P$$

For example:

$$P = \{a,b,c,d,e,f\}$$

$$A = \{(a;0{,}8), (c;0{,}3), (d;0{,}3), (f;0{,}5)\}$$

$$B = \{(b;1), (c;0{,}4), (d;0{,}5), (e;0{,}2)\}$$

$$A + B = \{(a;0{,}8), (b;1), (c;0{,}58); (d;0{,}65), (e;0{,}2), (f;0{,}5)\}$$

4.3.3 And operator

The 'and' operator is defined as the intersecting set.

Minimum function

In fuzzy logic, the 'and' operator is normally represented by the minimum function (also called Mamdani implication). As an example:

$$P = \{a,b,c,d,e,f\}$$

$$A = \{(a;0{,}8), (c;0{,}3), (d;0{,}3), (f;0{,}5)\}$$

$$B = \{(b;1), (c;0{,}4), (d;0{,}5), (e;0{,}2)\}$$

Elements with a degree of truth $\mu = 0$, are not listed. When applying the minimum operator, we get

$$A \cap B = \{(c;0{,}3); (d;0{,}3)\}$$

Note: Elements not existing in any of the sets A and B have a degree of truth of $\mu = 0$ for the intersecting set. Hence, when applying the minimum operator, these elements are not listed.

Product function

An alternative to the sum operator is the product operator. Then we have

$$\mu_{A \cap B}(x) = \mu_A(x) \cdot \mu_B(x)$$

For example:

$$P = \{a,b,c,d,e,f\}$$

$$A = \{(a;0{,}8), (c;0{,}3), (d;0{,}3), (f;0{,}5)\}$$

$$B = \{(b;1), (c;0{,}4), (d;0{,}5), (e;0{,}2)\}$$

Hence

$$A \cap B = \{(c;0{,}12); (d;0{,}15)\}$$

4.3.4 Be part of operator

If A is a part of B, this corresponds to $\mu_A(x) \leq \mu_B(x)$. An example is:

$$P = \{a,b,c,d,e,f\}$$

$$A = \{(a;0{,}8), (c;0{,}3), (d;0{,}3), (f;0{,}5)\}$$

$$C = \{(a;0{,}4), (c;0{,}3)\}$$

$$D = \{(a;0{,}7), (b;0{,}1)\}$$

Set C is a part of set A, but set D is not part of set A.

4.3.5 Modifier

Modifiers can be used to slightly change the degree of truth. In spoken language, this corresponds to terms such as 'very' and 'more or less'.

Very

The modifier 'very' can be expressed mathematically by taking the square of the membership function of the truth value. For complex problems, there are also a range of so-called compensating operators to express 'and' and 'or' but they are only rarely used. A comprehensive list can be found in Zimmermann (2001).

More or less

The modifier 'more or less' can be expressed mathematically by extracting the square root of the membership function or of the truth value. An example is:

Expensive

More or less expensive = $\sqrt{\text{expensive}}$

Very expensive = expensive^2

Not very expensive = 1 – very expensive = 1 – $expensive^2$

It is also possible to define separate membership functions.

4.4 Fuzzy logic versus probabilistic logic

The truth values in fuzzy logic are not probabilities. A probability is a measure for the likelihood of a certain event to happen. It does not give any information about the actual condition. In contrast, the degree of truth represents the real condition of a certain element. A degree of truth of $\mu_A(x) = 0.8$ means that the element x belongs with a degree of 80% to the imprecise set A and with a degree of 20% to other sets. It does not mean that the probability of x belonging to A is 80%.

Probability $p_A(x) = 0.8$: when considering x ten times, it will belong on average eight times to A and twice not. This also means that in the future, the truth value (probability) of x belonging to A is 80%.

Degree of truth $\mu_A(x) = 0.8$: when considering x, each time x will belong to the set A by a degree of 80%. This is valid for every single consideration. In the future, x will always meet the required properties to belong to set A by a degree of 80%.

4.5 Ultra-fuzzy logic

So far, x was attributed to the imprecise set A with a degree of truth of $\mu_A(x)$. The degree of truth itself was a precise value (e.g. 0.2). It is also possible to define the degree of truth imprecisely, e.g. within an interval. This can be helpful when taking into account rules described by different people. As an example, depending on personal preferences, the term 'hot' and the respective boundary values separating it from the next set can be defined differently. Also the degree of truth can be different for different points of view. Hence, when taking into account expert knowledge of different people, ultra-fuzzy can be a powerful and very effective tool to overcome this difficulty.

4.6 Fuzzy control

The main area of application of fuzzy logic is control engineering. As mentioned previously, when using fuzzy logic, a clearly defined mathematical model describing the process is no longer necessary. More or less simple rules that describe the relation between the variables in question are fully sufficient. Hence, these systems are often named 'expert systems' as the knowledge of experts is taken into account when defining these rules. As depicted in Fig. 4.2, fuzzy control can be divided into three stages:

1. Fuzzification (assigning of values to imprecise sets).

2. Inference (defining and linking of rules).
3. Defuzzification (calculation of final result).

4.6.1 Fuzzification

During the fuzzification stage, the problem at hand is assigned to imprecise sets using membership functions and truth values. Hence, it is determined by which degree an element (e.g. measurement value) possesses a certain property (e.g. temperature, price). See Fig. 4.5. The procedure is as follows:

1. Define imprecise sets.
2. Define membership functions.
3. Determine degree of truth values.

As an example, the water temperature in a colouring bath is to be controlled. The imprecise sets of the water temperature are 'cool', 'warm' and 'hot'. The membership functions are defined in such a way that the sum of the degrees of truth for each variable is always $\Sigma\mu_i = 1$. Hence, the temperature $\vartheta = 92\,°C$ belongs with a degree of truth of 60% to 'warm' ($\mu_{\text{warm}} = 0.6$) and with a degree of truth of 40% to 'hot' ($\mu_{\text{hot}} = 0.4$) and not at all to 'cool' (Fig. 4.6).

4.6.2 Inference

During this stage, rules have to be defined to process the imprecise data. These are normally formulated in the form 'if…then'. The rules are then combined and applied to the data. This results in degrees of truth to the previously defined output sets as shown in Fig. 4.7. Hence, we have the following procedure:

1. Define the processing rules.
2. Define the operators 'and', 'or', etc.
3. Calculate the degree of truth values of the output set values.

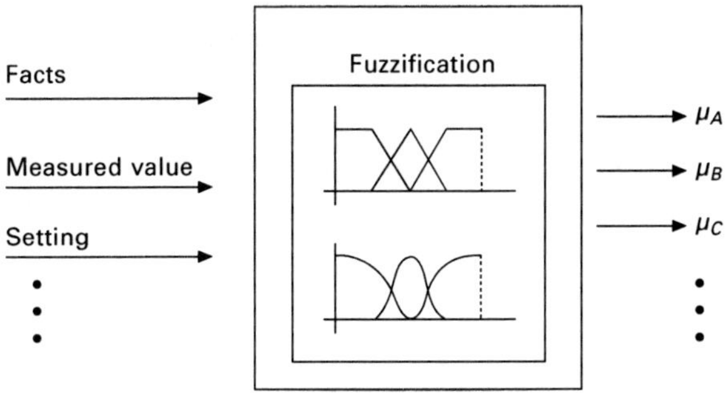

4.5 Principle of fuzzification.

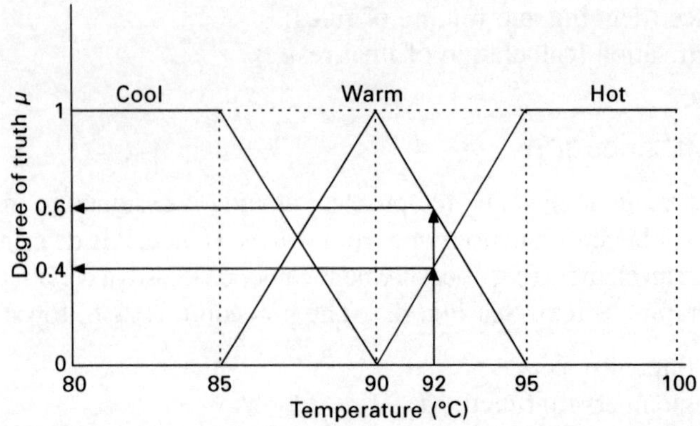

4.6 Determination of degrees of truth for a bath temperature of ϑ = 92 °C.

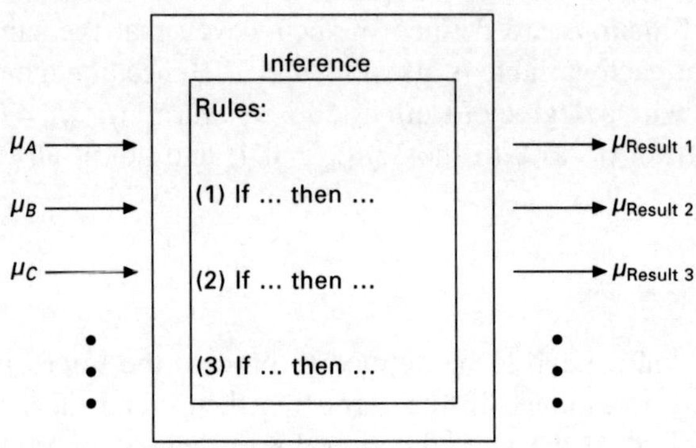

4.7 Principle of inference.

As an example, the temperature of water entering the colouring bath shall be controlled depending on the temperature of water already in the bath.

1. Definition of a simple set of rules:
   a. If water in bath is cool, then influx water is hot.
   b. If water in bath is warm, then influx water is warm.
   c. If water in bath is hot, then influx water is cool.
   d. If water in bath is warm or hot, then influx water is cool.

   It is of crucial importance to have at least one rule for each condition that may occur during the process! Otherwise, a un-defined state would result which in turn could lead to completely wrong output with catastrophic consequences in the worst case.
2. Definition of the operators 'and', 'or', etc. When selecting proper operators for 'and' and 'or', there are many possibilities of which some were described earlier. Normally, minimum and maximum operators

are used as the respective calculations are simple and do therefore not require much computing time.

3. Calculation of the truth values of the output sets. From the rules mentioned earlier, only the rules b, c and d apply. Hence:

Rule b: $\mu_{\text{inf lux,water,warm}} = \mu_{\text{warm}} = 0.6$

Rule c: $\mu_{\text{inf lux,water,cool}} = \mu_{\text{hot}} = 0.4$

Rule d: $\mu_{\text{inf lux,water,cool}} = \max\{\mu_{\text{warm}}; \mu_{\text{hot}}\} = \max\{0.6;0.4\} = 0.6$

In this example, different rules lead to different results, which is not uncommon in fuzzy logic. In the case of a similar result (in this case the result 'cool' according to rules c and d), normally the degrees of truth are connected by either taking their maximum or their arithmetic mean value. Another possibility is to summarize the two imprecise sets according to

$$\mu_{A+B} = \mu_A + \mu_B - \mu_A \cdot \mu_B \qquad [4.2]$$

The resulting sum is always greater than the greatest single degree of truth. The various degrees of truth are used in the next step to determine an area which is then used during defuzzification to calculate the final result.

Max/Min method

The membership functions of each imprecise set are clipped at the respective degree of truth parallel to the *x*-axis. This results in several areas that are then superimposed. As an example, we have:

$\mu_{\text{influx, water,cold}} = 0$

$\mu_{\text{influx, water,cool}} = 0.4$

$\mu_{\text{influx, water,warm}} = 0.6$

$\mu_{\text{influx, water,hot}} = 0$

The result area can then be determined according to Fig. 4.8.

Max/Prod method

The membership function of each imprecise set is multiplied with the respective truth value. The resulting areas are again superimposed (Fig. 4.9). As an example:

$\mu_{\text{influx, water,cold}} = 0$

$\mu_{\text{influx, water,cool}} = 0.4$

$\mu_{\text{influx, water,warm}} = 0.6$

$\mu_{\text{influx, water,hot}} = 0$

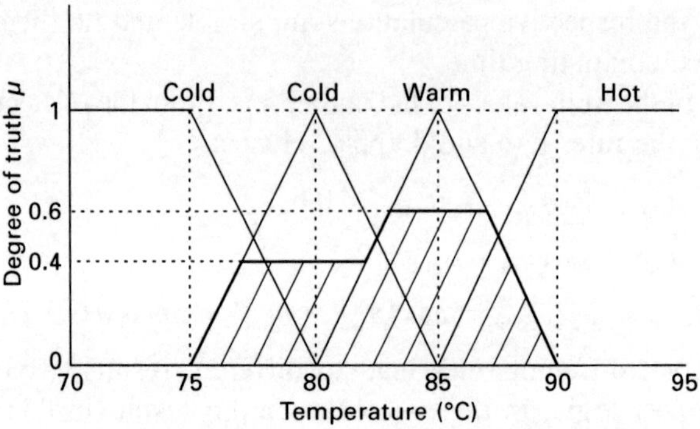

4.8 Determination of resulting areas with Max/Min method.

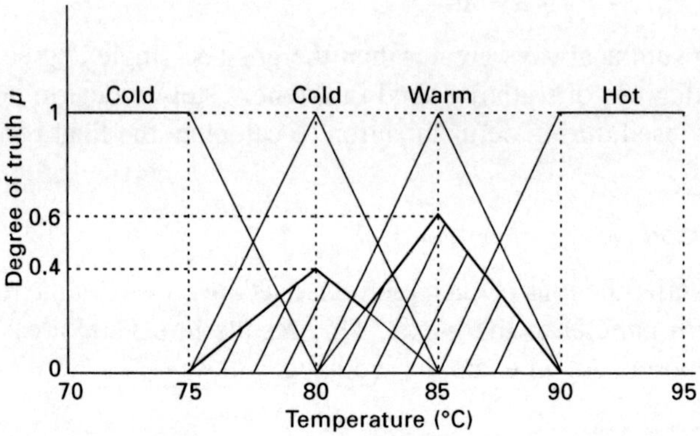

4.9 Determination of result areas with the Max/Prod method.

The actual temperature of the influx water can then be determined in the last stage, the defuzzification.

4.6.3 Defuzzification

The imprecise result of the inference is transformed into an exact value. This leads to clearly defined instructions, such as 'set temperature of influx water to 37.8 °C'. The method of defuzzyfication (Fig. 4.10) to be used depends on the problem. There are a wide range of possibilities of which only a few are described here. The principal approach is divided into two steps:

1. Determination of the defuzzification method.
2. Determination of result (output).

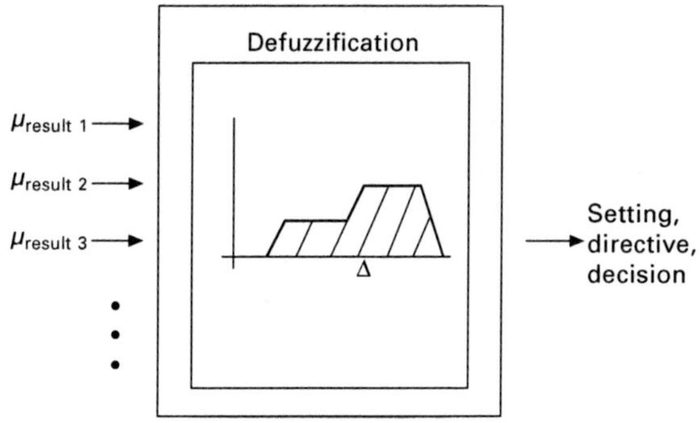

4.10 Principle of defuzzification.

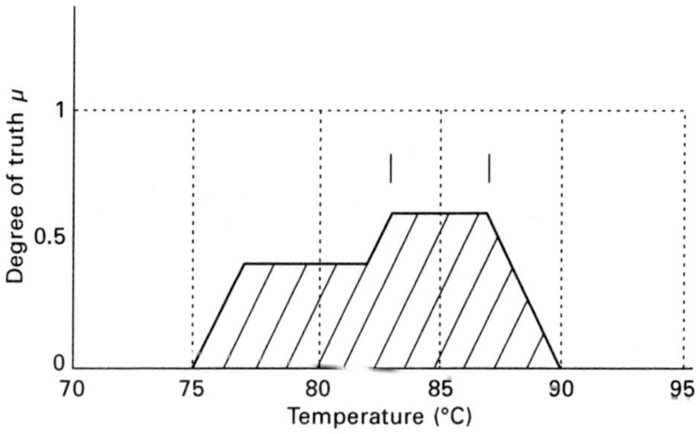

4.11 Mean-of-maximum method.

Mean of maximum

This method is suitable only for rough estimations. The mean x-value of the maximum value of the output set is taken as the value for the output variable as the shown in Fig. 4.11. The maximum value in this example is $\mu = 0.6$ and the mean x-value of this membership function is 40 °C, which is then taken as output value.

Centre of gravity

This method is more accurate than the mean-of-maximum method but also more complex. The x-value of the centre of gravity of the result area is taken as the output value. This can be done graphically as shown in Fig. 4.12. The output value for the influx water in this example is $\vartheta = 37\,°C$. A numerical solution can be derived with

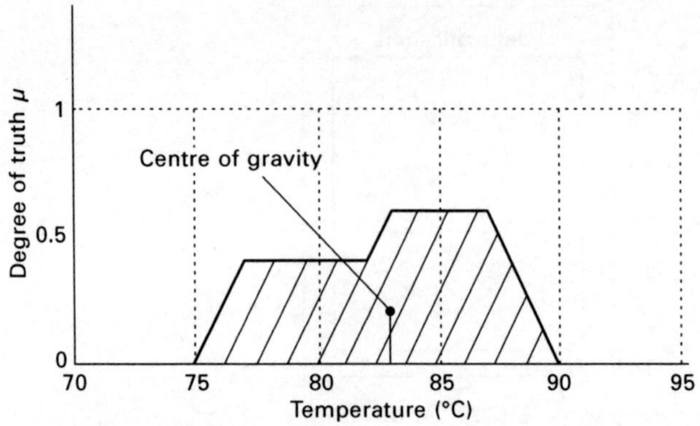

4.12 Centre of gravity method.

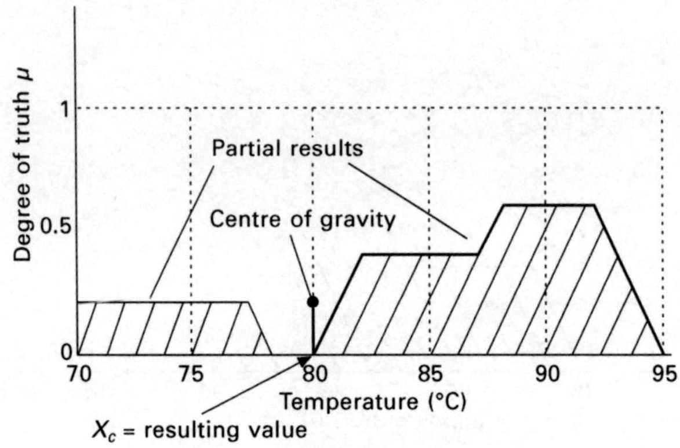

4.13 Centre of gravity outside result areas.

$$x_c = \frac{\int_{x_A}^{x_E} x \cdot f(x)\,\mathrm{d}x}{\int_{x_A}^{x_E} f(x)\,\mathrm{d}x} \qquad [4.3]$$

where

x_c = x-coordinate of the centre of gravity
x_B = starting value x of the area
x_E = end value x of the area
$f(x)$ = function (boundary curve) of the area

In Fig. 4.12, we have $x_B = 75$, $x_E = 90$ and $x_c = 82.8$. Overlapping sections of the result areas are only counted once. If the centre of gravity is located outside all of the result areas, the principal approach is unchanged as is shown in Fig. 4.13. For many applications, it is not overly important to have an exact output value as the fuzzy control loop is calculated frequently thus always leading to a stable condition within a very short period of time.

4.7 Fuzzy control with four input and one output variable

To illustrate the advantages of fuzzy logic to solve engineering problems, we now look at a typical example in ring spinning.

4.7.1 Fuzzification

Assume we want to predict the evenness value of a ring spun yarn depending on the fineness of the flyer roving, the preliminary draft, the yarn count and the number of operating hours of the traveller. The respective triangular membership functions are shown in Figs 4.14–4.18.

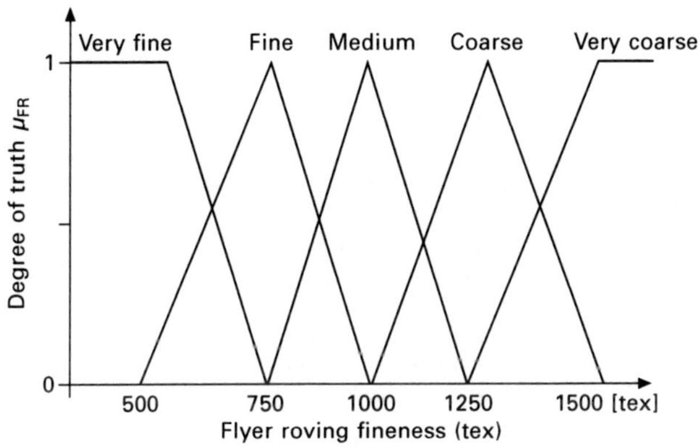

4.14 Membership functions for flyer roving fineness.

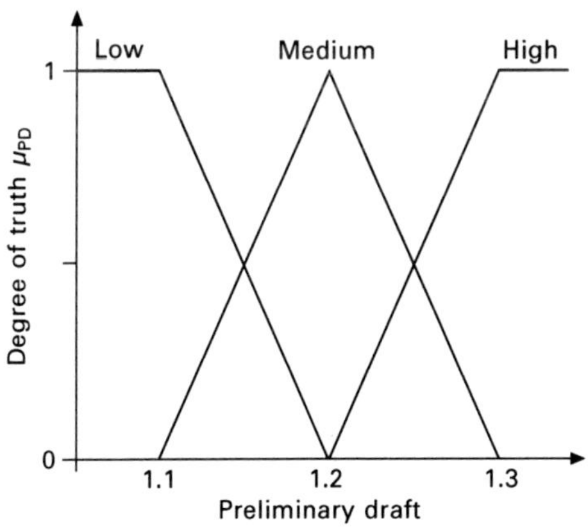

4.15 Membership functions for preliminary draft.

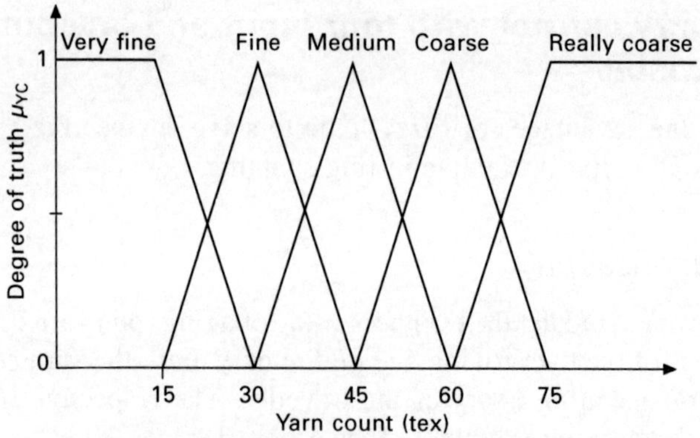

4.16 Membership functions for yarn count.

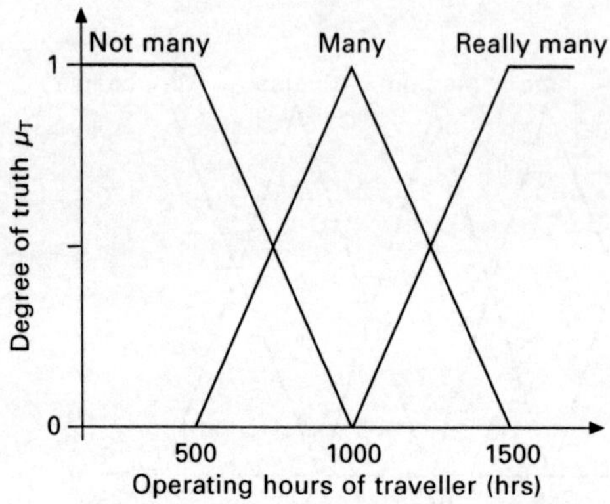

4.17 Membership functions for operating hours of traveller.

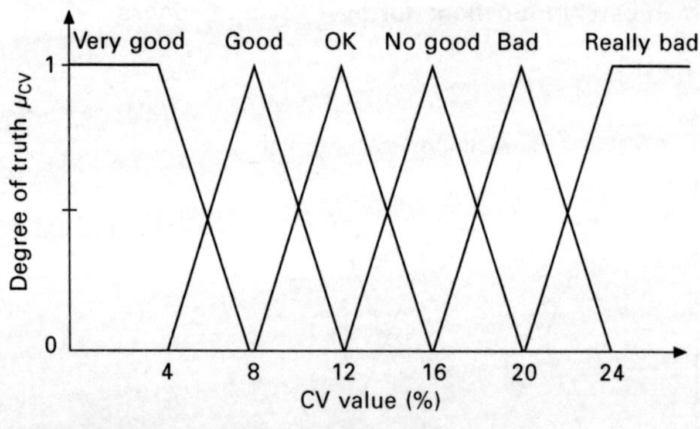

4.18 Membership functions for CV-value.

4.7.2 Inference

A set of rules for this example is defined as shown in Table 4.1. In this example, only a few rules are formulated. In general, all possible situations

Table 4.1 Set of rules

		Flyer roving fineness		Preliminary draft		Yarn count		Operating hours of traveller		CV
1	If	vf	AND	l	AND	(vf OR f)	AND	rm	then	bad
2	If	(f or m)	AND	l	AND	(f OR m)	AND	(m OR rm)	then	no good
3	If	(m or c)			AND	(m OR c OR vc)			then	good
4	If	C	AND	H	AND	(c OR vg)	AND	(nm OR m)	then	good
5	If	vc			AND	(c OR vc)	AND	nm	then	very good
6	If	m	AND	m	AND	m	AND	rm	then	no good
7	if	m	AND	m	AND	m	AND	m	then	ok

that may occur must be assigned to a rule. Using the operator ‘or’ does significantly reduce the number of necessary rules as is shown in Fig. 4.19. In this example, when using the operator ‘and’, 18 rules are necessary. Instead, when using the operator ‘or’, only 3 rules are required to describe the whole universe of discourse.

In order to illustrate the usefulness of fuzzy logic systems, typical data for the example above are given in Table 4.2. As is obvious, for example (a), only rules 2, 3 and 7 of Table 4.1 apply. For example (b), only rules 2, 3, 6 and 7 apply and for example (c), only the rules 3 and 4 apply. The inference results for (a) are hence:

$$\text{Rule 2: } \mu_{CV}(\text{no good}) = \min\{\max[\mu_{FR}(f); \mu_{FR}(m)]; \mu_{PD}(l); \max[\mu_{YC}(f);\mu_{YC}(m)];\mu_{T}(m)\} = 0.4$$

	Parameter B		
Parameter A	b1	b2	b3
a1	AND	AND	AND
a2	AND	AND	AND
a3	AND	AND	AND
a4	AND	AND	AND
a5	AND	AND	AND
a6	AND	AND	AND

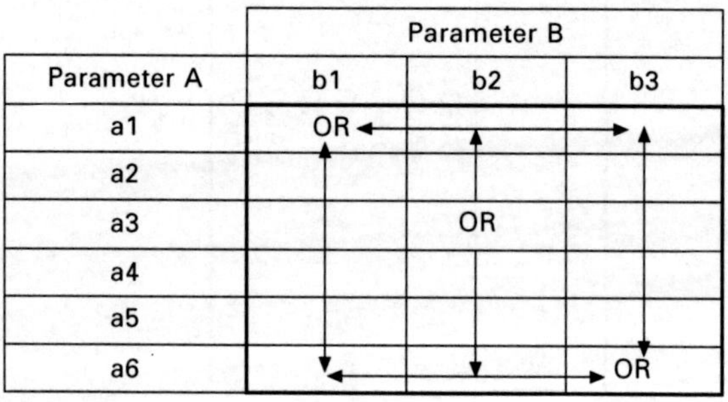

4.19 Reducing the number of necessary rules to describe the universe of discourse by using the operator ‘or’ (bottom) instead of ‘and’ (top).

Table 4.2 Example data

	Flyer roving fineness (tex)	Preliminary draft	Yarn count (tex)	Operating hours of traveller	CV (%)
a	800	1.15	35	700	
b	800	1.15	35	1400	
c	1300	1.25	50	1200	

Rule 3: $\mu_{CV}(\text{good}) = \min[\mu_{FR}(m); \mu_{YC}(m)] = 0.2$

Rule 7: $\mu_{CV}(\text{ok}) = \min[\mu_{FR}(m); \mu_{PD}(m); \mu_{YC}(m); \mu_{T}(m)] = 0.2$

4.7.3 Defuzzification

When the Max/Min method is used for the defuzzification as depicted in Fig. 4.20, we get as the final result CV = 13.5. In a similar way, we get the results for (b) CV = 14.5 and for (c) CV = 8. It is apparent from Fig. 4.20 that a slight inaccuracy of the rules and the respective μ-values does not affect the final result considerably. This shows the great advantage of fuzzy logic. If this method is used as a real-time controller, the final result is recalculated permanently thus leading quickly to a stable process even if the single parameters that are taken into account consist of inaccurate values. As can be seen from Fig. 4.21, is also possible to include the component content of yarns into such a system. This kind of control mechanism can

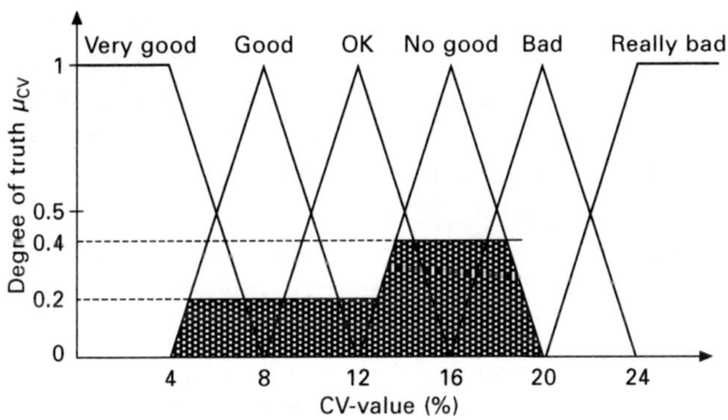

4.20 Defuzzification using the Max/Min method.

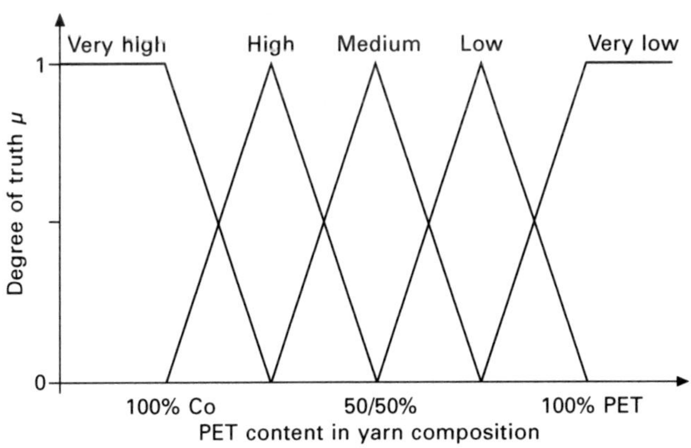

4.21 Membership function for yarn composition.

easily be implemented into a control circuit and is very fast, making it often superior to classic control approaches.

4.8 Applications of fuzzy logic in textile technology

Fuzzy control systems have been widely used since the 1990s. Many applications have set standards in control technology. In general, fuzzy logic systems offer a promising approach where expert knowledge is available and formulated in rules and a classic PID control systems is too complex to use. For an easier handling of fuzzy controllers, there is a wide range of software available on the market which simplifies its usage considerably. A selection is shown at the end of this chapter.

4.8.1 Intelligent diagnosis system for fabric inspection

Lin *et al.* (1995) developed a system to determine the breakdown causes of fabric defects of woven fabrics. Based on the experience of weavers, they defined rules relating 18 symptoms (e.g. broken end, pick-out mark, slack weft, hole) with 44 causes (e.g. poor quality of warp yarn, improper let-off, low weft tension). The causes comprised electrical problems, the weaving mechanism as such and quality grading aspects. For each different rule, sets were defined. A typical rule for a defect caused by the weaving process was 'if (fabric defect is incorrect let-off) and (weaving density of fabrics is big) then (decrease the let-off length of fabrics)'.

The operator fed the symptoms that were observed into the fuzzy system, which then came up with a range of up to three probable causes based on the predefined fuzzy rules. When the operator then inspected the machine and found additional defects, these could also be fed into the system to narrow the breakdown cause down. This system can also be used to teach untrained personnel the relation between fabric defects and improper machine function. In a later paper (Lin, 2009), a genetic algorithm according to Holland was employed to help find the best solution quicker.

4.8.2 Modelling of colour yield in polyester (PET) dyeing

Tavanai *et al.* (2005) developed a fuzzy system to model the colour yield (k/s) of polyester (PET) yarns as a function of time, temperature and disperse dye concentration. They applied triangular as well as exponential coefficients, and both led to acceptable results. A typical rule for the colour yield was 'if (temperature is low) and (time is low) and (concentration is low), then (colour yield is very low)'. Only a small number of rules was necessary to create a viable model. This paper also compares the derived models with the classic approach of statistical regression and clearly shows in this example that fuzzy inference systems (FIS) are superior. Nasiri and Berlik (2009) further improved the model by applying evolutionary algorithms to further tune the fuzzy sets. This led to a slight improvement of the prediction with regard to colour yield (see Fig. 4.22).

In a later paper, Nasiri *et al.* (2011) developed this model further and compared knowledge-driven versus data-driven and rule-based versus tree-based fuzzy systems. They developed a system that offers a compromise between the accuracy of classic Takagi–Sugeno systems and the interpretability of Mamdani models.

4.8.3 Modelling of false-twist texturing yarn

Nasiri (2005) developed a fuzzy regression model to predict the percentage of yarn contraction of false twist textured polyester yarns depending on the texturing temperature. She proved that this approach is superior to conventional statistical regression when only a small number of data sets is available.

4.8.4 Classification of fabric lustre

Hadjianfar and Semnani (2010) propose a fuzzy system to classify the lustre of textile fabrics. They use triangular membership functions for all fuzzy sets. The input consisted of a lustre index which is determined using digital image processing. Thirteen rules converted this input into one of six output classes ranging from 'very matt' to 'very bright'. A typical rule was as follows: 'if (lustre index is in interval 2) then output is 'very matt' to a

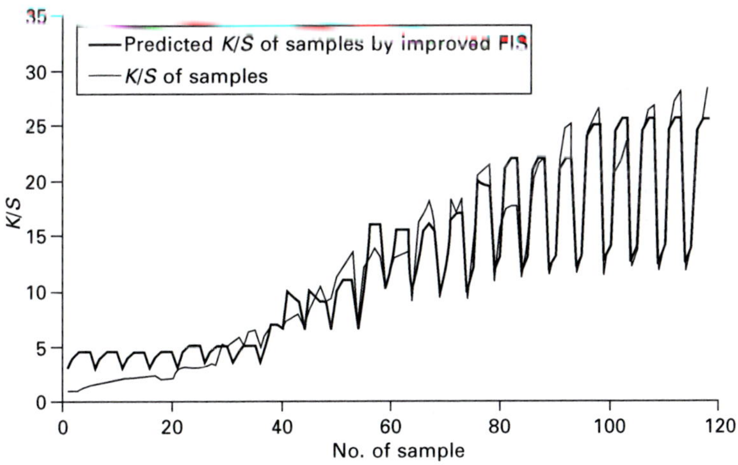

4.22 Predicted vs. actual colour yield for PET fabrics (adapted from Nasiri and Berlik, 2009).

degree of truth of 0.05'. The classification results of the fuzzy system were compared with the classification of 10 observers. As Fig. 4.23 shows, the correlation between reality and model is excellent.

4.8.5 Determination of warp tension in weaving

A simple model to determine the warp tension during weaving was proposed by Dayik *et al.* (2008). They used the shed height (5 – 60 mm), yarn elasticity (0–20%) and weft yarn density (10–60 cm^{-1}) to accurately predict warp tension (35–60 cN). All membership functions were triangular and only 10 rules were needed, e.g. 'if (shed height is very low) and (yarn density is very low) and (yarn elongation is normal) then (warp yarn tension is very low)'. A comparison between calculated and measured warp tension in this example showed an excellent correlation (Fig. 4.24). This system can be used to determine optimized settings for these three parameters quickly without carrying out time-consuming trials. The idea is related to a concept derived 10 years earlier and published in Osthus (1997).

4.8.6 Classifiying dyeing defects

Huang and Yu (2001) used a fuzzy approach to improve the classification of dyeing defects when using a neural network. After extracting the critical data that define a dyeing defect with digital image processing, these values were converted into fuzzy sets. These were fed into a neural network (backpropagation) thus increasing the dimensions of the feature space. The

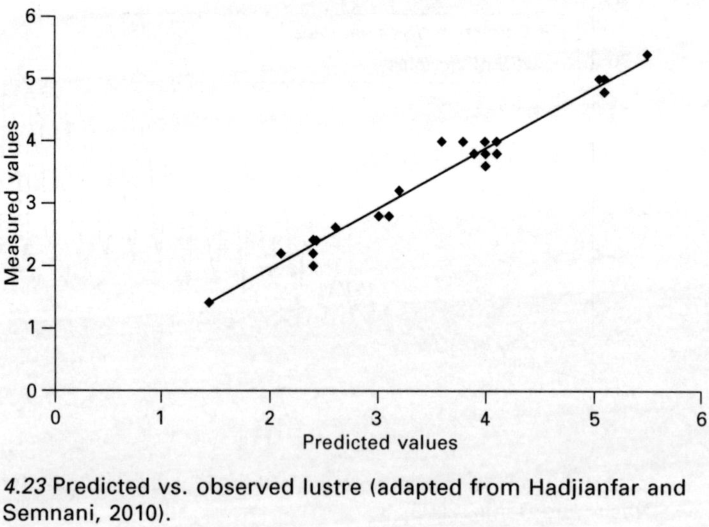

4.23 Predicted vs. observed lustre (adapted from Hadjianfar and Semnani, 2010).

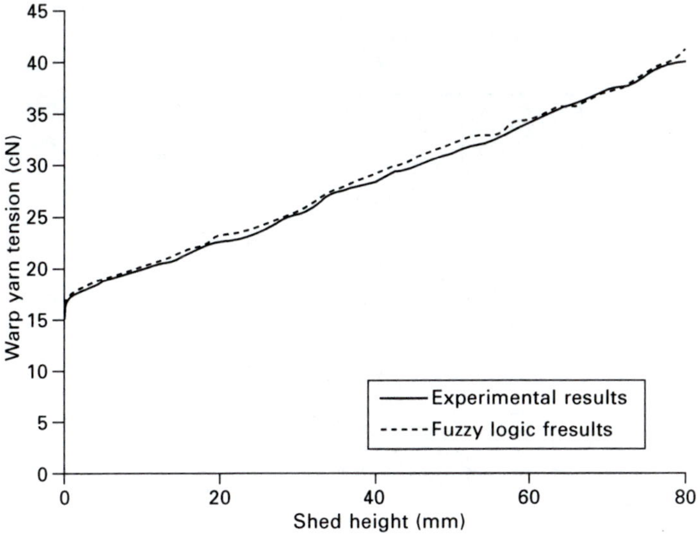

4.24 Predicted vs. measured warp tension (adapted from Dayik *et al.*, 2008).

output of the neural network was then converted into the number of the defect class, e.g. filling band in shade, dye and carrier spot, oil stain.

4.8.7 Predicting garment drape

Fan *et al.* (2001) developed a neuro-fuzzy system applying a classic approach to predict the garment drape of women's dresses. The input of the backpropagation network consisted of fabric properties and garment feature dimensions, the output was the expected drape chosen from an available range of images that were taken from actual garments. The fuzzy module was designed to choose the correct drape image from those that were calculated by the neural network as principally possible. A comparison between predicted and actual drape image shows that this idea does basically work.

This approach can be used to improve the output of a pure ANN considerably, thus making the results more reliable. Figure 4.25 shows the principle for the classification of input data which was developed by Hormes and Bergmann and successfully applied to the classification of trash particles in carded webs (Veit *et al.*, 1996). There, the output of the ANN is fed into a fuzzy system. In most cases, the output of an ANN as shown in Fig. 4.25 would not consist of a vector such as, for example, (1 0 0 0 0) but rather a vector such as (0.5 0.2 0.1 0.1 0.1). The output of the ANN is therefore not clearly class #1. The fuzzy system is able to 'translate' this fuzzy output into an accurate classification, e.g. class #1. The system used in Fan *et al.* (2001) and Su *et al.* (2008) is very similar.

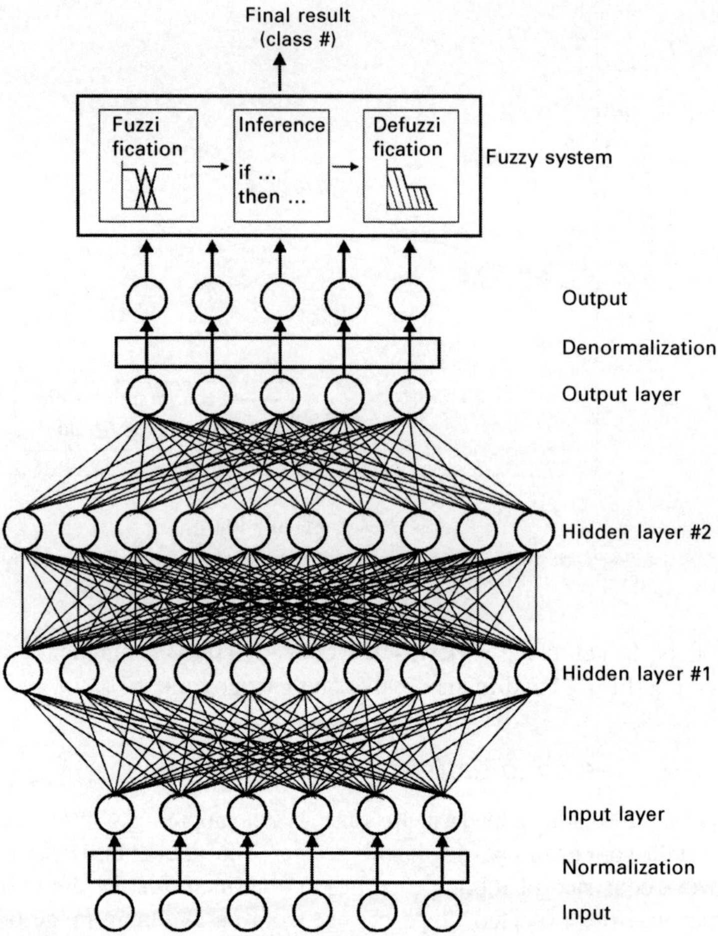

4.25 Principle of neuro-fuzzy system to classify input data.

4.8.8 Prediction of yarn properties in melt-spinning

Kuo *et al.* (2004) designed a fuzzy system that could determine tensile strength and yarn count of as-spun polypropylene fibres in the melt-spinning process from extruder screw speed, gear pump speed and winder speed. Although this approach is questionable as there are many more parameters influencing the tensile strength of melt-spun fibres, this system worked very well. They used triangular membership functions to describe the input and output parameters as shown in Fig. 4.26 for extruder screw speed and winding speed. Although the exact linguistic expressions and fuzzy rules are not elaborated, this approach appears worth pursuing further by including other possible relevant parameters in order to predict the characteristics of melt-spun filament yarns in general.

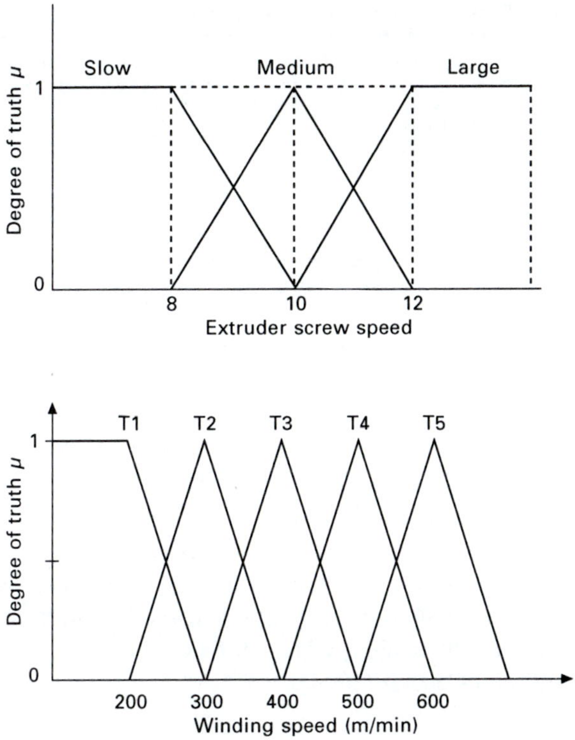

4.26 Membership functions for extruder screw speed and winding speed (adapted from Kuo *et al.*, 2004).

4.8.9 Prediction of fibre and yarn relationships

A very interesting approach to relate cotton fibre properties to yarn characteristics was designed by Admuthe and Apte (2010). They used a neural network-type system that contained adaptive nodes assigning membership function values to input data and propagating output to the subsequent layer. After a training with 450 data sets from 30 different varieties of Indian cotton, this system was able to determine, for example, yarn tenacity very accurately and even better than a conventional ANN as is shown in Table 4.3.

4.8.10 Other applications

Vachtsevanos *et al.* (1994) and Vachtsevanos and Sungshin (1999) describe the classic application of a fuzzy-based control system to improve several textile processes in a weaving mill. They describe the principle behind the system in detail and show that the use of a fuzzy control improves the overall performance of the weaving mill. This paper is recommended reading for everyone interested in applying fuzzy control to textile machines.

Table 4.3 Correlation between actual and predicted values

Yarn properties	ANN	ANN with fuzzy module
U (%)	0.93	0.97
Thick places	0.96	0.98
Thin places	0.94	0.99
Neps	0.87	0.91
Tenacity	0.89	0.97
Elongation	0.87	0.98

Source: adapted from Admuthe and Apte, 2010.

Other interesting papers include Huang and Chang (2001) who designed a fuzzy system to smooth the linear density deviation in a drawing frame. Liu *et al.* (2010) developed a neuro-fuzzy system to predict the fabric hand. An approach to determine the uniformity of nonwovens using a fuzzy neuro-system was published by Liu *et al.* (2010). A comparatively early paper on a fuzzy application was published by Zhao *et al.* (1996). They designed a tool to help with cotton allocation: if a certain type of cotton was running out, this system helped to decide by which type of cotton to replace it by keeping quality high and costs low. A comprehensive literature review on other fuzzy applications in textiles, which is recommended reading, was published by Cimilli and Candan (2008).

4.9 Practical advice in applying fuzzy logic

In general, the performance of a fuzzy system is only as good as the rules on which it is based. In order to be able to describe a process accurately with rules, it is therefore highly recommended to base the model not only on statistical data gathered through laboratory trials but especially on the knowledge of experts, preferably from industry. As explained above, it is an absolute requirement that to all possible states of the system at least one rule must be assigned. The membership functions also have an influence on the result, but are not crucial in terms of whether the fuzzy controller works in the first place. Fuzzy modules can be used to further improve the output of an artificial neural network, especially for classification problems. In publicly available literature, there are few papers on fuzzy applications for textile machine control, although the system is perfectly suited for this kind of task. It remains to be seen whether this changes in the near future as fuzzy logic experiences a renaissance in science.

4.10 Web references for software tools

There is a wide range of software tools and general information on the world wide web. The following links are a selection of useful websites.

Fuzzy software

http://www.freedownloadmanager.org/downloads/fuzzy_logic_software

Free fuzzy software

http://fuzzy.cs.uni-magdeburg.de/nefcon/
http://www.trollynet.de/en/open-source/fuzzy-software-samt-fuzzy
http://sourceforge.net/projects/jfuzzyqt/

Commercial fuzzy software (selection)

http://www.fuzzytech.com/
http://www.mathworks.com/help/toolbox/fuzzy/rn/br5s6fn-1.html

Online tutorials

http://www.jimbrule.com/fuzzytutorial.html
http://www.calvin.edu/~pribeiro/othrlnks/Fuzzy/home

4.11 References

Admuthe, L.S., Apte, S. (2010), 'Adaptive neuro-fuzzy inference system with subtractive clustering: a model to predict fiber and yarn relationship', *Textile Research Journal, 80 (9)*, 841–846.

Cimilli, S., Candan, C. (2008), 'Fuzzy logic and its usage in textile', *Proceedings of the 4th International Textile, Clothing and Design Conference*, 724–729.

Dayik, M., Kayacan M.C., Acar, M., Calis, H. (2008), 'Modeling of warp tension variation during shedding operation using fuzzy logic', *Journal of the Textile Institute*, *99 (6)*, 505–514.

Fan, J., Newton, E., Au, R. (2001), 'Predicting garment drape with a fuzzy-neural network', *Textile Research Journal*, *71 (7)*, 605–608.

Hadjianfar, M., Semnani, D. (2010), 'Fuzzy classification of luster of textile fabrics', *Textile Research Journal 2011*, *(81)*, 649–656.

Hayashi, Y. (1994), 'Neutral expert system using fuzzy teaching input and its application to medical diagnosis', *Information Sciences Applications*, *1*, (*1*), 47–58.

Huang, C.-C., Chang, K.-T. (2001), 'Fuzzy self-organizing and neural network control of sliver linear density in a drawing frame', *Textile Research Journal*, *71 (11)*, 987–992.

Huang, C.-C., Yu, W.-H. (2001), 'Fuzzy neural network approach to classifying dyeing defects', *Textile Research Journal*, *71 (11)*, 100–104.

Kuo, C.-F. J., Hsiao, K.-I., Wu., Y.-S. (2004), 'Using fuzzy theory to predict the properties of a melt spinning system', *Textile Research Journal*, *74 (3)*, 231–235.

Lin, J.-J. (2009), 'A genetic algorithm-based solution search to fuzzy logical inference for breakdown causes in fabric inspection', *Textile Research Journal*, *79*, 394–409.

Lin, J.-J., Lin, C-H., Tsai, I-S. (1995), 'Applying expert system and fuzzy logic to an intelligent diagnosis system for fabric inspection', *Textile Research Journal, 65*, 697–709.

Liu, J., Zuo, B., Vromen, P., Rabenasolo, B., Zeng, X. (2010), 'Identification of nonwoven uniformity using generalized Gaussian density and fuzzy neural network', *Journal of the Textile Institute, 101, (12)*, 1080–1094.

Mauer, G. (1995), 'Fuzzy logic controller for an ABS braking system', *IEEE Transactions on Fuzzy System, 3 (4)*, 381–388.

Meier, W., Weber, R., Zimmermann, H.-J. (1994), 'Fuzzy data analysis. Methods and industrial applications', *Fuzzy Sets and Systems, 61 (1)*, 19–28.

Nasiri, M. (2005), 'Fuzzy regression modelling of false twist texturing yarn', *Proceedings of International Fuzzy Systems Association World Congress 2005*, 142–146.

Nasiri, M., Berlik, S. (2009), 'Modeling of polyester dyeing using an evolutionary fuzzy system', *Proceedings of IFSA-EUSFLAT 2009*, 1246 –1251.

Nasiri, M., Hüllermeier, E., Senge, R., Lughofer, E. (2011), 'Comparing methods for knowledge-driven and data-driven fuzzy modeling: a case study in textile industry', *Proceedings of International Fuzzy Systems Association World Congress/Asian Fuzzy Systems Society 2011*, p. 103.

Nguyen, H.T., Walker, E.A. (2005), *A first course in fuzzy logic*, CRC Press Inc., 3rd edition.

Nitta, T. (1993), 'Applications of neural networks to home appliances', *Proceedings of the International Joint Conference on Neural Mateworks (1)*, 1056–1060.

Osthus, T. (1997), 'Prozessoptimierung in der Weberei durch automatische Einstellung von Streichbaum und Kettwächterkorb', Dissertation, RWTH Aachen University.

Ross, T. (2010), *Fuzzy Logic with Engineering Applications*, John Wiley & Sons, 3rd edition.

Stadnik, O.E. (1986), 'Fuzzy decision rules in medical diagnosis and prognosis', *Automation and Remote Control 47, (3) pt.* 2, 406–411.

Su, T.-L., Kung, F.-C., Kui, Y.-L. (2008), Applicatoin of back-propagation neural network fuzzy clustering in textile texture automatic recognition system', *Proc. of International Conference on Wavelet Analysis and Pattern Recognition*, 46–49.

Tavanai, H., Taheri, S.M., Nasisi, M. (2005), 'Modelling of colour yield in polyethylene terephthalate dying with statistical and fuzzy regression', *Iranian Polymer Journal, 14 (11)*, 954–967.

Vachtsevanos, G.J., Sungshin, K. (1999), An intelligent approach to integratoin and control of textile processes', *Information Sciences, 123 (2000)*, 181–199.

Vachtsevanos, G.J., Dorrity, J.L., Kumar, A., Kim, S. (1994), 'Advanced application of statistical and fuzzy control to textile processes', *IEEE Transactions on Industry Applications, 30, (3)*, 510–516.

Veit, D., Hormes, I., Bergmann, J.; Wulfhorst, B. (1996), 'Image processing as a tool to improve machine performance and process control', *International Journal of Clothing Science and Technology, 8*, (1/2), 66–72.

Yazicioglu, Y. (2008), 'A fuzzy logic controlled anti-lock braking system (ABS) for improved braking performance and directional stability', *International Journal of Vehicle Design, 48, (3–4)*, 299–315.

Zadeh, L. A. (1965), 'Fuzzy sets', *Inf. Control, 8*, 338–353.

Zhao, L., Peng, C.-S, Cheng, S.-C. (1996), 'An application of fuzzy dynamic integrated judgement to cotton allocation', *Information Sciences, 94*, 277–290.

Zimmermann, H.-J. (2001), *Fuzzy Set Theory – and its applications*, Springer, 4th edition.

Zimmermann, H.-J. (2010), 'Fuzzy set theory', *Wiley Interdisciplinary Reviews: Computational Statistics, 2, (3)*, 317–332.
Zimmermann, H.-J., Sebastian, H.-J. (1994), 'Fuzzy design – integration of fuzzy theory with knowledge-based system-design', *Proceedings of the IEEE International Conference on Fuzzy Systems,* 1, 352–357.

5

Computational fluid dynamics (CFD) and its application to textile technology

P. JUNGBECKER and D. VEIT, RWTH Aachen University, Germany

Abstract: Computational fluid dynamics (CFD) deals with numerical flow simulation. By applying CFD, even complex problems can be solved using a numerical iteration process. A commonly used method involves the Navier–Stokes equations which represent a system of equations for continuity, impulse and energy conservation of a fluid. Typical examples for the simulation of complex flows are the air current around aircrafts and inside cars, and the modelling of heat transfer for a wide range of applications, including in textile processing.

Key words: computational fluid dynamics (CFD), simulation, modelling, measuring techniques, textile technology.

5.1 Introduction

All programs that simulate currents are based on equations describing fluid mechanics. Those date back to Roman times when canals, baths and aqueducts were built using laws of fluid mechanics. Leonardo da Vinci (1452–1519) used fluid mechanics for various applications, e.g. to describe water passing obstacles. Important contributions were made between the seventeenth and the nineteenth centuries by Isaac Newton (1643–1727) with the three axioms of classic mechanics, by Jakob I. Bernoulli (1655–1705) with the one-dimensional energy equation and by Leonhard Euler (1707–1783) who described the law of impulse conservation for frictionless currents.

Two of the most important researchers in the field of fluid mechanics were Claude Navier (1785–1836) and George G. Stokes (1819–1903). Their findings allowed the extension of Euler's equation using the Navier–Stokes equations by taking into account the influence of friction. Although originally regarded as of not much use, today the Navier–Stokes equations are the basis for most computational fluid dynamics (CFD) software. Because of its complexity, CFD can only be solved using an iterative, numerical approximation, which still requires a powerful computer. Calculation time varies, depending on the problem at hand, between fractions of seconds and months.

More recent contributions include Rodi and Spalding (1970) who implemented several algorithms that are still used today. Another important contribution was made by Baliga and Patankar (1980). The first commercial CFD software became available in the 1980s. It was firstly used by the aircraft industries. Since the 1990s, CFD simulations have been widely used, e.g. in the automotive and the textile industries. Today, CFD is an accepted part of computer-aided engineering (CAE). Good introductions into CFD and the theory behind it can be found in Abbot and Basco (1989), Bhaskaran and Collins (2008), Cebeci *et al.* (2005), Hirsch (2007) and Mason (1998).

5.2 From reality to simulation

Today, a wide range of commercial CFD software is available (see Section 5.9). Nevertheless, it is of crucial importance that the user is familiar with simulation in general and also with the physical laws that apply in order to select a suitable model and the respective algorithms. This is a major difference to many other simulation tools (e.g. neural networks). This is illustrated in Fig. 5.1.

Starting from reality, the first step includes abstracting and creating a model which should be kept as simple as possible. This already incorporates the possibility of an error which should not be underestimated. A thorough analysis of the laws governing the problem and hence a suitable model is crucial for the success of the simulation. In subsequent steps, applying the model first leads to the physical relations between the parameters, then a numerical representation, subsequently to the actual simulation and final results which are to be verified. During each step, the user is advised to check for errors so as not to distort reality beyond acceptable limits or to achieve a completely wrong solution.

Common sources of error of a CFD simulation are:

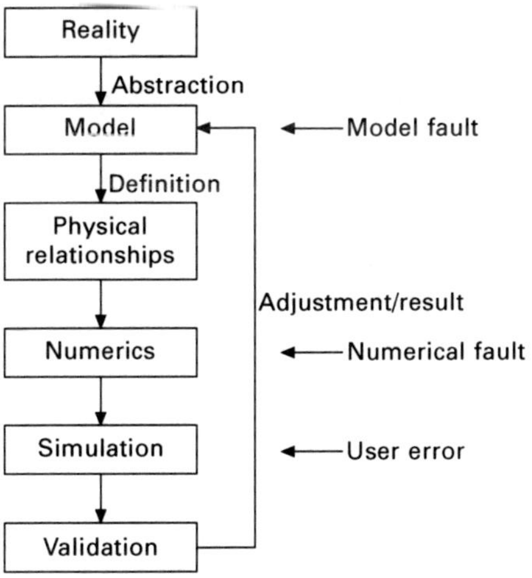

5.1 From reality to simulation.

- simplification of reality, hence not all aspects and details are taken into account;
- inaccuracies of the solution due to rounding errors depending on the numerical algorithms which are used;
- wrong boundary conditions;
- wrong net structure.

As Fig. 5.2 shows, choosing a suitable number of nodes to represent the current flow profile is a decisive step. In this example, the parabolic velocity profile can be approximated by three nodes with a trapezoidal cross-section which represents an easy-to-calculate solution. However, using 15 nodes leads to a solution which coincides much better with the original parabolic profile. In the end, every simulation solution is only as good as its validation which is thus an indispensable part of the process. It must be noted though that it is very difficult to measure currents without affecting the current itself and also that measuring inaccuracies are not uncommon.

5.2.1 Reality

Most phenomena involving the flow of a fluid are very complex in reality. In order to describe them, a certain degree of abstraction is necessary which still takes into account all relevant parameters and conditions. For example, one milligram of water consists of approximately 3.3×10^{19} molecules with different velocities. It is impossible to create a model which incorporates each molecule separately. Hence, water is normally regarded as a continuum having a constant density and temperature if only the macroscopic, integral forces are of interest.

5.2.2 Model

As described above, a model normally only describes the influence of those parameters that are considered to be of importance. If non-relevant parameters are not taken into account, the model becomes more complex without improving the final results. If, in contrast, relevant parameters are neglected, the model will often not mirror reality to the desired extent. Typical examples of models for a CFD simulation are:

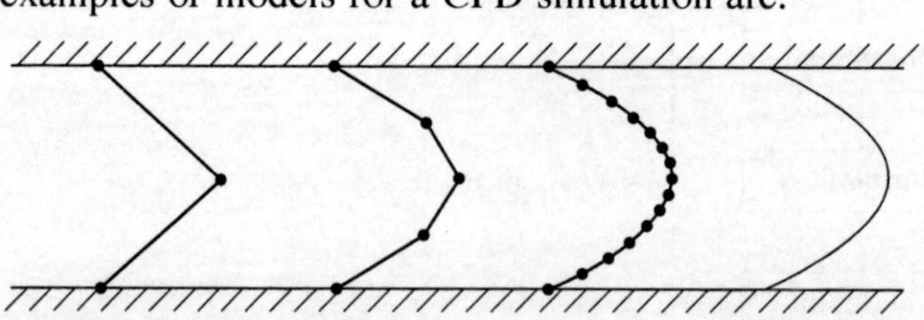

5.2 Influence of the number of nodes on the model.

- *Water waves*: in order to design a model that simulates wave propagation, water is regarded as a continuum. Friction effects in the water and between water and air are neglected. Also temperature, water colour and the number of living beings within the observed volume are not relevant. Salt content is represented by the average density. Important parameters include the distance between wave crests (wave length), wave amplitude and gravitational force.
- *Perfusion of fabrics*: a fabric made up of staple fibre yarns represents a very complex geometrical problem for a flow simulation. It consists of a large number of intersecting yarns which in turn are built up of single fibres. For an outdoor application where the fabric is subjected to wind, an intimate knowledge of the current around each single fibre is not relevant. Hence, applying an air resistance factor of the fabric relative to the air velocity would be sufficient to model the fabric as a flat area.
- *Polymer flow in melt spinning*: melt distribution inside a spin pack can be simulated by regarding the temperature as isothermal despite its high value. However, density and viscosity of the melt must be adapted to the spin temperature.

In order to create a model for fluids, there is a wide range of possibilities, namely:

- *Lagrange approach*: a fluid particle's (molecule, atom, agglomerate) trajectory is observed and its properties are recorded. As a result, the time-dependency of the parameters is recorded but not their dependency on a position. A typical example are currents of granulate or sand.
- *Euler approach*: the flow field is regarded as a continuum which allows to determine the parameters depending on their location.
- *Euler–Lagrange approach*: this is a combination of both approaches and is especially suitable for the description of single particles in a continuous medium, e.g. the movement of fibres or yarns in a current.

To illustrate these approaches, one can make a thought experiment: if a continuous flow is recorded for a given point in time, this results in a description according to Euler. If a fluid particle in this field is marked repeatedly at given time intervals, this results in a series of Euler fields. By connecting the marked particle hence obtaining their trajectory, one obtains a description according to Lagrange.

5.2.3 Physical relations

After determining the parameters of the model by abstracting reality, they need to be connected applying the laws of physics. Newton fluids, e.g. air and water, can be described according to Euler's approach using the

Navier–Stokes equations, provided that the laws of continuum mechanics apply. They make up the all-embracing model for the description of fluids, and consist of a system of non-linear, partial differential equations of the second order which describe a continuous flow, taking into account friction and heat conduction.

The conservation equation for mass is given by:

$$\frac{\partial v}{\partial t} + \nabla \cdot \rho \cdot \vec{v} = 0 \quad [5.1]$$

According to equation [5.1], the sum of time-dependent change of the density ρ and the three-dimensional change of the current density $\rho\vec{v}$ is nil. The mass of the current is preserved.

The conservation equation for impulse is given by:

$$\frac{\partial \rho\vec{v}}{\partial t} + \nabla \cdot (\rho\vec{v}\vec{v} + \bar{\bar{\sigma}}) = \vec{f}_{\text{Vol}} \quad [5.2]$$

The term $\vec{v}\vec{v}$ represents the dyadic product of the velocity, with

$$\vec{v}\vec{v} = \begin{pmatrix} u^2 & nv & uw \\ vu & v^2 & vw \\ wu & wv & w^2 \end{pmatrix} \quad [5.3]$$

which, when multiplied with the density, results in $\rho\vec{v}\vec{v}$. The term $\bar{\bar{\sigma}}$ stands for the tension tensor. The areal change of the sum of impulse flow and tension, added by the time-dependent change of the flow density, results in the volume force $\vec{f}_{\text{Vol}}$, which can be equal to, for example, the gravitational force, hence we have $\vec{f}_{\text{Vol}} = \rho \cdot \vec{g}$.

The conservation equation for energy is given by:

$$\frac{\partial \rho E}{\partial t} + \nabla \cdot (\rho E\vec{v} + \bar{\bar{\sigma}}\vec{v} + \vec{q}) = \vec{f}_{\text{Vol}} \cdot \vec{v} \quad [5.4]$$

The time-dependent change of the energy per volume unit ρE, added by the areal change of the sum of the energy flow $\rho E\vec{v}$, the power of the tensions $\bar{\bar{\sigma}}\vec{v}$ and the heat flow $\vec{q}$ is equal to the power of the volume forces $\vec{f}_{\text{Vol}} \cdot \vec{v}$. The conservation equations for mass, impulse and energy also include deduction after time. These can be neglected for stationary processes so that only the stationary conservation equations remain.

5.2.4 Numerics

Numerics are a subarea of classic mathematics. The respective approaches are mainly used to solve equations, in most cases with a computer. Some physical relations can be determined analytically. Others, such as systems of partial equations, are normally solved numerically via an iteration process. The equations are then discretized into algebraic equations.

Discretization

By applying a time discretization, the process is analysed at different points in time. Similarly, a local discretization implies the division of the geometry into small parts which are then analysed separately. According to Löhner (2008), this results in either structured or unstructured cross-linkings, which can be defined as follows:

- micro-structured: each node within the net is surrounded by the same amount of adjacent nodes;
- micro-unstructured: each node within the net is surrounded by a different amount of adjacent nodes;
- macro-unstructured, micro-unstructured: each net consists of sub-meshes which are micro structured.

Figure 5.3 shows the cross-linking of a volume with a structured (left) and an unstructured (right) cross-linking.

The discretized equations are then linearized in order to transform them into a numerically solvable system of equations which represents the whole flow field. According to Peiró and Sherwin (2005), the most important methods of discretization are:

- finite differences method;
- finite element method (see Chapter 6); and
- finite volume method.

When using the finite differences method, the current is represented by single points. This comparatively simple method is especially suitable for structured grids and one-dimensional problems. This method can cause convergence problems for a steep gradient.

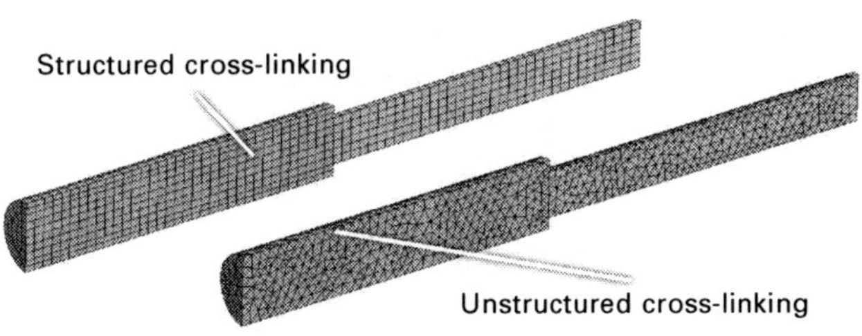

5.3 Volume element with structured (left) and unstructured (right) cross-linking.

When using the finite volume method, the current is converted into basic functions with several degrees of freedom. Their programming is complex but this method allows the simulation of any unstructured grid. Steep gradients can be taken into account with special algorithms. The finite element method is a direct approximation of the integral conservation equations. It is common to solve problems related to flow simulations. Mass, impulse and energy are preserved. Owing to their integral form, the equations can be used for discontinuous problems as well as for unstructured grids. Figure 5.4 shows a comparison of the three methods.

The discretized systems of equations can be approximated using a range of approaches, all striving to achieve convergence via iterative approaches. In general, there are the following methods:

- *One-step method*: data about the solution curve are used for only one time step in the interval $[t_{n-1}, t_1]$. Hence, this method takes data into account only of point in time t_{n-1}, but not those of previous steps. The timespan Δt determines the congruence behaviour of the solution. It is sufficient to set the starting point, hence the method is called 'self-starting'.
- *Multi-step method*: this method also uses data prior to t_{n-1}, hence taking several steps into account. It is not suitable to numerically solve complex, dynamic systems as the determination of data sets prior to the current set is a time-consuming process.
- *Explicit method*: this method is the most simple time discretization process. The new values of the parameters at point in time t_{n+1} can directly be deducted from those at point in time t. Its main advantage is the easy and hence quick dissolution of the system of equations. Its drawback is the limitation of the increment due to numerical instability.
- *Implicit method*: this method does not have that restriction, but the time required to calculate the values for each time step is much greater. Hence, this method is preferred for applications with large increments, especially when convergence for a stationary flow is an issue.

Turbulence models

Turbulence in the current can have a direct effect on the flow field. It is often caused by surface irregularities or barriers blocking the current. Normally,

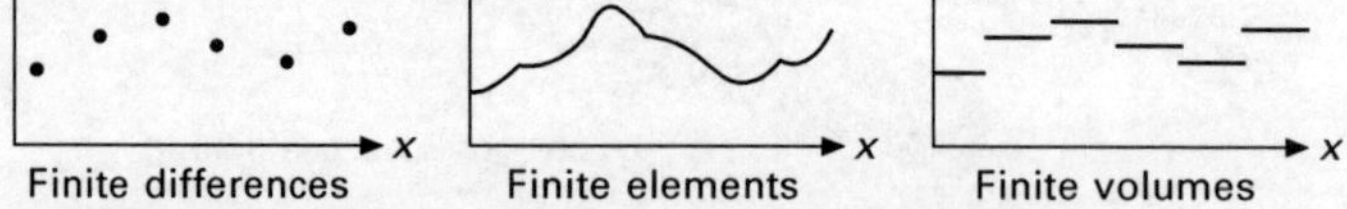

5.4 Comparison between finite differences, finite element and finite volume method (adapted from Roller *et al.*, 2005).

the average flow velocity is superimposed by a time-dependent velocity oscillation. In contrast to a laminar flow, this can cause a mixing of the various flow layers. The current velocity of a turbulent flow is given by

$$v(t) = \vec{v} + v' \quad [5.5]$$

where $\vec{v}$ stands for the average velocity over time and v' represents the turbulence component of the velocity.

A turbulent flow is characterized by the degree of turbulence *Tu*. It represents the size of the time-dependent oscillations of the flow velocity relative to the average velocity. For a three-dimensional flow field, it can be calculated as follows:

$$Tu = \frac{1}{v} \cdot \sqrt{\frac{1}{3} \cdot (\overline{u}'^2 + \overline{v}'^2 + \overline{w}'^2)} \quad [5.6]$$

where u', v' and w' stand for the oscillation components of the flow speed.

Figure 5.5 shows the time-dependent oscillations of a turbulent flow for high and low values of Tu. The time-dependent oscillations of a turbulent flow can be explained by vortex-like structures within the current. The so-called eddies are three-dimensional structures which can be as large as the current itself and as small as a single fluid particle. Whereas the large eddies cause an exchange of impulses within the current, the small eddies transform turbulent energy into heat through dissipation.

Several models are used to simulate a turbulent flow.

Reynolds-averaged Navier–Stokes equations (RANS)

The Reynolds-averaged Navier–Stokes equations (RANS) are most commonly used in engineering. They do not disintegrate turbulent structures with regard to time and location but model the relevant parameters through a turbulent kinetic energy approach. This is a comparatively quick method which normally leads to good results.

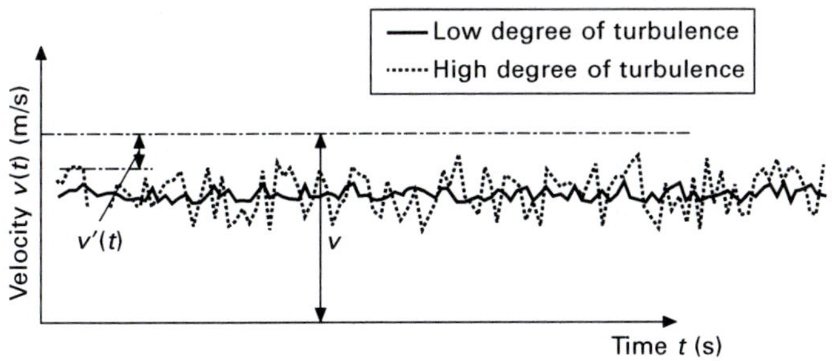

5.5 Comparison between turbulent currents with high and low degree of turbulence.

Standard k–ε turbulence model

This model was developed by Launder and Spalding (1974) and describes the development of the turbulent, kinetic energy k and the isotropic dissipation rate ε. It leads to viable results for those areas away from the surrounding wall. The isotropic dissipation rate ε represents the kinetic energy of the turbulent flow which is either transformed or dissipated into inner energy (heat) per time and mass unit. Energy is extracted from the main current by large turbulence elements and passed on to smaller elements.

K–ω turbulence model

Salari *et al.* (1994) developed another model which can also be used to describe the flow field close to the surrounding wall and in shear layers. It describes the relation between kinetic energy k and the specific, isentrope dissipation rate ω according to:

$$\omega = \frac{\sqrt{k}}{l_t} \qquad [5.7]$$

where l_t represents the eddies' length. This model is very sensitive to small changes of the turbulence parameters at the edges of the flow field.

Shear–stress transport $k - \varpi$ model

The so-called shear–stress transport (SST) model was invented by Menter (1994). It combines the advantages of the aforementioned models, shows a higher accuracy and is more reliable while still requiring the same computing time. In areas close to the wall, turbulence is described applying the k–ω model, in areas away from the walls, the k–ε model is used. A transition function is defined which assumes values between 0 and 1. This value is multiplied with the respective function and both functions are then added. The distinguishing feature of this model is how the flow at the wall is simulated. If the node which is nearest to the wall is further away from it than a defined distance, the wall function of the k–ε model is used. If it is closer, then the Low-Reynolds equation (for currents with low Re numbers) is gradually taken into account. This meets on the one hand the general demand for a sufficiently high number of nodes in order to describe the flow with a high degree of accuracy. On the other hand, the automatic switch to the Low-Reynolds equation ensures that, independent of the grid resolution, the flow close to the wall is always described properly.

5.2.5 Simulation

During the simulation, numerical methods are applied in order to solve the equations representing the governing laws of physics. In CFD, this is normally a finite element method and hence an iterative process. Thus, in the beginning, to each cell, a starting value for velocity, turbulence and pressure is allocated. This initial solution is then converted to the actual solution step-by-step. This convergence process can be supervised by determining the residues. They represent the difference between the current solution and the true solution. This can be achieved by observing the relevant parameters, e.g. maximum flow velocity or temperature at the outlet. If both convergence criteria are met, the simulation can be terminated as the flow field will not change any more.

If the initial solution is too far away from the true solution, the process can be divergent which requires ending the simulation. In this case, it can be useful to adjust the boundary conditions step-by-step to their final values. A typical example is a supersonic current where the simulation starts with a low pressure which is then gradually increased to the maximum value. Plate I (see colour section between pages 152 and 153) shows a typical example of the development over time of the residuals.

5.2.6 Validation

After the simulation, each cell is described by values that characterize the state of the cell within the current. From them, the necessary information to describe the flow must be extracted. One possibility is to export the numerical result as single values or in tables, another their summary in diagrams for velocity and pressure over time, yet another the three-dimensional depiction of the flow. The main difficulty is the presentation of the results in such a way that users receive the information they want.

A plausibility test is an indispensable part of every validation. This can simply be done by checking whether the current flows into the right direction and whether all boundary conditions are met. In addition, sophisticated measuring technology can be employed to validate the simulation in certain areas (see Section 5.4).

5.3 Simulation of currents with fibres, yarns and textiles

For many applications in textile technology, apart from the movement of the flow, the movement of fibres and yarns inside the flow is also of great interest. This leads to coupled simulations where fibres and yarns move freely through a current and are in turn affected by it. In most cases, the preferred model to tackle this problem is the finite volume method. Fibres and yarns are then not regarded as a continuum but as single objects and hence treated separately.

If fibres and yarns are to interact with the current, there must be an interface at which an exchange of impulse can take place. Fibres and yarns are normally longer than they are wide; typical ratios are in the range of 1000:1–100 000:1. Besides, the geometrical dimensions of the current are in most cases much greater than the fibre or yarn diameter. Magnitude differences between the smallest (fibre) and the largest (machine) dimension can reach 1 000 000 and more.

One possibility to exchange impulses between yarn and current is the body-fitted-mesh method (Fig. 5.6, top). There, the yarn is completely integrated into the grid that describes the current. This implies an enormously high number of nodes and the calculation step in each time interval is very time consuming. The great advantage of this method is though that the exchange of impulses can be directly determined via turbulence models as described above.

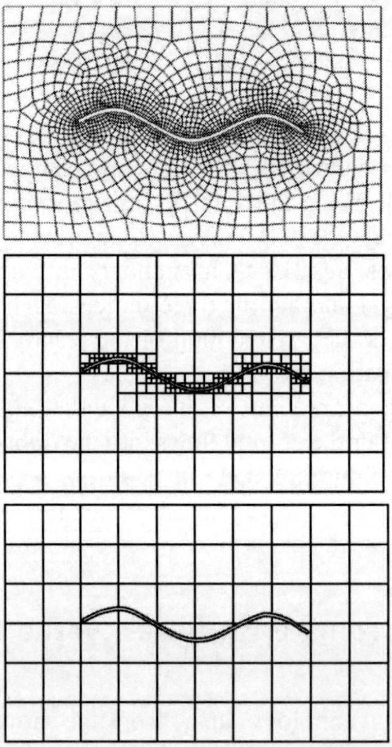

5.6 Different methods to model a fibre or yarn in a current (top: body-fitted mesh; middle: immersed-boundary-layer; bottom: unchanged net).

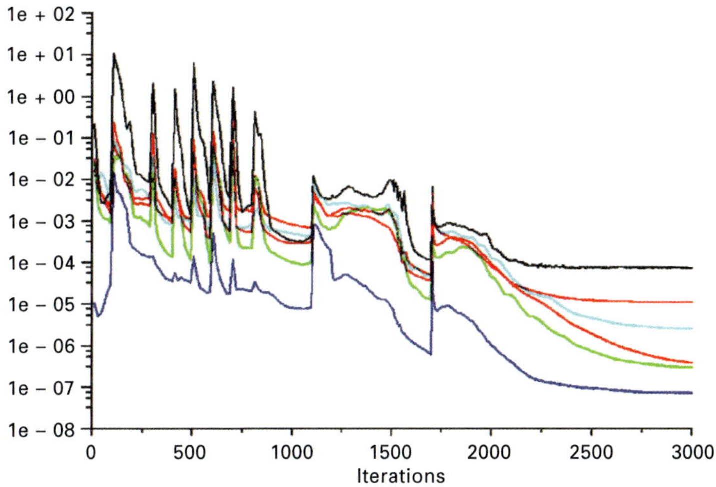

Plate I Development over time of the residuals for a typical CFD simulation.

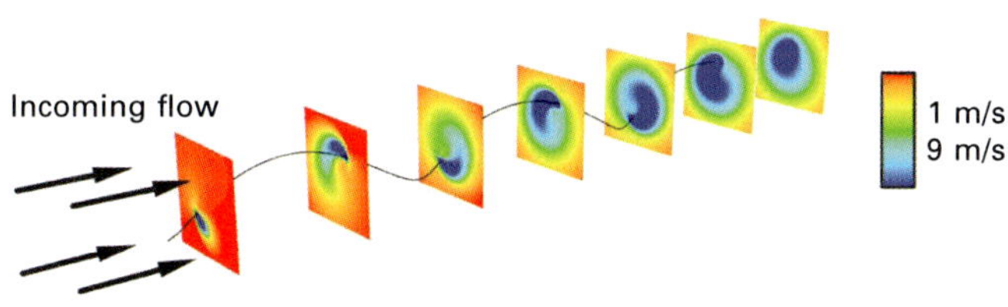

Plate II Boundary layer around a crimped fibre in an air flow.

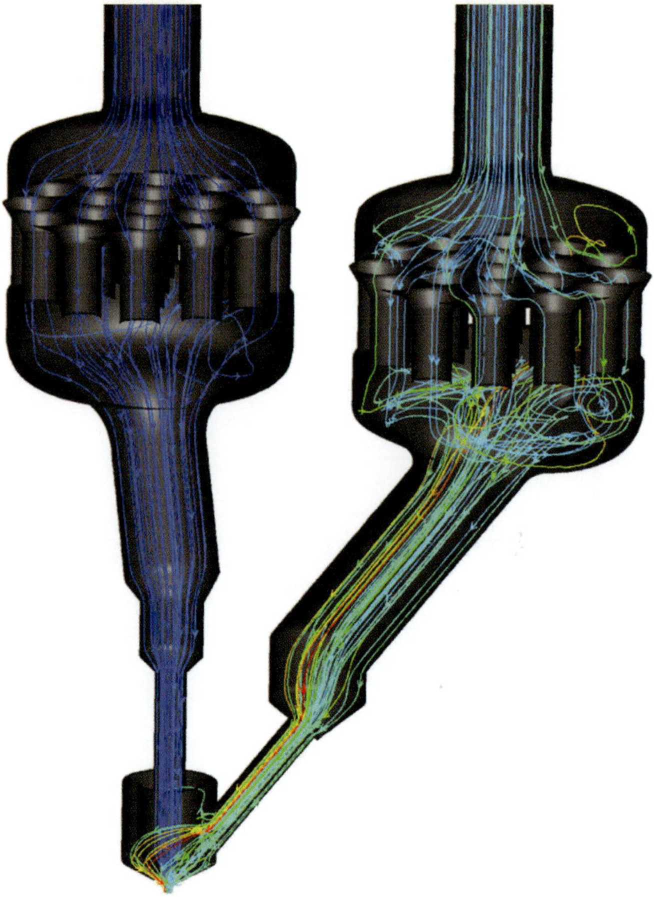

Plate III Flow lines of the polymer inside a bicomponent spin pack.

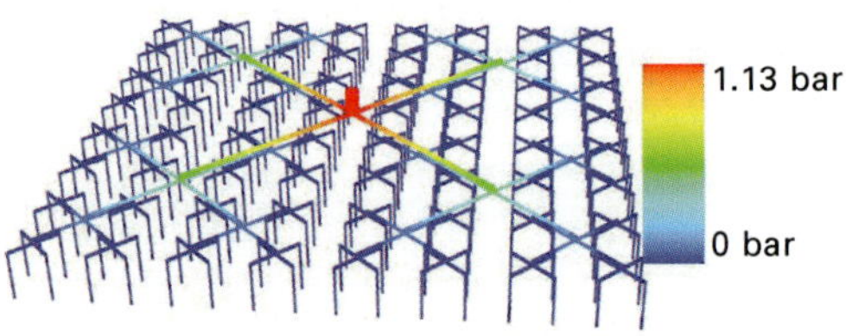

Plate IV Simulation of pressure distribution in spin pack.

Plate V Flow in the redesigned drafting zone.

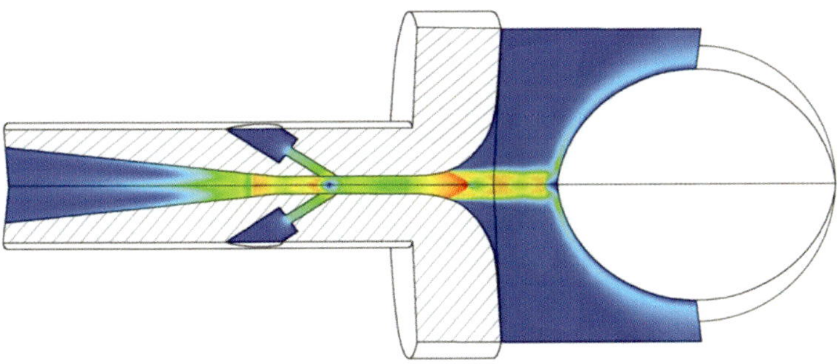

Plate VI Flow inside and outside the air-jet.

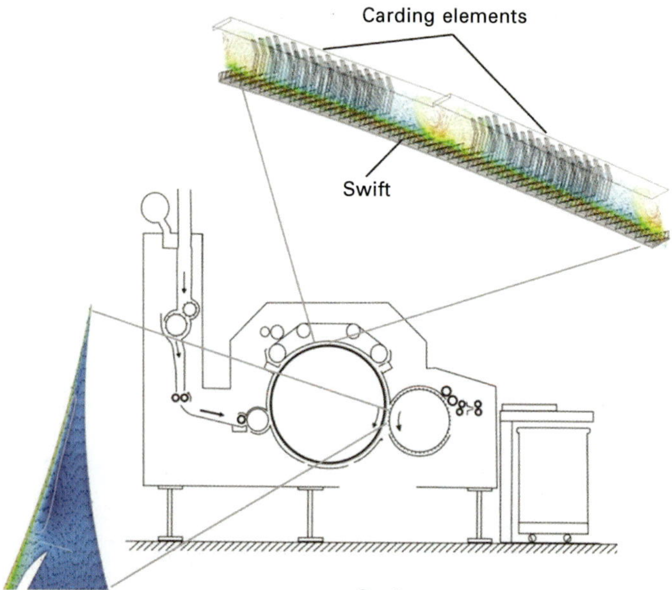

Plate VII Air flow inside a card (left: between swift and doffing cylinder, top: between carding elements and swift).

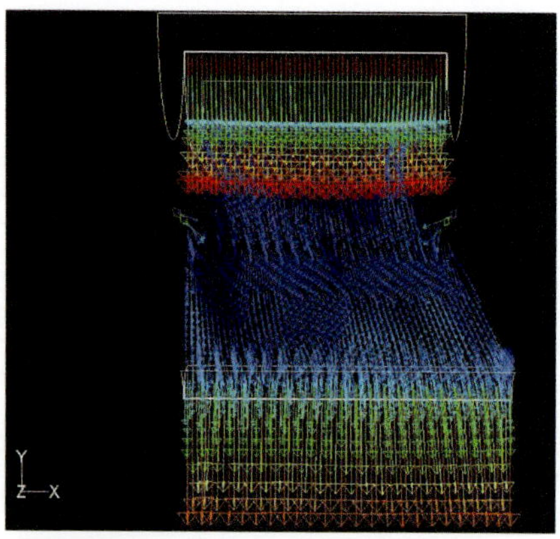

Plate VIII Air-flow profile in the duct with one slot at 45° on each side (courtesy of Professor A. Das, IIT Delhi).

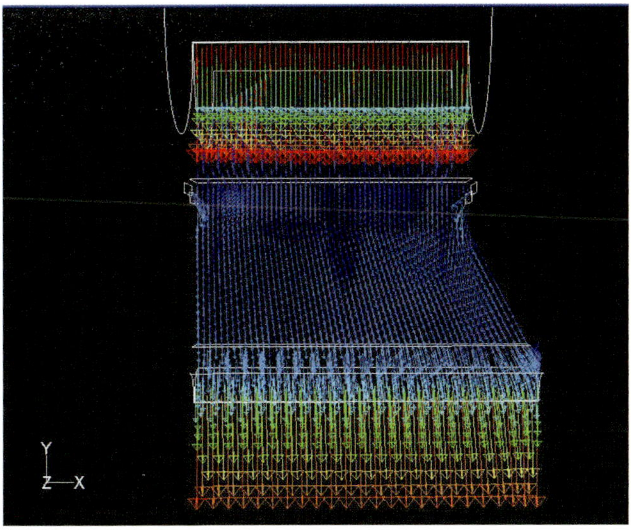

Plate IX Air-flow profile in the duct with one slot at 30° on each side (courtesy of Professor A. Das, IIT Delhi).

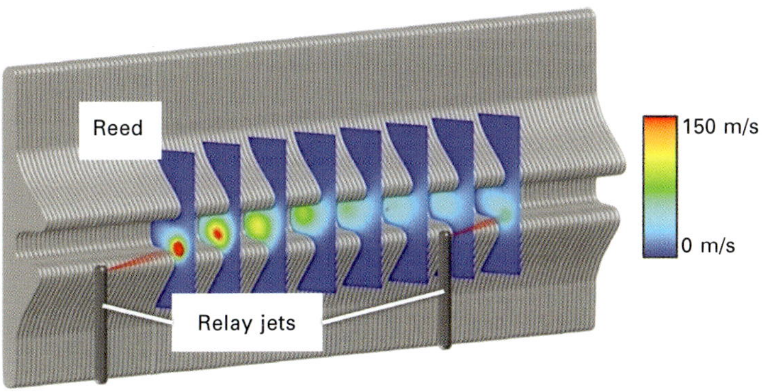

Plate X Three-dimensional flow in the reed channel.

Plate XI Flow through a multi-layered braiding.

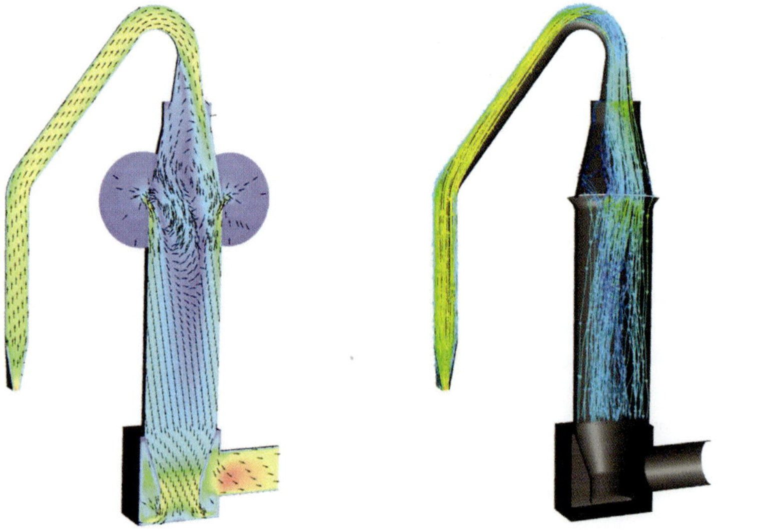

Plate XII Simulated air flow (left) and fibre trajectories (right) for an aerodynamical nonwovens process.

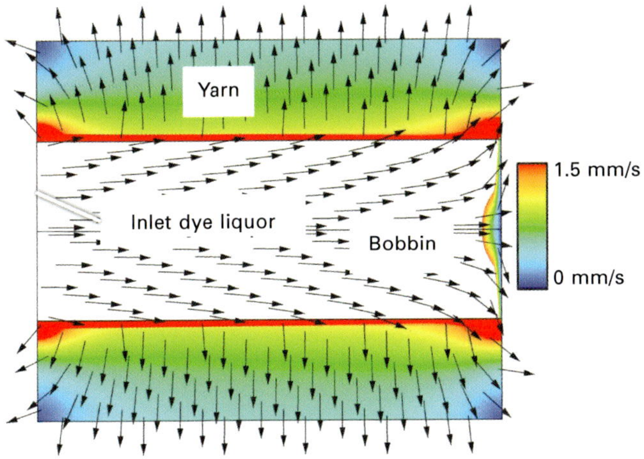

Plate XIII Flow velocities inside a dye bobbin.

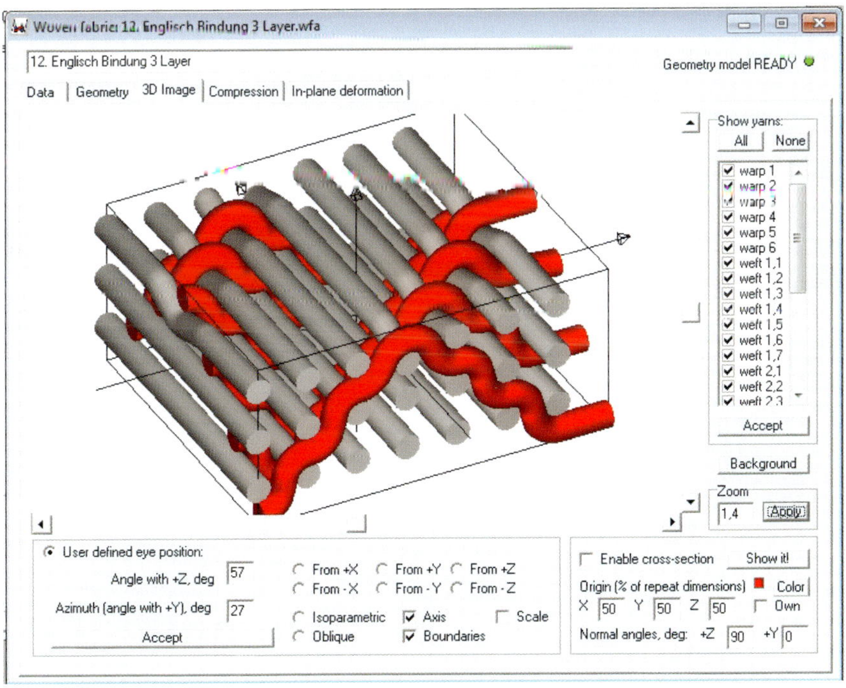

Plate XIV A multilayer woven structures (WiseTex Software).

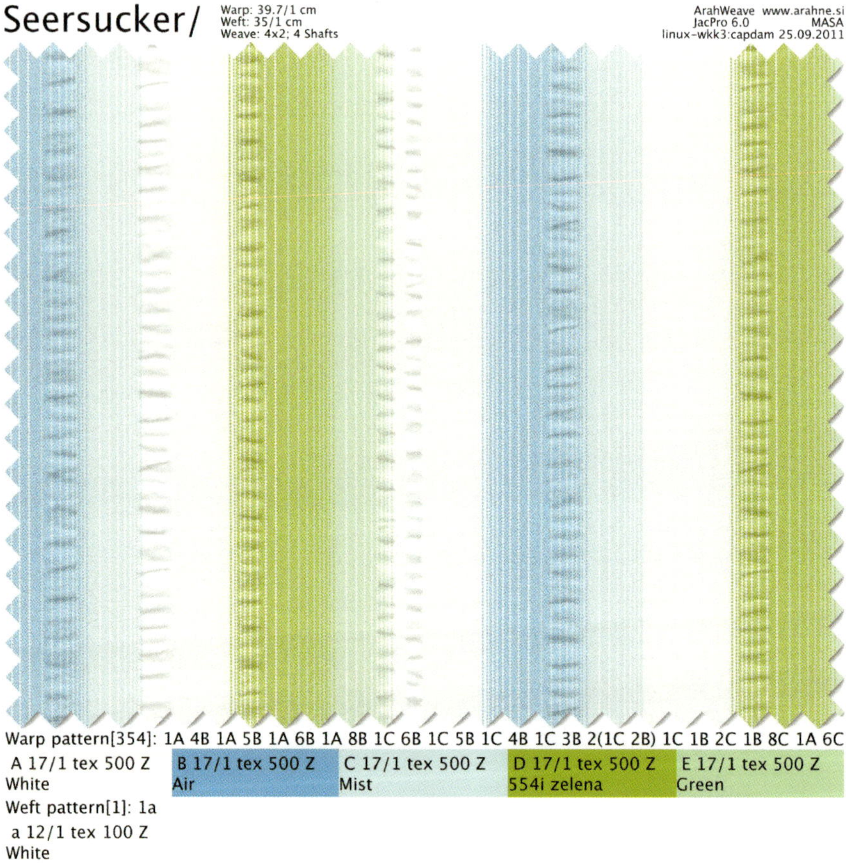

Plate XV Simulated fabrics by Arahne (Arahne, 2011). Courtesy of Arahne company http://www.arahne.si/.

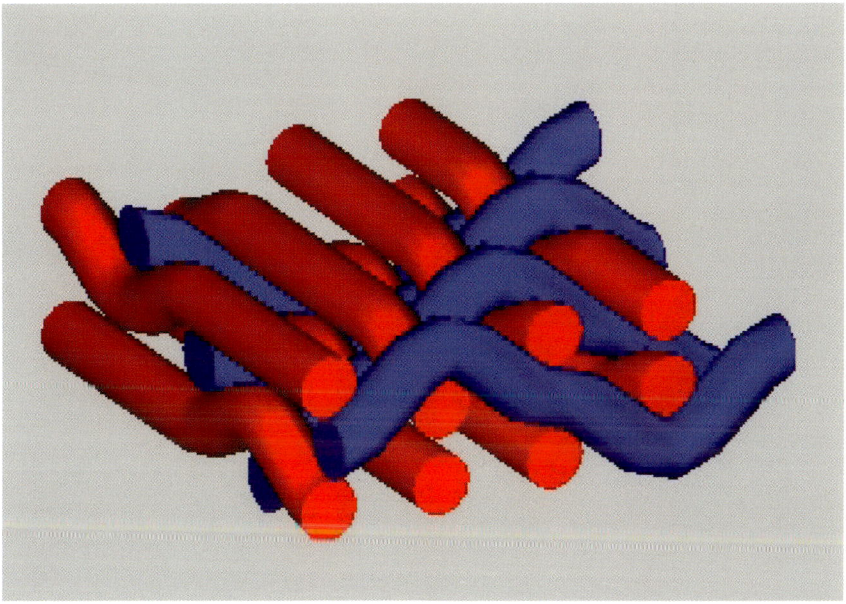

Plate XVI Explanation of the coding (WiseTex Theory Manual).

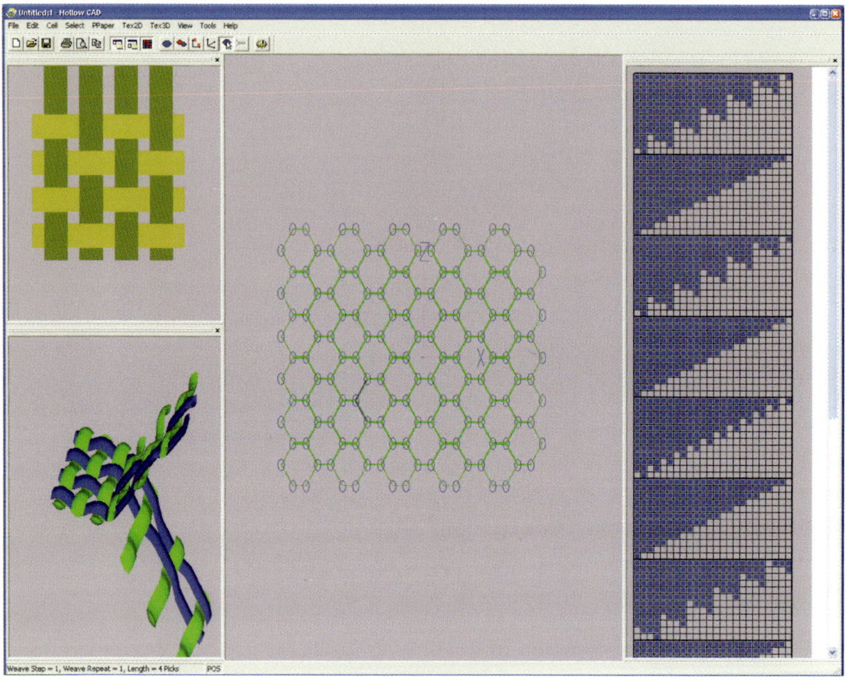

Plate XVII Interface for 3D hollow fabrics with the patterning for the weaving machine.

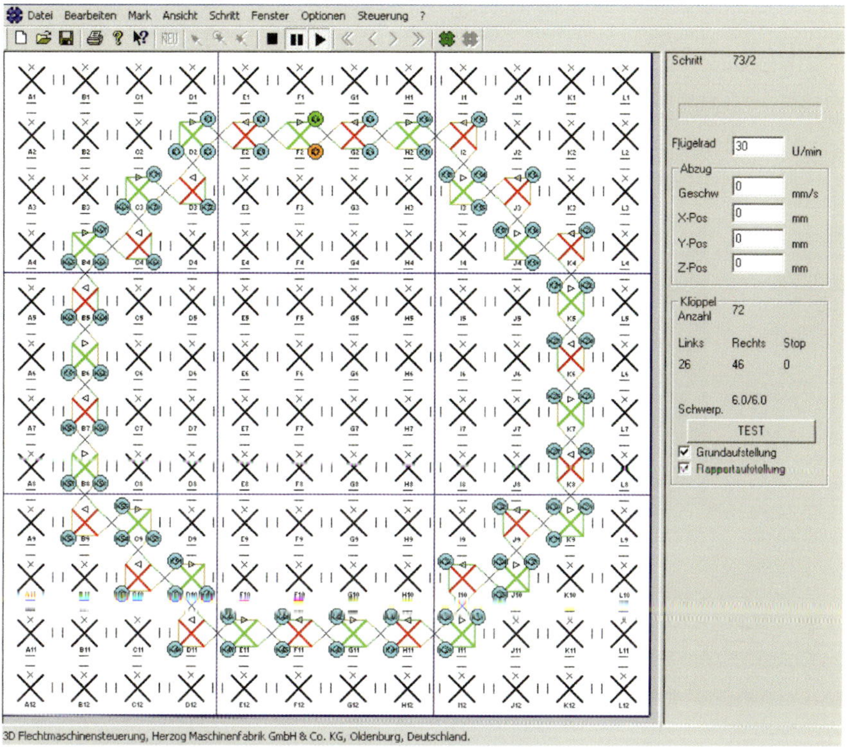

Plate XIII Software for simulation and control of the braiding process on a 3D braiding machine of August Herzog Maschinenfabrik, used by ITA of RWTH Aachen University (Stueve and Gries, 2009).

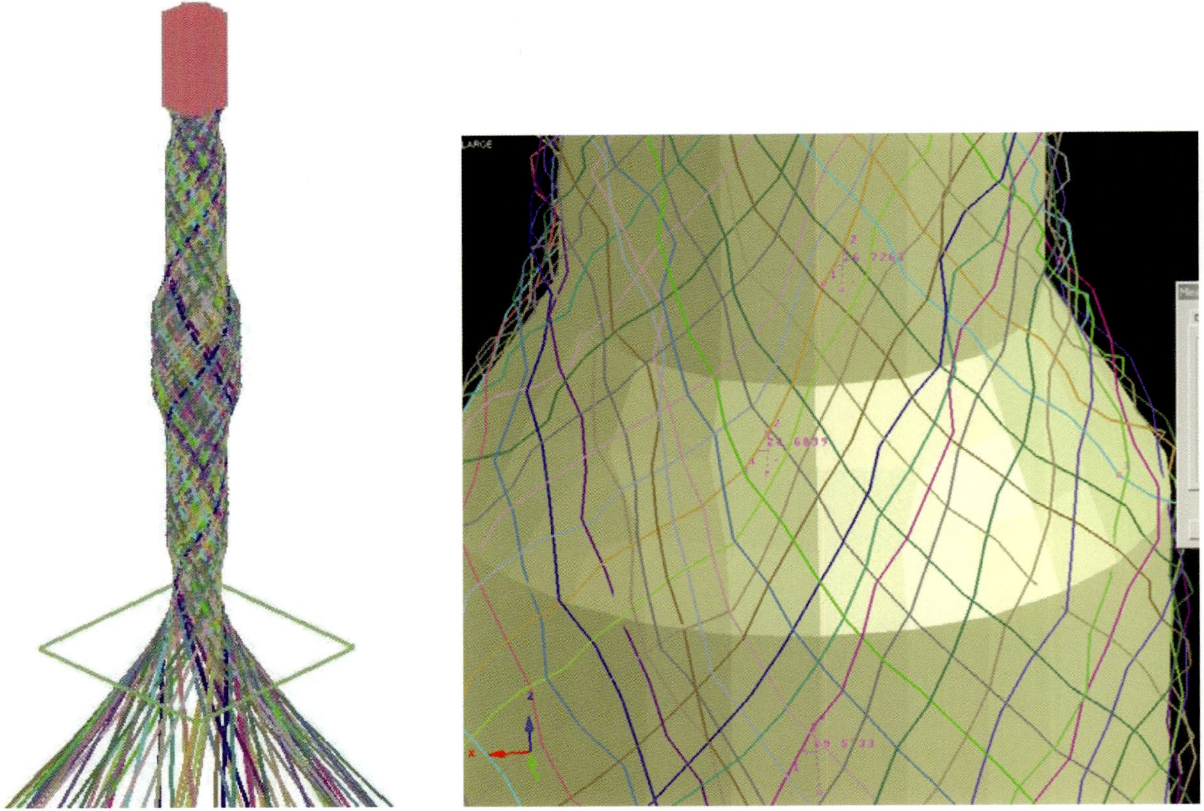

Plate XIX Simulation of the overbraiding of a body using explicit FEM Software PAM-CRASH (Stueve and Gries, 2009).

An alternative is the immersed-boundary-layer method as described by Lai and Peskin (2000). In a cartesian mesh, those nodes surrounding the fibre and the yarn are distributed until the desired mesh fineness is achieved. The other nodes are distributed evenly (Fig. 5.6, middle). The major advantage of this method is the considerably lower number of nodes required and hence a much shorter computing time. However, in order to represent the exchange of impulse for turbulent flows, separate models need to be defined.

A third possibility as shown in Fig. 5.6, bottom, is an unchanged net. This approach is similar to the immersed-boundary-layer method but the mesh is not denser close to the fibre or the yarn. This method can then be applied when fibres or yarns cover more than one cell which makes a refinement of the mesh to describe the different flow forces onto the fibre or the yarn dispensable.

When using the last two approaches, separate methods to describe the exchange of impulse must be defined by applying resistance factors. A good summary is published in Marheineke and Wegener (2009). A major challenge are fibres or yarns that flow against the current at angles smaller than 5°. Plate II shows the simulation result of a single, crimped fibre in an air flow and illustrates the complex conditions.

Besides the interactions of the fibre and the yarn with the fluid, interactions with walls and other fibres and yarns must also be taken into account. Another difficulty is the representation of elasticity in such a simulation. When considering this parameter, the time increment becomes very small as the computing algorithm otherwise becomes unstable. According to Courant *et al.* (1928), the maximum time increment can be calculated by

$$\Delta t \leq \sqrt{\frac{\rho}{E}} \cdot \Delta l \qquad [5.8]$$

Equation [5.8] implies that the maximum time increment increases with density ρ and segment length Δl and decreases with the E-modulus of the fibre or the yarn.

5.4 Validation methods

Apart from the plausibility tests as described above, the validation of the simulation results by measuring the actual current is very important to assess the quality of the model and the results. In the following, three commonly used methods to measure the flow velocity are described briefly.

5.4.1 Laser Doppler anemometry (LDA)

Here, the inference pattern of two coherent laser beams is used to determine the speed of particles within a current. The laser beam is first separated by a beam divider. The two phase-delayed beams are then brought into focus using a lens so that an interference pattern made up of bright and dark spots arises as shown in Fig. 5.7. Tracer particles brought into the current move through this light pattern and start to light up in a velocity-dependent frequency. This frequency is detected and the respective value converted into a velocity. The main advantage of this method is that the current is not affected as this is a purely optical technique. The main disadvantage is the extraordinary high price for the necessary equipment which can easily reach € 250000. An interesting paper on the subject was published by Lehmann (1988).

5.4.2 Particle image velocimetry (PIV)

The principle of Particle image velocimetry (PIV) is based on the time-dependent measurement of the position of particles (so-called seeds) moving through the current. One layer of the current is subjected to two very brief light impulses (duration in the range of nanoseconds) within microseconds. The displacement of the particles over time is recorded via CCD or CMOS camera for the points in time t and t'. Applying a cross-correlation, the particle distribution in this layer at t and t' can then be detected. Normally, 10000 velocity vectors can be calculated for one frame. By averaging up to several hundred measurements, a velocity chart can be drawn. Figure 5.8 shows the principle of PIV.

When using a second PIV system with another camera, it is possible to realize a stereo measurement. Hence, the velocity vectors can be determined in three dimensions. This principle is applicable for small and large volume flows and for a wide range of flow velocities ranging from a few mm/s up to 1 km/s. The main advantage of this method is that the result is a two-

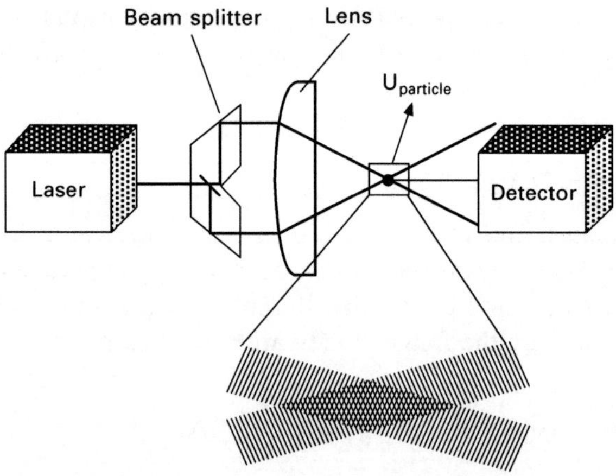

5.7 Principle of laser Doppler anemometry (LDA).

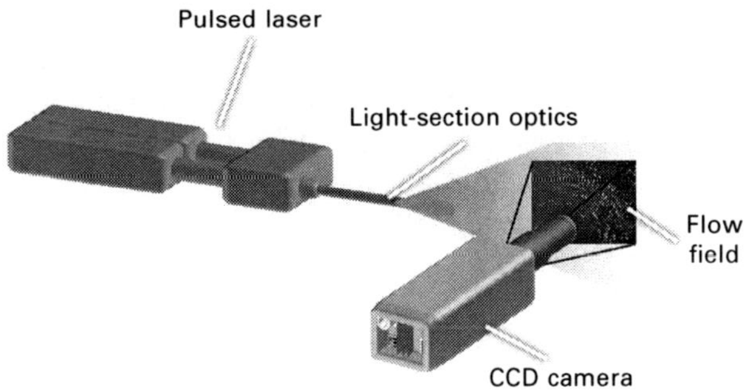

5.8 Principle of particle image velocimetry (PIV).

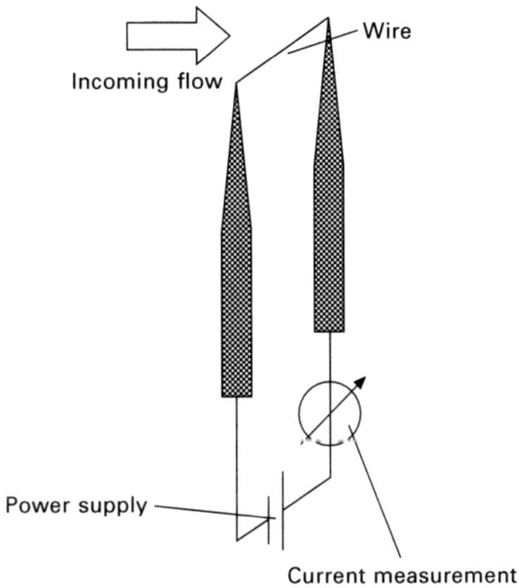

5.9 Principle of hot-wire anemometry.

dimensional vector field which mirrors the flow. This makes it easy to determine areas of turbulence and laminar flow which is very difficult with LDA measurements. A detailed explanation of the measuring principle can be found in Raffel *et al.* (2007).

5.4.3 Hot-wire anemometry (HWA)

With this method, air flow velocities can be measured. The principle is based on the cooling effect of an air flow on a hot wire which depends directly on the flow velocity (Hussein, 1990). By keeping the hot wire at a constant temperature, a dissipation loss can be recorded which depends on the fluid velocity. Air temperature and humidity at the beginning of the measurement have a direct effect on the result. Hence, the hot-wire sensor must be calibrated before the actual start of the measurement. Figure 5.9 shows the principle of hot-wire anemometry (HWA). The main advantage of this method is its easy applicability in contrast to LDA and PIV for which the technical equipment is much more complicated. Another advantage is the possibility to record fluid velocity continuously at a frequency of several kHz. This allows the measurement of laminar and also turbulent flows.

Besides these three methods, there are other techniques, such as measuring

- air speed (impeller);
- pressure (barometer, ionization vacuum-meter),
- wall shear stress (scales, hot film);
- density of the fluid (striation optics);
- flow rate (volumetric, Venturi tube, ultrasonic);
- temperature (thermometer, thermoelement); and
- concentration (ultrasonic).

5.5 Economic aspects

The simulation of currents is one of the most expensive methods to optimize a process. The usage of the costly software requires a thorough knowledge in fluid dynamics, different modelling techniques and the algorithms used to solve the problem. This makes CFD simulations different from the methods described in Chapters 2–4 where even a little knowledge can lead to viable results. Hence, in many applications, prior to receiving the first usable results, high costs can arise from personnel and software. In some cases it is thus easier and cheaper to achieve results by using empirical methods or by simply measuring the currents. But for complex problems, CFD often is the only possible way to tackle them.

An empirical approach is the most simple and obvious method to optimize a process: based on existing knowledge and results of trials, dependencies between the interesting parameters are derived. If the number of influencing parameters is small and their relations are clear, an empirical approach often leads quickly to first results, thus allowing a process optimization at little cost.

By using measuring instruments, the influencing parameters can also be quantified so allowing conclusions to be drawn of the fundamental laws governing the process. Typical examples are temperature sensors or position measuring devices. The necessary instrumentation can become very complex depending on the problem to be solved (Ramakers, 2005; Ramakers *et al.*, 2006). An important aspect when using measuring instruments is the determination of the measuring range and accuracy which in turn has a direct effect on the costs.

In many cases, an empirical approach in combination with measuring techniques is limited due to boundary conditions. Then, CFD is an interesting alternative although it requires a considerable amount of preparation: the geometrical conditions must be fed into the model, the boundary conditions must be defined and suitable models must be chosen in order to achieve usable results. Besides, it is of crucial importance with regard to costs and computing time to define the required accuracy of the model at a very early stage.

The major benefits of CFD modelling versus empiricism are hence:

- currents can be depicted in three dimensions and are time-dependent;
- details inside the current can be shown which could not be detected by pure measurement;
- geometries can be analysed that only exist in virtual reality;
- parameter values can be analysed within a wide range without the need to build expensive prototypes;
- complex interrelations can be shown.

Empiricism, measuring technology and simulation are compared qualitatively in Fig. 5.10 by showing costs of process optimization versus process quality.

By using empiricism, it is possible to achieve improvements with comparably small investment by changing parameters, e.g. of a machine, ever so slightly until a certain criterion is met. The curve in Fig. 5.10 shows a slow increase which levels out after some time. From this point on, it is very hard to further improve the process. The main reason for this is the missing understanding of the process and the interactions between the different parameters.

When using measuring technology to optimize a process, costs arise due to the initial investment. This leads with a delayed increase of the process quality compared with empiricism. The curve then increases quicker and reaches a higher level, owing to a better understanding of the process itself. At a certain point, the curve levels out as not all interactions between the process parameters can be determined.

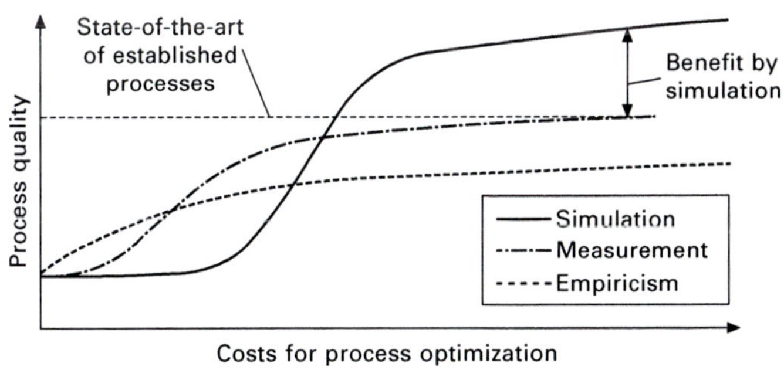

5.10 Costs of process optimization vs. process quality for empiricism, measuring technology and simulation.

For a typical CFD simulation, the curve initially does not show any improvement as investments into personnel and software are needed. In addition, during the time which is required to prepare the actual simulation (e.g. determination of geometry, boundary conditions and models to be used), the process quality is not improved. As soon as the simulation leads to the first results, the curve rises sharply. With relatively small further expenses, the optimization develops quickly and, owing to a good understanding of the process and the possibility of varying the parameters within a wide range, the process quality increases and exceeds that of the other two methods.

To illustrate this, Fig. 5.11 shows a typical example for cost effectiveness through simulation. At the beginning of the simulation, additional costs arise when applying simulation tools. The development time, however, is much shorter. Hence, the product can be brought to market much quicker. When using conventional methods, additional costs can occur before the product reaches marketability owing to design faults which have not been noticed prior to producing the first prototypes. A simulation program would normally show these faults at a much earlier stage so that they can be eliminated before they cause costs. This results in the product developed by simulation normally reaching the break-even point earlier than its competitor which was developed by conventional methods. A good overview of economic aspects is given in Lemon (2011).

5.6 Applications of computational fluid dynamics to textile technology

CFD is used in a wide range of applications, including the following:

- Air- and spacecraft: new aircraft designs are modelled in order to decrease fuel consumption.

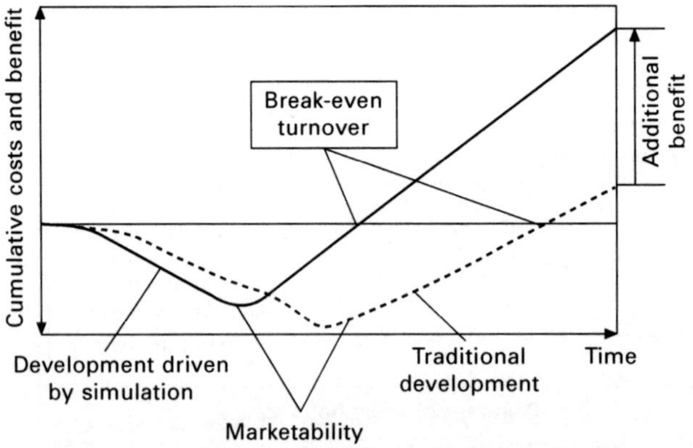

5.11 Advantage of process development by simulation vs. conventional methods (adapted from Lemon, 2011).

- Automotive: improvement of the aerodynamical properties of new car bodies, simulation of fuel injection and modelling of car interior to increase passenger comfort.
- Medical: inhalation processes are analysed in order to minimize the effect of surgical operation on the patient, flow simulation of heart pumps show areas where unwanted depositions can occur which can be fatal.
- Chemical: processes in reactors are optimized with regard to reaction speed and energy consumption.
- Air conditioning: comfort of human beings in a room can be enhanced by optimizing heat distribution and air humidity.
- Process engineering: improving mixing processes, e.g. homogenization and deposition of substances.
- Maritime: development of new propeller systems to reduce energy consumption.
- Energy: increasing efficiency by better design of, e.g. windmill blades, combustion chambers.
- Sports: effects of biological examples are transferred into textiles, e.g. shark skin effect on a swim-suit, knob-structured golf balls.

In textile technology today, computational fluid dynamics is widely used to analyse processes where fast moving fibres and yarns or fabrics are involved (Martens *et al.*, 2009). Typical examples are air-jet texturing, air-jet spinning, air-jet weaving and a range of finishing processes. Most of these processes were developed well before CFD could be used. Hence, it is now possible to directly improve existing technologies by literally taking a closer look which was unthinkable only 15 years ago. Figure 5.12 shows that, prior to 1998, there were no papers on record in the compendex database, but there has been a steady increase in the number of publications over the last 5 years. However, the overall number is still comparatively low compared with other simulation techniques described in Chapters 2–4. In this section, if not otherwise stated, all results were achieved by or under the supervision of the main author of this chapter, Ph. Jungbecker.

5.6.1 Melt-spinning

Melt-spinning is widely used to produce synthetic fibres and yarns, e.g. from polyester and polyamide. Granulate is molten by an extruder, the molten mass pumped through the spin pack, distributed onto a large number of capillaries and spun into single filaments before they are quenched, drafted and wound up onto bobbins. CFD can be applied to several components of this process. A typical application is the spin pack, where the current of the polymer can be analysed in order to avoid dead water areas in which the polymer flow is subjected to different and hence undefined conditions. This could lead to

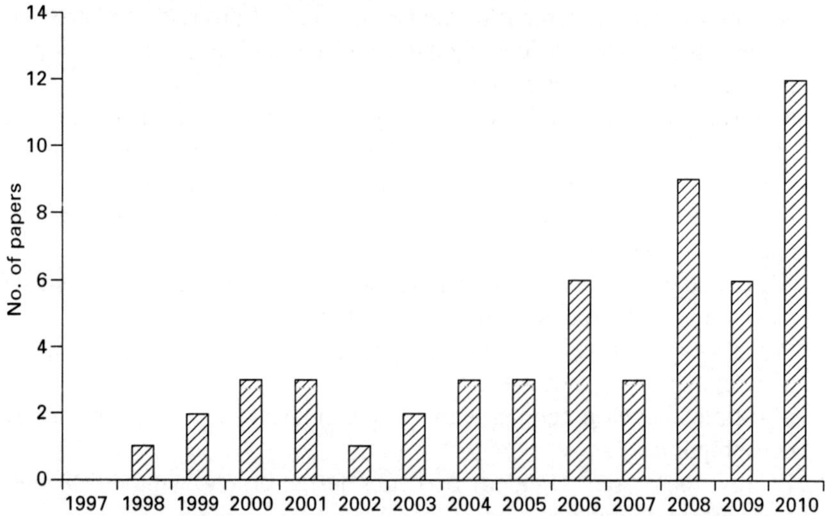

5.12 Number of publications on CFD in compendex database ('CFD' and 'textile').

different properties of the filaments which may go unnoticed up until the final stages of the textile process chain, e.g. finishing and dyeing. In the production of spun-bonded nonwovens, several thousands of filaments are produced by one spinneret. Modelling the behaviour of the polymer flow inside the spin pack can significantly reduce costs by increasing quality of the product (Jungbecker *et al.*, 2009). Plate III shows the flow lines of a polymer inside a bicomponent spin pack. The distribution of the polymer inside the capillaries can clearly be seen.

Owing to the visco-elastic properties of the polymer, a widening and thus a distortion of the filaments can occur when the polymer leaves the spinneret. A modelling of this process using CFD can help to achieve the desired filament cross-sections. The quenching process of the filament process is determined by heat transfer inside the yarn and heat transport outside the yarn. By applying CFD, it is possible to analyse this process and thus to optimize it, which cannot be achieved by pure measurement as this does directly effect the quenching itself.

5.6.2 Melt-electrospinning

Electrospinning has been used since the 1920s but has not yet been commercialized on an industrial scale. A polymer is melted, filled into a syringe and injected into an electrical field where the fibres are drafted. With this process, fibre diameters below 500 nm can be realized which is not possible with conventional processes. A CFD simulation was employed in order to upscale the process to commercial throughput (Hacker *et al.*, 2009). Based on these results, a new spin pack design was realized which had an even distribution of the polymer melt and low residence times. This improved the fibre quality considerably. By analysing the temperature gradient directly underneath the spinneret, it is now possible for the first time to produce nano-fibres in a continuous process. Plate IV shows the pressure distribution inside the spin pack (left) which led to the design of a new spin pack.

5.6.3 Drafting of filaments

Drafting processes are common in many fibre and yarn manufacturing processes. The filaments are normally heated to temperatures above glass transition point, thus allowing easy drawing. By applying CFD to analyse the heat loss in draft zones, it was possible to identify the weak spots in a typical drafting machine and hence to redesign it in order to reduce energy consumption and to equalize the drafting conditions. The new design was put into practice (Plate V) and measurements showed that the energy consumption could be reduced by up to 30% owing to the new design.

5.6.4 Air-jet texturing

Air-jet texturing is a process to improve the handling of flat yarns made of synthetic fibres. The yarn is guided through ceramic jets and subjected to pressured air. This results in a supersonic flow inside the jet and an impulse wave at the outlet. CFD was used to analyse the current, which provided an insight into the flow inside the jets which is not possible with conventional measuring technology. Plate VI shows a typical flow inside and outside of the actual air-jet.

5.6.5 Card

In a card, fibres are parallelized, cleaned and a sliver is formed. The main components of a card are the licker-in drums, the swift, the carding elements and the doffing cylinder. Owing to the small distances between those fast rotating elements, there are strong air flows by which the fibres are affected considerably during carding. Since space is limited, the use of measuring devices is confined to certain areas where they do not have an impact on the actual process. The application of CFD to model the currents inside the card is therefore an interesting approach to analyse and hence improve the carding process (Seide *et al.*, 2008). Plate VII shows the drag flow between swift and the slowly rotating doffing cylinder. The simulation results show that this current is directly affected by the distance between swift and doffing cylinder, the carding cloth and the velocities of the components. By adjusting the respective parameters, it is hence possible to improve the carding process.

Han *et al.* (2009) analysed the air flow between back plate and swift by applying CFD. They found that the static pressure increases to its maximum value near the inlet and decreases towards the outlet along a zone between back plate and swift. A comparison of the simulation results with experimental data showed a good correlation. A practical application was not given.

5.6.6 Staple fibre spinning

Zou *et al.* (2009) analysed the air flow in the condensing zone of a compact spinning device with a perforated drum (Fig. 5.13). By modelling the trajectory of single fibres applying CFD, it was possible to determine the parameters affecting the condensing mechanism. The trajectories were calculated by having single fibres starting from different positions of the last drafting roller. Firstly, the forces on an individual fibre were calculated. This led to the accelerations of the fibres and the respective velocities. They were

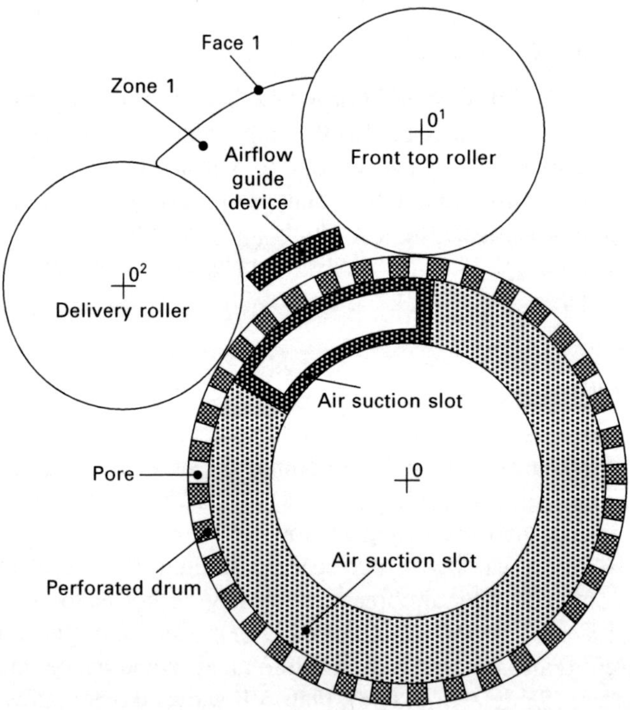

5.13 Condensing zone of a compact spinning device (adapted from Zou *et al.*, 2009).

used to calculate the displacements in all three dimensions and hence the trajectories of the fibres. For the model, it was assumed that the air flow is viscous and incompressible and the process adiabatic.

Rengasamy *et al.* (2008) investigated the influence of an air-jet on ring spinning with regard to hairiness. They assumed the air flow to be viscous and compressible and the process to be adiabatic. The results in terms of air velocity and drag forces acting on the fibres showed the influence of the axial angle of the air inlets of the nozzles on yarn hairiness. They correlated well with measured values.

Zeng and Yu (2003) analysed the air flow in the nozzle of an air-jet spinning machine. They determined the resulting air flow when changing nozzle pressure, jet orifice angle and jet orifice position. A higher nozzle pressure led to higher axial and tangential velocities of the fibres and subsequently to better tensile properties. The influence of the jet orifice angle is complex and was not clearly visible. The position of the jet orifice did not considerably affect the air flow pattern. In a follow-up paper, they used the results to simulate the motion of fibres inside the nozzle using a two-dimensional model (Zeng and Yu, 2004). A comparison with photographs taken by a high-speed camera showed a reasonable correlation to the theoretical results. Zhu *et al.* (2008) developed a three-dimensional CFD model of the flow field inside the nozzle of a murate vortex spinning machine. No verification with experimental data was given.

In a paper by Morimoto *et al.* (2010), the currently used design of a nozzle for Murata vortex spinning is analysed applying a two- and a three-dimensional CFD approach. Different nozzles were investigated and produced so that theoretical and practical results could be compared. It was found that the currently used nozzles are suitable to produce vortex yarns which did not come as a surprise.

An interesting paper on friction spinning was published by Ishtiaque *et al.* (2010). The authors design a CFD model to analyse the air flow inside the transport duct of a DREF-III friction spinning machine. After specifying the boundary conditions, e.g. geometry of inlet, outlet and moving wall, pressure and velocity distribution were calculated by applying the k–ε model for turbulent flow. After a thorough investigation of the influencing parameters, they simulated the effect of additional slots in the duct on air pressure and velocity in order to reduce the turbulence and hence to improve the yarn quality. The results are shown in Plates XIII and IX. The effect of the slots on the turbulence of the flow is apparent. A practical test with a modified duct showed that the yarn properties could indeed be improved with this measure. This is an excellent example of a CFD simulation that was actually used to improve machine design.

5.6.7 Fabric production

For many kinds of fabrics, air-jet weaving is a very economical process due to its high production speed. A major disadvantage is the high cost for the pressured air which is required for the weft insertion. The instationary flow field during weft insertion was analysed using CFD by Jungbecker *et al.* (2011). The results of the analysis can then be used to significantly decrease air consumption without affecting the machine efficiency. The weft insertion components comprise main jets in tandem configuration, a large number of relay jets, e.g. 52, and the reed which in turn consists of, for example, 4200 dents.

The geometric complexity makcs it impossible to simulate one single CFD model to calculate the instationary flow field. Hence, developing a simplified model which does not affect the simulation results too much is unavoidable. In a first step, main jet, relay jets and reed are considered independently of each other which allows designing a highly accurate model for each component. In a second step, these models are simplified and new models are designed. The results of the respective simulations are then compared with the results of the detailed simulations of the first step until they coincide sufficiently well. In a third step, the whole weft insertion process is simulated taking into account the findings of the previous steps.

The results of the flow simulation show that a major part of the usable flow energy is lost during the detonation shocks of the relay jets. Air arriving at the reed channel flow backwards out of the channel and is hence lost. In other simulations, the yarn behaviour during weft insertion was analysed based on the previous findings. By changing the relay jets' turn-off times from a linear to a parabolic profile where the relay jets in the middle section are turned off earlier, it was possible to reduce pressured air consumption by 30% for a prototype weaving machine and by 10% for a commercially used weaving machine. Plate X shows the three-dimensional air flow in the reed channel.

Belforte *et al.* (2009) studied the flow field inside an air-jet loom main nozzle by applying a two-dimensional CFD model. They designed different models and compared the simulation results with experimental data in order to find the best-fit model. Then they analysed the influence of the length, shape and size of the acceleration tube on the yarn drag force. This led to an optimum value of the acceleration tube length for which the drag force on the yarn can be maximized.

5.6.8 Permeability of fabrics

The permeability of textiles is of great interest for a wide range of applications where air or fluids flow through textile structures. Typical examples are outdoor clothing and textile preforms for composite production. A CFD simulation can be used to predict how the infusion resin spreads inside the structure and whether air inclusions occur. Plate XI shows a typical example for a multi-layered braided structure which is used for a ceramic-fibre reinforced aluminum part.

The flow of the resin through a carbon fibre fabric for use in composites was analysed by Zeng *et al.* (2010). They designed a unit cell to represent the three-dimensional woven fabric geometry from a set of experimentally determined geometric parameters. The calculated through-thickness and in-plane flow was then compared with actual results. The correlation was found to be satisfactory. Then, the influence of fabric geometry on permeability was analysed and it was found that the weft yarn cross-section has the largest effect.

5.6.9 Heat resistance of fabrics

A mathematical model to determine the thermal resistance of woven fabrics was designed by Kothari and Bhattacharjee (2008). They regarded the yarns, intersections and air pores of a fabric as a system of resistances. The authors assumed the woven fabric to have a cellular geometry through which conductive heat transfer occurs from the air pores as well as the yarns. Hence, all basic weaves can be regarded as modifications of a simple geometry of air pores, yarn sections and intersections. The yarns were considered to be porous and comprising infinite cylindrical fibres and air. Heat transfer through convection was neglected but was included in the CFD model in part 2 (see below). Heat transfer by radiation is considered. The underlying model is explained in great detail and could possibly serve as an example for other similar applications. The predicted and the measured values coincided very well which proves the viability of this approach.

In part 2 of the publication (Bhattacharjee and Kothari, 2008), this model was used as the basis of a CFD simulation which also included heat transfer through convection and a wind tunnel simulation. Forced convection was accurately represented by the model, natural convection could not be modelled to the same extent. This publication is recommended reading for any user wishing to simulate heat transfer through textiles by applying CFD. Another CFD-based simulation of the permeability of multifilament woven fabrics was developed by Wang *et al.* (2006). The results are compared with those from another analytical model and show good correlation.

Hjohlman and Andersson (2009) used a CFD simulation to predict the flame spread in textile materials. One model was used to analyse the pyrolysis of the material, the other to simulate flame spread over a surface. The results were then compared with experimental data. Although there was a large difference, especially in the calculated and the actual heat release rate, the model could still accurately predict the rating of the textiles according to the European building product classification system EN 13501.

5.6.10 Nonwovens

Nonwovens made of staple fibres can be manufactured using an aerodynamic process which dissolves the material to single fibres. They are transported in an air flow onto a sieve where they form a web. The air flow is a crucial component with regard to the quality of the web. Due to their inertia and turbulences, the fibres follow the air current only partially, resulting in different trajectories of the fibres and hence in an uneven web. A model of the air flow was created without fibres which formed the basis for a simulation of the fibres. For a high fibre content (exceeding 5%), the air current can be modelled in connection with the fibre trajectories. This allows analysing the repercussions of the fibre movement onto the air flow. A simple and effective approach to simulate the fibre bundles is to consider them as spheres. Diameter and density of the spheres can be adjusted so that their aerodynamical properties resemble those of fibres. Plate XII shows the air flow and the fibre trajectories for this aerodynamical nonwovens process.

The mixing process used to disperse synthetic fibres in the wet-lay process was modelled by Ramasubramanian *et al.* (2010). A multiple reference frame model and a standard k–ε turbulence model were employed. After obtaining a converging solution for the mixing tank with water, a discrete phase model was designed by injecting spherical particles into the flow. The results obtained from the simulation and measured values coincided very well qualitatively. The CFD model can hence be used to optimize the mixing tank design for this process.

5.6.11 Finishing

Schmidt *et al.* (2009) analysed the commonly used blade coating process with a CFD simulation. One of the main findings was that pressure decreased with viscosity in the coating material in the area underneath the blade's edge. This is due to the geometry of the blade and can be used to optimize it. Hence, a deeper understanding of the rheological behaviour of the coating material can help to improve this process that is normally used to produce technical textiles. Other interesting papers include Menshutina and Kudra (2001), which presents a CFD based model for textile drying processes and Scharf *et al.* (2002), which deals with textile dyeing processes.

Yarns can be dyed in a continuous process or discontinuously on bobbins. In the latter case, the yarn is wound onto perforated plastic cones and the dye liquor flows through the material. Flow velocity decreases from inside to outside as the diameter increases. This effect was analysed using a CFD model. This approach can also be used to determine the perfusion through the bobbin for different yarn densities across the cone. As shown in Plate XIII, especially the areas close to the edges are critical as yarn density is higher and the dye liquor leaves the bobbin tangentially.

With rising energy costs and the demand for an environmentally friendly production of textiles, an efficient drying of textiles becomes an increasingly important issue. Textiles can be dried using convection, e.g. in a stentering frame, infrared radiation and microwaves, or by contact drying with rollers. Especially for convective drying processes, a simulation of the flow can help to homogenize the air flow and thus increase the process efficiency. Not only flow velocities can be determined but also relative air humidity and water vapour intake of the textiles across the dryer's cross-section. This can lead to lower energy consumption and to better product quality.

5.7 Practical advice in applying computational fluid dynamics

- *Aim of the simulation*: it is important to define the aim of the simulation before designing the respective model as a model can never mirror reality in all aspects. By describing the target, both, researcher and customer, know what to expect. If certain effects are to be included, the respective models need to be activated in the software.
- *Boundary conditions*: the result of the simulation is strongly affected by the chosen boundary conditions. Choosing wrong or imprecise boundary conditions can lead to large differences between simulation result and reality. If that happens, a sensitivity analysis of the boundary conditions is recommended. It will show whether and to what extent an error in the initial conditions does affect the simulation result.
- *From simple to complex model*: it is strongly recommended to begin with simple models before proceeding to more complex algorithms. If too many parameters are changed at once, it is hard to tell the effects of each parameter. A coarse discretization keeps computing time at a minimum. Hence, effects of different models on the result can easily and quickly be recognized and one does not lose precious time. Models based on discretization accuracy, e.g. for turbulent currents, are exceptions: here. It can be sensible to first test the respective models on a simple geometry before applying them to more complex geometries.
- *Model simplification*: simple models are preferable over mirroring a process in all details. A typical example are textiles: in a detailed simulation, the perfusion can be analysed for a small area on microscopic scale. The resulting dependency between mean velocity and pressure loss can be represented by a porosity model according to Darcy (1856). For the subsequent simulations on macroscopic scale, the textile can then

be regarded as an unstructured area with a velocity-dependent impulse sink.

- *Cross-linking*: if possible, the use of hexaeder cells is highly recommended. Having the same edge length as tetraeder cells, 87.5% fewer cells are needed. The relation between cross-linking effort and simulation costs should be considered. With powerful hardware and a simple model, in many cases, a fine cross-linking (small mesh size) should be chosen. Especially for instationary processes and parameter studies, it can pay off quickly if the cross-linking parameters are adjusted manually.
- *Validation*: it is advised to check the results for plausibility already after only a few iteration steps. In particular, the magnitude order and the correct flow direction should be checked.
- *Final analysis*: a final analysis and interpretation of the results should never be carried out before the results are converging.

5.8 Summary

Computational fluid dynamics is a very powerful tool to simulate currents with fibres and yarns as well as flows through and around textiles. In contrast to the methods described in Chapters 2–4, however, it is essential to acquire a deeper insight into the theoretical background behind the various mathematical models, e.g. fluid mechanics, before starting to create CFD-based models and simulations. Existing software is often very complex to use and hence requires extensive experience to apply it properly. CFD is therefore not a quick-and-dirty approach to solve fluid-related problems but often a time-consuming and thus costly process. Nevertheless, CFD simulation can lead to an insight into the relation of parameters that could otherwise not be analysed. Today, its use is wide spread and has led to a considerable number of machine optimizations in textile technology ranging from fibre and yarn manufacturing to fabric production and finishing.

5.9 Web references for software tools

A site dedicated to collecting and sharing information, successful applications and software tools about CFD is:

http://www.cfd-online.com

In the 'Codes' section of the wiki on this website, interested readers can find an extensive list of free and commercially available software packages to carry out CFD simulations. This includes, besides the actual modelling and simulation software, also grid generation as well as visualization tools. As far as the authors know, this is the most comprehensive list in this respect on the internet. Hence, there is no need to give a separate list of recommended software in this section.

5.10 References

Abbot, M.B.; Basco, D.R. (1989), *Computational Fluid Dynamics: An introduction for engineers*, Wiley.

Baliga, B.R., Patankar, S.V. (1980), 'New finite-element formulation for convection-diffusion problems', *Numerical Heat Transfer 3:4*, p. 393–409.

Belforte, G., Mattiazzo, G., Viktorov, V., Visconte, C. (2009), 'Numerical model of an air-jet loom main nozzle for drag forces evaluation', *Textile Research Journal 79*, p. 1664–1669.

Bhaskaran, R., Collins, L. (2008), *Introduction to CFD Basics*, Cornell Univ., URL: http://courses.cit.cornell.edu/fluent/cfd/intro.pdf (accessed 20 April 2011).

Bhattacharjee, D., Kothari, V.K. (2008), 'Prediction of thermal resistance of woven fabrics. Part II: heat transfer in natural and forced convective environments', *Journal of the Textile Institute 99:5*, p. 433–449.

Cebeci, T., Shao, J.P., Kafyeke, F., Laurendeau, E. (Eds., 2005), *Computational Fluid Dynamics for Engineers: From panel to Navier–Stokes methods with computer programs*, Springer.

Courant, R., Friedrichs, K., Lewy, H. (1928), 'Über die partiellen Differenzengleichungen der mathematischen Physik', *Mathematische Annalen 100:1*, p. 32–74.

Darcy, H. (1856), *Les Fontaines Publiques de la Ville de Dijon*, Dalmont.

Hacker, C., Jungbecker, P., Gries, T., Thomas, H. (2009), 'Mehrdüsen-Elektrospinnen aus der Polymerschmelze – Die Entwicklung zum Upscaling', *Proceedings of 12. Chemnitzer Textiltechnik-Tagung*, p. 32–35.

Han, X.-G., Sun, P.-Z., Zhao, Y.-P. (2009), 'Analysis of static pressure in area between back plate and cylinder of a carding machine with CFD', *Journal of Donghua University*, p. 242–246.

Hirsch, C. (2007), *Numerical Computation of Internal and External Flows: Fundamentals of computational fluid dynamics*, vol. 1, 2nd ed., Elsevier/Butterworth-Heinemann.

Hjohlman, M., Andersson, P. (2009), 'Flame spread modelling of complex textile materials', *Fire Technology 47*, p. 85–106.

Hussein, H.J. (1990), *Measurements of Turbulent Flows with Flying Hot-wire Anemometry*, American Society of Mechanical Engineers, Fluids Engineering Division (Publication) FED 97, p. 77–80.

Ishtiaque, S.M., Singh, S.N., Das, A., Mittal, S., Dave, V. (2010), 'Optimisation of fluid flow phenomena inside transport duct of a DREF-III friction spinning machine. Part III: CFD approach', *Journal of the Textile Institute 101:10*, p. 906–916.

Jungbecker, P., Hacker, C., Gries, T., Thomas, H., Möller, M. (2009), 'Melt-electrospinning of nano-nonwovens', *Chemical Fibers International 59:4*, p. 228–229.

Jungbecker, P., Schenuit, H., Holtermann, T., Seide, G., Gries, T. (2011), 'Reduction of energy consumption in air-jet weaving', *Melliand International 17:1*, p., 32–33.

Kothari, V.K., Bhattacharjee, D. (2008), 'Prediction of thermal resistance of woven fabrics. Part I: mathematical model, *Journal of the Textile Institute 99:5*, p. 421–432.

Lai, M.-C.; Peskin, C.S. (2000), 'An immersed boundary method with formal second-order accuracy and reduced numerical viscosity', *Journal of Computational Physics 160:2*, p. 705–719.

Launder, B.E., Spalding, D.B. (1974), 'Numerical computation of turbulent flows', *Computer Methods in Applied Mechanics and Engineering 3:2*, p. 269–289.

Lehmann, B. (1988), 'Large-Eddy interpretation of LDA results obtained by conditional seeding in a circular jet', *Advances in Turbulence 2*, p. 102.

Lemon, J. (2011), 'Why simulation should drive product development', *http://www.iti-global.com/Education/Articles/SimLedDev.htm* (accessed: 20 April 2011).

Löhner, R. (2008), '*Applied CFD Techniques: An introduction based on finite element methods*, 2nd ed., Wiley.

Marheineke, N., Wegener, R. (2009), 'Modeling and validation of a stochastic drag for fibers in turbulent flows', *Berichte des Fraunhofer ITWM 172*.

Martens, S., Jungbecker, P., Seide, G., Gries, T. (2009), 'Fostering energy efficiency in textile machinery by flow simulation', in: Küppers, B. (Ed.): *Proceedings of the 3rd Aachen–Dresden International Textile Conference*, p. 1–7.

Mason, W.H. (1998), 'Introduction to computational fluid dynamics', *Applied Computational Aerodynamics, 2:8*.

Menshutina, N.V., Kudra, T. (2001), 'Computer aided drying technologies', *Drying Technology 19:8*, p. 1825–1849.

Menter, F.R. (1994), 'Two-equation eddy-viscosity turbulence models for engineering applications', *AIAA Journal 32:8*, p. 1598–1605.

Morimoto, K., Tada, Y., Takashima, H., Minamino, K., Tahara, R., Konishi, S. (2010), 'Micro spinning nozzle having 3D profile for fiber generation with spiral air flow', *proceedings of IEEE International Conference on Micro Electro Mechanical Systems (MEMS)*, p. 244–247.

Peiró, J., Sherwin, S. (2005), 'Finite difference, finite element and finite volume methods for partial differential equations', in Yip, S. (Ed.): *Handbook of Materials Modeling*. Vol. 1.: *Methods and models*. Springer, p. 1–32, http://www2.imperial.ac.uk/ssherw/spectralhp/papers/HandBook.pdf (accessed: 21 April 2011).

Raffel, M., Willert, C., Wereley, S., Kompenhans, J. (2007), *Particle Image Velocimetry – A practical guide*, Springer.

Ramakers, R. (2005), 'Systematische Entwicklung von sensorbasierten Online-Überwachungssystemen für die Filamentgarnverarbeitung', doctoral thesis, RWTH Aachen, Shaker, 2005.

Ramakers, R., Besen, A., Gries, T. (2006), *Konstruktionskatalog Sensortechnologie für die Online-Überwachung von Produktionsprozessen*, Shaker.

Ramasubramanian, M.K., Shiffler, D.A., Jayachandran, A. (2010), 'A computational fluid dynamics modeling and experimental study of the mixing process for dispersion of synthetic fibers in wet-lay forming', *Tappi Journal*, p. 6–13.

Rengasamy, R.S., Patanaik, A., Anandjiwala, R.D. (2008), 'Simulation of airflow in nozzle-ring spinning using computational fluid dynamics: study on reduction in yarn hairiness and the role of air drag forces and angle of impact of air current', *Textile Research Journal 78*, p. 412–420.

Rodi, W., Spalding, D.B. (1970), 'Two-parameter model of turbulence, and its application to free jets', *Waerme-Stoffuebertrag Thermo-Fluid Dynamics* 3:2, p. 85–95.

Roller, S., Ferch, M., Munz, C.-D., Dumbser, M. (2005), 'Numerische Grundlagan: Finite Differenzen, Finite Elemente, Finite Volumen', Höchstleistungrechen-zentrum Stuttgart. https://fs.hirs.de/projects/par/par_prog-ws/2005D/cfd-05-be.pdf.

Salari, K., Roache, P.J., Wilcox, D.C. (1994), *Numerical Simulation of Dynamic Stall Using the Two-equation k-ω Turbulence Model*, American Society of Mechanical Engineers, Fluids Engineering Division FED 196, p. 1–7.

Scharf, S., Cleve, E., Bach, E., Schollmeyer, E., Naderwitz, P. (2002), 'Three-dimensional flow calculation in a textile dyeing process', *Textile Research Journal 72*, p. 783–788.

Schmidt, M., Schloßer, U., Schollmeyer, E. (2009), 'Computational fluid dynamics investigation of the static pressure at the blade in a blade coating process', *Textile Research Journal 79*, p. 579–584.

Seide, G., Jungbecker, P., Gries, T. (2008), 'Simulation of staple fiber kinematics in a turbulent air stream', *Melliand International 14:1*, p. 22–23.

Wang, Q., Mazé, B., Vahedi Tafreshi, H., Pourdeyhimi, B. (2006), 'A note on permeability simulation of multifilament woven fabrics', *Chemical Engineering Science 61*, p. 8085–8088.

Zeng, Y.C., Yu, C.W. (2003), 'Numerical simulation of air flow in the nozzle of an air-jet spinning machine', *Textile Research Journal 73:4*, p. 350–356.

Zeng, Y.C., Yu, C.W. (2004), 'Numerical simulation of fiber motion in the nozzle of an air-jet spinning machine', *Textile Research Journal 74:2*, p. 117–122.

Zeng, X.S., Endruweit, A., Long, A.C., Clifford, M.J. (2010), 'CFD flow simulation for impregnation of 3D woven reinforcements', *Proceedings of the 10th international conference on textile composites*, p. 455–462.

Zhu, Y.D., Zou, Z.-Y., Wu, J.-M., Xue, W.-L., Cheng, L.-D. (2008), 'Characterizing and analyzing the airflow field inside the nozzle block of murata vortex spinning', *Journal of Donghua University*, p. 670–675.

Zou, Y.Z., Cheng, L.D., Hua, Z.H. (2009), 'A numerical approach to simulate fiber motion trajectory in an airflow field in compact spinning with a perforated drum', *Textile Research Journal 80:5*, p. 395–402.

6
The finite element method (FEM) and its application to textile technology

Y. K. KYOSEV, Hochschule Niederrhein – University of Applied Sciences, Germany

Abstract: This chapter describes the basics of the finite element method (FEM) and its application for modelling of textile structures. It gives a brief introduction to the basics of the modelling of mechanical systems in the static and dynamic cases, some of the ideas behind the finite element method and a short overview of FEM software.

Key words: finite element method (FEM), modelling, textile structures.

6.1 Introduction

This chapter describes the basics of the finite element method (FEM). Since the development of the method and the publishing of the first book about it by Zienkiewicz and Cheung (1967), more than one hundred books have been published about the FEM, mostly written especially for engineers, mathematicians, programmers and 'dummies'. The aim of this chapter is not to summarize the contents of these books or to cover all the aspects they cover but to give a brief introduction to FEM for textile engineers and students. They may have learned about the FEM in lectures on statics, but do not know how to use it in dynamics. Alternatively, they may have studied chemistry, fluid dynamics or physics, but not enough continuum mechanics. As a result they may need to model a textile structure with FEM software, with not much knowledge of what is hidden behind the windows and buttons. For experienced FEM users, there could be some interesting points mainly in the comparison of explicit/implicit method and in some of the applications at the end of this chapter.

Engineers and scientists in general often analyse objects and processes by creating simplified mathematical models. The goal is to understand these objects in order to be able to predict their characteristics and in the ideal case to be able to design objects with improved or special characteristics. Every material object has three geometrical dimensions (x_1, x_2, x_3). The geometry and the properties of this object can be defined to be a function $f(x_1, x_2, x_3)$ of these three dimensions. The geometry and the investigated property can change during the timespan t. This function thus depends on four variables $f(x_1, x_2, x_3, t)$. Applying the laws of nature about mechanics, thermodynamics, acoustics, etc. to this property, a set of equations for the investigated object can be defined. They describe the dependencies between its variables and the object parameters p_1, p_2, etc.:

$$f(x_1, x_2, x_3, t, p_1, p_2, \ldots) = 0 \qquad [6.1]$$

Normally, in equation [6.1], there are several terms consisting of the partial derivatives $\partial f/\partial x_1$, $\partial f/\partial x_2$ of f with respect to the independent variables x_1, x_2, e.g. the coordinates in the principal direction. Such equations are named *partial differential equations* (PDE) and they are used to describe problems in the statics and dynamics of elastic bodies, propagation of sound or heat or fluid flow, electrostatics and dynamics and a wide range of other phenomena. In general, it is not possible to find an exact analytical solution of the partial differential equations. The FEM is an numerical method for solving PDE, developed over the last 60 years, owing to the increasing computer power which is required for matrix calculations (Fig. 6.1).

The idea of the FEM is that the investigated body is split (*meshed*) into a set of small parts, named *elements* with known finite size (*finite elements*) for which keypoints (*nodes*) the equations have an analytical solution (Fig. 6.1b). The nodes at the edges of one element are simultaneously nodes of some of the neighbour element. Because of this, the solutions of problems for both neighbour elements have to have the same values at these nodes. In order to ensure this, a system of equations for all the nodes is solved. Due to the linear elastic bodies, this system of equations is linear and can thus be solved within a very short period of time using a computer and, thus, a solution for a complex geometry can be found quickly. A big advantage of today's FEM software for engineers is that all the numerical methods for the solution of the PDF for each element and for solving the whole linear system remain hidden from the user. The user has to prepare the geometry,

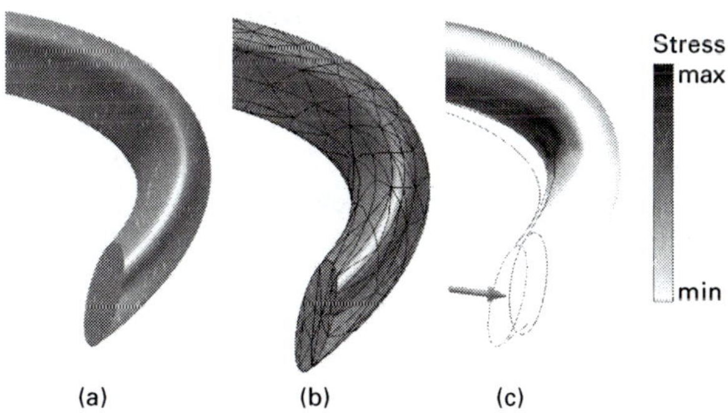

6.1 Part with nonlinear geometry: (a) CAD view; (b) the same part meshed with regular finite elements; (c) deformed part with the simulated Von Mises stresses.

the material, constraints and loads for the system and then the software more or less automatically looks for a viable solution. Because of this, this method is very popular, especially for applications where a stable automatic solution can be defined.

6.2 Modelling of mechanical systems

The modelling process of mechanical systems consists of several steps (Fig. 6.2). The real system has to be idealized into a physical model, then a mathematical model is created and solved. This section discusses important issues of this modelling process in general.

6.2.1 The modelling process

During the modelling with FEM, the real system (Fig. 6.3a), like a tree with a ladder, has to be idealized first in order to achieve a somewhat simplified physical system. This first step is the building of the physical model, where certain assumptions have to be made. For example, it could be assumed that all steps consist of the same material, they have the same geometry and

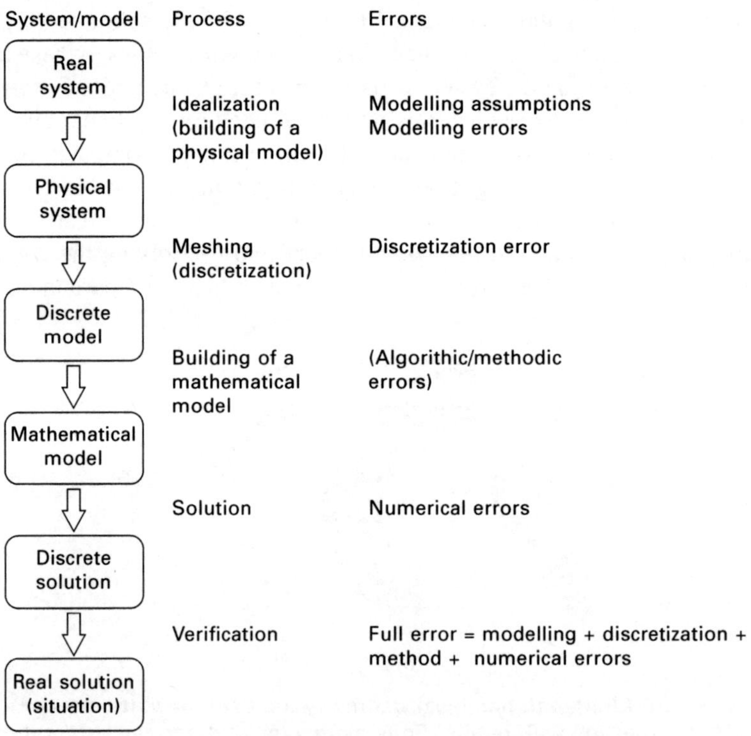

6.2 Steps during the modelling of mechanical systems.

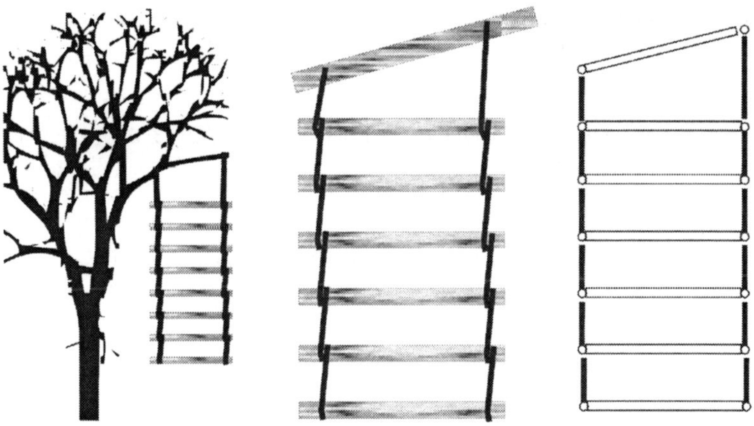

6.3 The way between the physical object and FEM model: (a) real object – ladder on a tree; (b) physical model – consists of equal steps; (c) discretized model – the steps are represented as a beams and ropes, connected at the nodes.

mechanical properties, and that the connections between them are the same. This idealization step generates the first source for errors, e.g. influences from the differences of the weight and the geometry of the steps are already excluded and will not be considered in the next step. The created physical model (Fig. 6.3b) represents the investigated system in a simplified way, so that some analytical or numerical method could be applied. During the FEM modelling, a discrete model of the system has to be created (Fig. 6.3c), which consists of beams and ropes, connected at the nodes. The chosen discretization method leads to a discretization error, e.g. using one beam element per step, so only a coarse approximation of the step deformation can be achieved, and the real deformation cannot be represented.

Having defined the elements and the nodes, as well as the problem type which has to be solved (static, dynamic, vibration, etc.), the equations for this system can be created and solved. These equations represent the mathematical model of the system. The advantage of using commercial or non-commercial, but well-tested and evaluated, FEM software is that at this step, the methods are well investigated and the risk of an additional error is minimized. In most standard load cases with standard elements, an error can be ruled out. This is not the case for the solution of the mathematical model, where the user still has the possibility to influence the step size during the integration, etc., so numerical errors are possible. In the case of FEM in mechanics, the calculated solution represents the displacements of the nodes. This is a solution of the discrete system and it differs from the real system – because of all previously described error sources. If the difference between this solution and the real situation of the system is less than the allowed error, it can be assumed that the modelling was successful and the model is verified.

6.2.2 Types of problem from the mechanical point of view

The aim of the FEM in statics and dynamics is to calculate the behaviour of mechanical bodies and systems. In the different subareas of classical mechanics, the objectives of the FEM differ. The equation of the dynamics equilibrium of an elastic body consists of different components:

$$K \cdot u + D \cdot \frac{du}{dt} + M \cdot \frac{d^2u}{dt^2} = F \qquad [6.2]$$

The first component, $K \cdot u$, describes the internal elastic forces of the body, which are proportional to its stiffness (matrix) K and the displacement vector u. The second element, $D \cdot \frac{du}{dt}$, describes internal viscous forces, which are proportional to the velocity of the nodes. These forces are important for polymer materials or can be used to describe air resistance. The third component, $M \cdot \frac{d^2u}{dt^2}$, represents the forces of inertia, which depend on the acceleration of the system and on its mass M. All external forces are represented by the vector on the right-hand side of the system – F. The solution of the system [6.2] represents the motion of the steps (Fig. 6.4) and their deformation due to the given loads.

If the deformations inside the steps are negligible compared with the displacements (to the motion), then the internal forces and the damping can be neglected and it is possible to reduce the system to:

$$M \cdot \frac{d^2u}{dt^2} = F \qquad [6.3]$$

The motion law can be derived by equation [6.3]. For the investigation of the vibrations of the body, the natural frequencies and forms have to be found. If the motion of the system with consideration of the elastic changes in the body has to be found, the system [6.2] has to be completely solved, with the right-hand side set equal to zero:

$$K \cdot u + D \cdot \frac{du}{dt} + M \cdot \frac{d^2u}{dt^2} = 0 \qquad [6.4]$$

These natural frequencies and forms then represent then the vibrations of the steps, if the branch is not moving (Fig. 6.5a). If the branch is swinging with a frequency ω, then the forced vibration of the system can be investigated by integrating equation [6.5]:

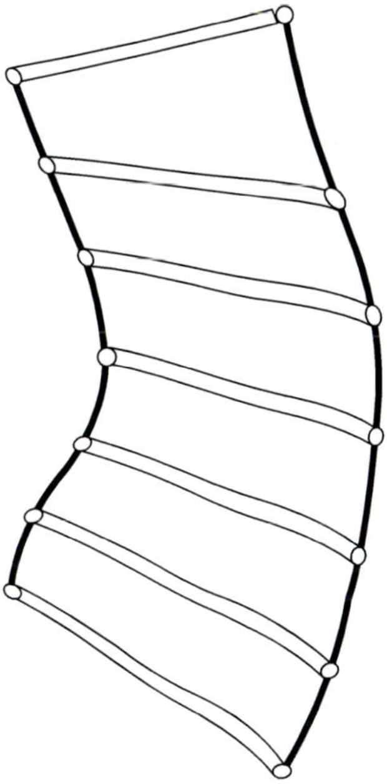

6.4 Dynamical calculation – the motion of the entire ladder is investigated.

(a) (b)

6.5 Frequencies: (a) the independent motion of the ladder; (b) forced vibration – the behaviour of the ladder if the branch of the tree is moving or vibrating.

$$K \cdot u + D \cdot \frac{du}{dt} + M \cdot \frac{d^2u}{dt^2} = F \cdot \cos(\omega t) \qquad [6.5]$$

All these cases consider the dynamics or the transient processes of the system. At present, the dynamic equations are solved usually, but not in all cases, using explicit numerical methods, where the displacements at the later time state are calculated using the state of the system at the present time. With implicit methods, the state of the system involving both the current and the subsequent state is calculated. They are usually used for static problems (Fig. 6.6). In statics there is no velocity and acceleration of the body and equation [6.5] becomes the simple form for linear (or nonlinear) systems:

$$K \cdot u = F \qquad [6.6]$$

In this case, the stiffness matrix K and the forces F are known at the current time step and the nodal displacements have to be found solving the system [6.6]. For this case, different discretizations are also possible, e.g. the steps, assuming a constant cross-section, can be represented by a single beam (Fig. 6.7b). A meshing of the steps with smaller elements requires more computational time, but allows a more precise solution and can be used even if the steps have a more complex geometry, which cannot be simply represented by a beam.

6.2.3 Elasticity theory for FEM calculations

For FEM calculations, three laws are required for the modelling of mechanical systems – kinematic, constitutive (material) and equilibrium.

Kinematic

The kinematic equations represent the relation between the nodal displacements and the internal deformations (strains) of the mechanical system. Figures 6.8

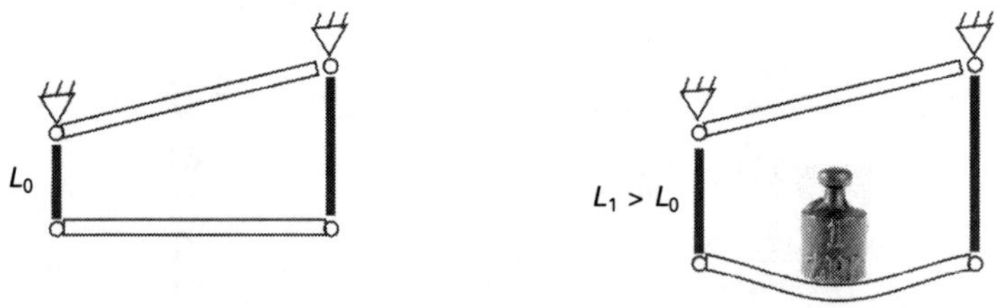

(a) (b)

6.6 Static calculation – the ladder and its components are not moving; their deformations and the forces at static equilibrium are investigated: (a) initial state; (b) deformed state.

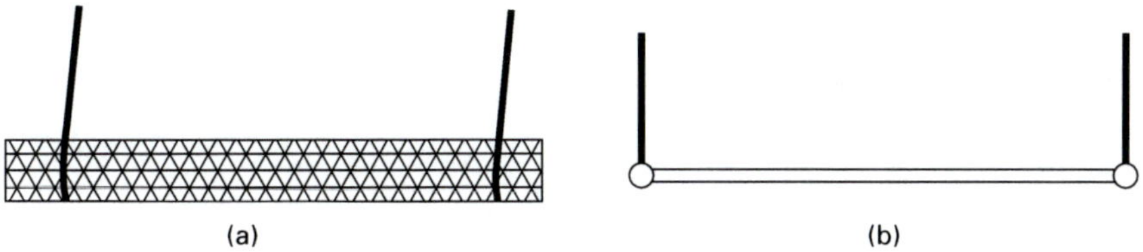

6.7 Modeling levels: (a) each step is presented as a solid body, discretized by small elements; (b) the entire step is presented as one beam element with nodes.

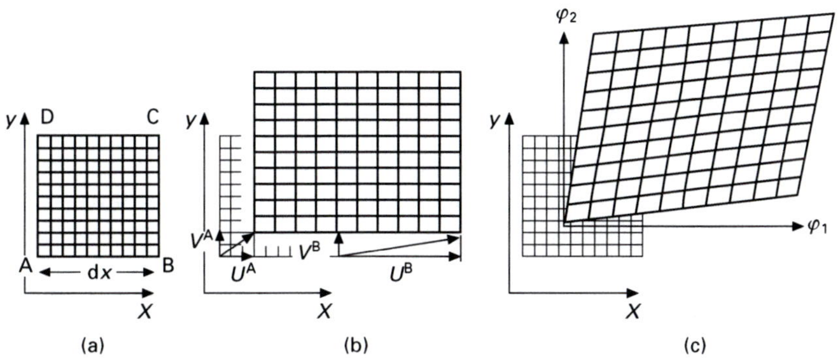

6.8 Kinematic relations during small deformations of elastic body: (a) initial state; (b) pure elongations without change of the angle between the sides (c) shear deformation.

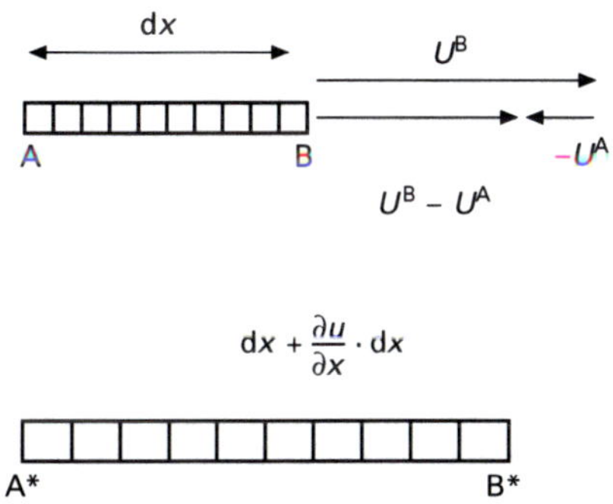

6.9 Elongation of line AB as relation of the displacements of both points A and B.

and 6.9 explain the fundamentals of the kinematic relations based on a square figure ABCD. Figure 6.8a shows a plate ABCD with an infinitessimal side d*x*. The mesh inside the figure is shown only to visualize the deformations inside the plate. When point A moves inside the *XY* plane, its displacement is given by

$$\mathrm{d}r = u + v \qquad [6.7]$$

The side AB, having an initial length dx will become longer (or shorter) with

$$u^{\mathrm{B}} - u^{\mathrm{A}} = \frac{\partial u}{\partial x} \cdot \mathrm{d}x \qquad [6.8]$$

Thus, the new length A*B* is given by

$$\mathrm{d}x + \frac{\partial u}{\partial x} \cdot \mathrm{d}x \qquad [6.9]$$

The linear elongation in X direction is the change of the length related to the initial length, so that:

$$\varepsilon_{xx} = \frac{\mathrm{d}u}{\mathrm{d}x} \approx \frac{u^{\mathrm{B}} - u^{\mathrm{A}}}{\mathrm{d}x} \qquad [6.10]$$

In the three-dimensional case, the linear elongations in the three main directions can be derived as follows:

$$\varepsilon_{xx} = \frac{\mathrm{d}u}{\mathrm{d}x} \qquad \varepsilon_{yy} = \frac{\mathrm{d}v}{\mathrm{d}y} \qquad \varepsilon_{zz} = \frac{\mathrm{d}w}{\mathrm{d}z} \qquad [6.11]$$

In addition to the linear elongations, the body can be deformed so that the angles of it change. In the case of the analysed plate ABCD (Fig. 6.8c), the angle ∠ BAD can change. This depends on the rotations of both sides AB and AD and can be represented as a sum of the angles φ_1 and φ_2.

It could be proven (for instance in Rieg and Hackenschmidt, 2003), that

$$\varphi_1 = \frac{\partial u}{\partial x} \qquad \varphi_2 = \frac{\partial v}{\partial y} \qquad [6.12]$$

and the change of the angle ∠ BAD in dimensionless form is

$$\gamma_{xy} = \frac{\mathrm{d}u}{\mathrm{d}y} + \frac{\mathrm{d}v}{\partial x} \qquad [6.13]$$

The variable γ_{xy} is called the engineering shear strain. According to the definition, we have:

$$\gamma_{xy} = 2 \cdot \varepsilon_{xy} \quad \gamma_{xz} = 2 \cdot \varepsilon_{xz} \quad \gamma_{yz} = 2 \cdot \varepsilon_{yz} \qquad [6.14]$$

Then, the entire strain matrix becomes a tensor

$$\varepsilon = \begin{bmatrix} \varepsilon_{xx} & \varepsilon_{xy} & \varepsilon_{xz} \\ \varepsilon_{yx} & \varepsilon_{yy} & \varepsilon_{yz} \\ \varepsilon_{zx} & \varepsilon_{zy} & \varepsilon_{zz} \end{bmatrix} \qquad [6.15]$$

where

$$\varepsilon_{xy} = \varepsilon_{yx} = \frac{1}{2} \cdot \left(\frac{dv}{dx} + \frac{du}{dy} \right)$$

$$\varepsilon_{xz} = \varepsilon_{zx} = \frac{1}{2} \cdot \left(\frac{dw}{dx} + \frac{du}{dz} \right)$$

$$\varepsilon_{yz} = \varepsilon_{zy} = \frac{1}{2} \cdot \left(\frac{dv}{dz} + \frac{dw}{dy} \right) \qquad [6.16]$$

Equations [6.11] and [6.16] represent the connection between the displacements of the nodes and the strain in the body in the linear case (small deformations) and build the kinematic equations. For larger displacements, the accuracy of these equations is not sufficient, since they are derived using only the first term of the Tailor development of the relation $u^B = f(u^A, dx)$. In such cases, more terms in the development have to be used or other equations for the relations between displacement and strain have to be derived. Since the nonlinearity in these equations is based on pure geometrical relations, the problems are known as 'problems with geometric nonlinearities'. For these, special options in the FEM software have to be activated to ensure that the proper equations are used. More details about geometrical nonlinearities are given in Section 6.5.

Constitutive equations

Constitutive equations represent the relations between the deformations of the system and the internal stresses. These are usually named material law, because they represent the behaviour of the material of the modelled object. In the simple case of a linear elastic material (e.g. a spring), the constitutive law is Hooke's law, here represented in a matrix form:

$$C \cdot \varepsilon = \sigma \qquad [6.17]$$

where the stresses in vector form are given by

$$\sigma = [\sigma_{xx}\, \sigma_{yy}\, \sigma_{zz}\, \tau_{xy}\, \tau_{yz}\, \tau_{zx}]^T \qquad [6.18]$$

The normal stresses result from tension or pressure applied on the working plane, for instance σ_{xx} (Fig. 6.10). The tangential τ_{xy} stresses are caused by shear forces. The first symbol shows the plane index, where the stresses are applied and the second symbol the direction of the forces. For example shear stress τ_{xy} is applied in the plane x, perpendicular to the x axis, and its direction is parallel to the y axis.

The strains here are in vector form:

$$\varepsilon = [\varepsilon_{xx}\, \varepsilon_{yy}\, \varepsilon_{zz}\, \varepsilon_{xy}\, \varepsilon_{yz}\, \varepsilon_{zx}]^T \qquad [6.19]$$

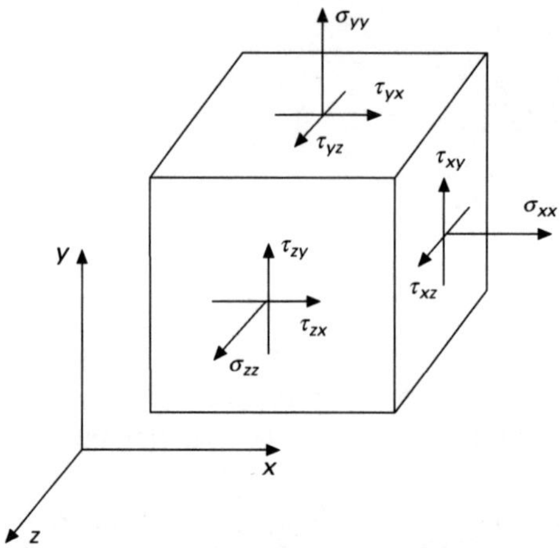

6.10 Directions of normal and shear stresses.

and the elasticity matrix is then given by

$$C = \begin{bmatrix} c_{11} & c_{12} & c_{13} & c_{14} & c_{15} & c_{16} \\ c_{21} & c_{22} & c_{23} & c_{24} & c_{25} & c_{26} \\ c_{31} & c_{32} & c_{33} & c_{34} & c_{35} & c_{36} \\ c_{41} & c_{42} & c_{43} & c_{44} & c_{45} & c_{46} \\ c_{51} & c_{52} & c_{53} & c_{54} & c_{55} & c_{56} \\ c_{61} & c_{62} & c_{63} & c_{64} & c_{65} & c_{66} \end{bmatrix} \qquad [6.20]$$

consisting, in the general case, of 21 independent constants, representing the material properties presented as a relation between stress in one direction and strain in the same or another direction.

The elasticity matrix is symmetrical, because coefficients with the same digits are equal, e.g. $c_{21} = c_{12}$. For most engineering materials, such as metals, this matrix is filled only by two independent constants – the elasticity modulus E and the Poisson ratio v:

$$\begin{bmatrix} \sigma_{xx} \\ \sigma_{yy} \\ \sigma_{zz} \\ \tau_{xy} \\ \tau_{yz} \\ \tau_{zx} \end{bmatrix} = \frac{E}{(1+v)\cdot(1-2v)} \cdot \begin{bmatrix} (1-v) & v & v & 0 & 0 & 0 \\ v & (1-v) & v & 0 & 0 & 0 \\ v & v & (1-v) & 0 & 0 & 0 \\ 0 & 0 & 0 & \frac{1}{2}(1-2v) & 0 & 0 \\ 0 & 0 & 0 & 0 & \frac{1}{2}(1-2v) & 0 \\ 0 & 0 & 0 & 0 & 0 & \frac{1}{2}(1-2v) \end{bmatrix} \begin{bmatrix} \varepsilon_{xx} \\ \varepsilon_{yy} \\ \varepsilon_{zz} \\ \varepsilon_{xy} \\ \varepsilon_{yz} \\ \varepsilon_{zx} \end{bmatrix}$$

[6.21]

In some special cases, where the Poisson ratio is equal to 0.5, the volume of the body does not change during deformation. In this case of incompressibility, nonlinear material models have to be used. Some more information about linear and nonlinear models is presented in Section 6.7. If the relation between the stresses in the body and its deformation is nonlinear, then we have

$$\sigma = f(\varepsilon) \tag{6.22}$$

This model is labelled 'model with material nonlinearity'. Such problems require the use of special numerical methods as standard models would fail.

Equilibrium equations

The equilibrium equation describes the static or dynamic equilibrium of all internal and external forces of the system. In the static case, the equilibrium equation is

$$K \cdot u = F \tag{6.23}$$

where K is the stiffness matrix of the system, u is the vector with the nodal displacements and F represents the external forces (Fig. 6.11). In the dynamic case, suitable equations from Section 6.2.2 are used instead of equation [6.23]. The left-hand side of equation [6.23] represents the internal forces, calculated on the basis of the displacements of the single nodes and the stiffness of all the elements. How the stiffness matrix K aggregates kinematic

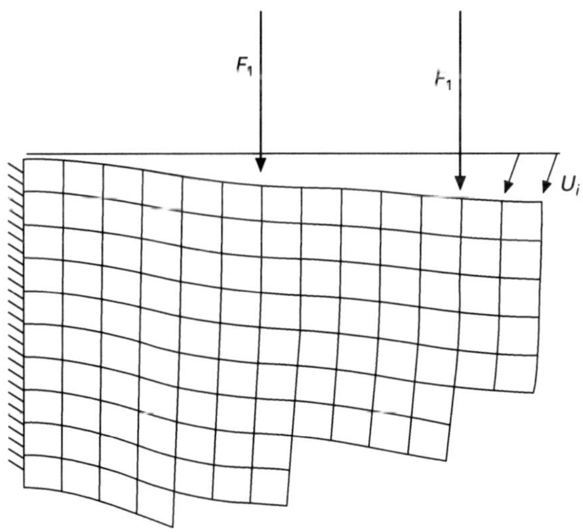

6.11 Example of deformed state under external forces and single displacements.

relations equations [6.11] and [6.16] with the constitutive equation [6.21] in one matrix is described in the next section.

6.2.4 Static equilibrium and element stiffness matrix

One of the ideas of the FEM is that the displacements at every point inside of one finite element $U(x, y) = [u(x, y)\ v(x, y)]^T$ can be represented by interpolation of the nodal displacements u_i using the known function $N_i(x, y)$

$$U(x, y) = \sum_i u_i \cdot N_i(x, y) \qquad [6.24]$$

The functions $N_i(x, y)$ are named shape functions. They can be linear, quadratic or from a higher power which determines the order of the element. If the functions are linear, the element is of first order, if quadratic-second order and so on (Fig. 6.12).

Some examples of shape functions are given in the next section, where the different elements are described. Since the shape functions are given and selected during the development of the FEM code and not during the modelling, the main operations, like differentiation over these can be performed analytically. This minimizes the numerical errors from numerical differentiations, and simplifies and speeds up computations. The strains ε_{ij} (here, only the strains for the two-dimensional case are taken) can be presented using a differential operator L and a displacement vector U as follows:

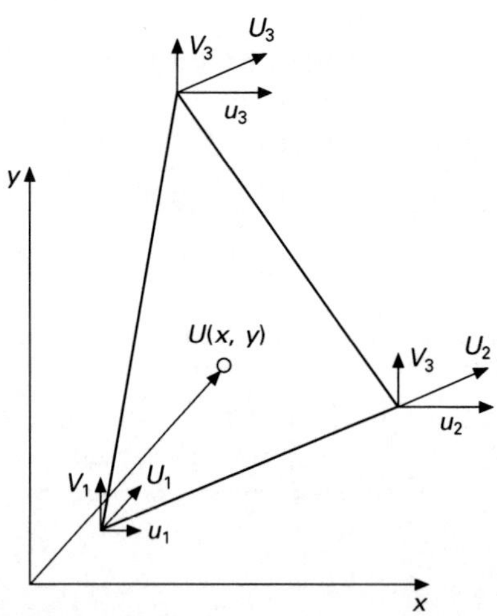

6.12 The displacement of any point inside the finite element is presented as a function of the nodal displacements.

$$\varepsilon_{ij} = \begin{bmatrix} \varepsilon_{xx} \\ \varepsilon_{yy} \\ \gamma_{xy} \end{bmatrix} = \begin{bmatrix} \frac{\partial u}{\partial x} \\ \frac{\partial v}{\partial y} \\ \frac{\partial u}{\partial x} + \frac{\partial v}{\partial y} \end{bmatrix} = \begin{bmatrix} \frac{\partial}{\partial x} & 0 \\ 0 & \frac{\partial}{\partial y} \\ \frac{\partial}{\partial x} & \frac{\partial}{\partial y} \end{bmatrix} \begin{bmatrix} u \\ v \end{bmatrix} = \mathrm{LU} \qquad [6.25]$$

Using the shape function for the displacements, these become

$$U = \begin{bmatrix} u(x, y) \\ v(x, y) \end{bmatrix} = \begin{bmatrix} N_1 \cdot u_1 + N_2 \cdot u_2 + N_3 \cdot u_3 \\ N_1 \cdot v_1 + N_2 \cdot v_2 + N_3 \cdot v_3 \end{bmatrix} \qquad [6.26]$$

which, in matrix form, with the displacements of the nodes leads to: $U_{(i)}$

$$U = \begin{bmatrix} N_1 & 0 & N_2 & 0 & N_3 & 0 \\ 0 & N_1 & 0 & N_2 & 0 & N_3 \end{bmatrix} \cdot \begin{bmatrix} u_1 \\ v_1 \\ u_2 \\ v_2 \\ u_3 \\ v_3 \end{bmatrix} = NU_{(i)} \qquad [6.27]$$

Hence, the strains from equation [6.25] become

$$\varepsilon_{ij} = LU = LNU_{(i)} = BU_{(i)} \qquad [6.28]$$

where the matrix B consists of the differentiated form functions. The relations for elements with other numbers of nodes in 2D and 3D space are similar.

At that point, the principle of the virtual displacements will be applied. If the system deforms with a virtual displacements $\delta\varepsilon$ the virtual internal work δW_i and the external work δW_e have to be equal:

$$\delta W_i = \delta W_e \qquad [6.29]$$

The virtual external work is due to the external forces and can be represented by

$$\delta W_i = \iiint_v \delta U^{\mathrm{T}} P \mathrm{d}V + \delta U_{(i)}^{\mathrm{T}} F_{(i)} \qquad [6.30]$$

where P stands for the distributed volume forces and $F_{(i)}$ represents the concentrated nodal forces.

The virtual internal work δW_i can be derived as the increase of the internal (potential) energy, or as a product of the displacements and the stresses (work is equal to the product of way $\delta\varepsilon^{\mathrm{T}}$ and the force) according to: $\sigma \cdot \mathrm{d}V$

$$\delta W_i = \iiint_v \delta\varepsilon^{\mathrm{T}}\sigma \mathrm{d}V \qquad [6.31]$$

The equilibrium of the internal and external work is then:

$$\iiint_v \delta\varepsilon^{\mathrm{T}}\sigma \mathrm{d}V = \iiint_v \delta U^{\mathrm{T}} P \mathrm{d}V + \delta U_{(i)}^{T} F_{(i)} \qquad [6.32]$$

Taking into account equations [6.28] and [6.27], we get:

$$\iiint_v (B\delta U_{(i)})^{\mathrm{T}}\sigma \mathrm{d}V = \iiint_v (N\delta U_{(i)})^{\mathrm{T}} P \mathrm{d}V + \delta U_{(i)}^{\mathrm{T}} F_{(i)} \qquad [6.33]$$

Applying the rules for matrix transpose $(AB)^{T} = B^{T}A^{T}$ that leads to

$$\iiint_v \delta U_{(i)}^{T} B^{T} \sigma \mathrm{d}V = \iiint_v \delta U_{(i)}^{T} N^{T} P \mathrm{d}V + \delta U_{(i)}^{\mathrm{T}} F_{(i)} \qquad [6.34]$$

The displacements $\delta U_{(i)}$ are not dependent on the volume and the equilibrium between the internal and external forces has to be true at every displacement, so we have:

$$\iiint_v B^{T} \sigma \mathrm{d}V = \iiint_v N^{T} P \mathrm{d}V + F_{(i)} \qquad [6.35]$$

According to equation [6.17], the stresses $\sigma = C \cdot \varepsilon$ are dependent on the material law and the strains on the displacements according to equation [6.28], hence: $BU_{(i)}$

$$\iiint_v B^{T} CB \mathrm{d}V \cdot U_{(i)} = \iiint_v N^{T} P \mathrm{d}V + F_{(i)} \qquad [6.36]$$

If the nodal forces on the right-hand side are not considered for simplification, then we get:

$$\iiint_v B^{T} CB \mathrm{d}V \cdot U_{(i)} = F_{(i)} \qquad [6.37]$$

Considering the element stiffness matrix as

$$K = \iiint_v B^{T} CB \mathrm{d}V \qquad [6.38]$$

the main equation of the FEM for the static problems in local, element coordinates is derived as:

$$K \cdot U_{(i)} = F_{(i)} \qquad [6.39]$$

In this equation, the displacements, the forces and the stiffness matrix are in the element (local) coordinate system. For the solution of the entire system of elements in one body, these have to be transformed to the global coordinate system.

It is important to note in this case that:

- equation [6.39] is linear and can be solved according to the unknown displacements using standard libraries and methods;
- the stiffness matrix is constant (for linear cases) and can be calculated on the basis of simple matrix operations using the form functions. These are defined before solving the problem and their differentiation can be done analytically, so that the matrix B is known during the building of the mathematical model at the moment and the appropriate element is selected.

The different element types are discussed in Section 6.3, and the main steps of the performing FEM simulation with available FEM software are explained in the next section.

6.2.5 Dynamic equilibrium – explicit versus implicit integration

In the same way as in the previous section, equation [6.39] can be derived in the case of additional forces from the acceleration and damping in the form (with global coordinates and the subscript (i) for the displacements is skipped for simplification):

$$M \cdot \ddot{U} + C \cdot \dot{U} + K \cdot U = F \quad [6.40]$$

where M, C and K are the mass, damping and stiffness matrices of the entire system, U is the displacements vector of all nodes, and F is the force vector with all acting forces in global coordinates. This equation consists of first and second order derivatives of the displacements and can be solved during implicit or explicit integration. The displacements U, velocities $\dot{U}$ and acceleration $\ddot{U}$ have to be calculated at every time step t_n. To solve it, the equation has to be transformed into an algorithm-dependent time-difference equation by quantizing and replacing

$$\dot{U}(t) = \lim_{dt \to 0} \frac{U(t + dt) - U(t)}{dt} \quad [6.41]$$

with some finite difference such as forward differences

$$\dot{U}(t) = \frac{U(t + dt) - U(t)}{dt} \quad [6.42]$$

or central differences

$$\dot{U}(t) = \frac{U(t + dt) - U(t - dt)}{2 \cdot dt} \quad [6.43]$$

In the explicit integration, the equation for accelerations is written in the same (analogous) way through the actual, known displacements.

The displacements U_n, velocities $\dot{U}_n$, accelerations $\ddot{U}_n$ and the forces F_n have to be calculated at every time step t_n. Actually, in order to calculate the accelerations and the velocities with some of the finite differences during the time t_n, the displacements and the next and/or previous time step must be known, depending on the type of the used finite differences. The problem is that the displacements in the next time step (in the future) are not known. At that point, depending on the method for the solution of this problem, two approaches can be used, which lead to two different groups of the FEM and software – explicit FEM and implicit FEM (Fig. 6.13).

For explicit integration, the next time point is calculated based on the information from the previous time point (a). During the implicit integration, a system of the unknown variables has to be solved at each time step, in order to find the proper solution for the next time point (b).

During explicit integration (Fig. 6.13a), the displacements at the next time step are calculated using the displacements of the previous time step and the velocity. This procedure is stable only for time steps smaller than a critical limit and can be applied only for some of the methods for the calculation of the velocity (for instance central differences). For instance, if the damping in equation [6.40] is neglected, for the time step t_n, we get

$$M \cdot \ddot{U}_n + K \cdot U_n = F_n \tag{6.44}$$

and the new point is extrapolated for the next time step $n + 1$.

Assuming that the displacement can be interpolated as a linear function of time, the velocity at one step after the current one is given by:

$$\dot{U}_n = \frac{1}{\Delta t}(U_{n+1} - U_n) \tag{6.45}$$

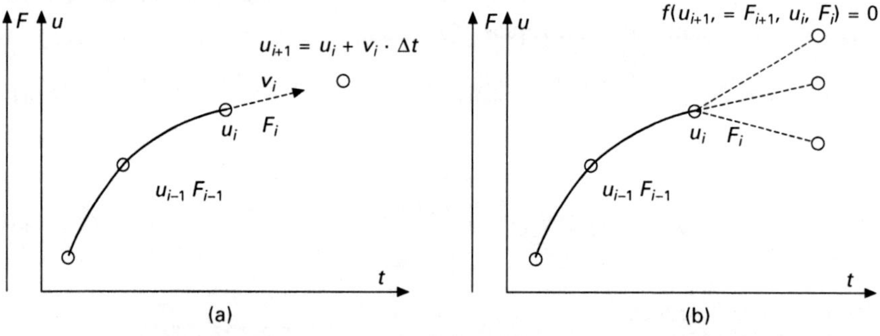

6.13 The ideas of the explicit (a) and implicit (b) integration of the differential equation. For the explicit integration the next time point is calculated based on the information from the previous time point (a). During the implicit integration, a system of the unknown variables have to be solved at each time step, in order the proper solution for the next time point to be found (b).

The acceleration at the current time step can then be calculated as:

$$\ddot{U}_n = \frac{1}{\Delta t}(\dot{U}_{n+1} - \dot{U}_n) \qquad [6.46]$$

where replacing [6.45]

$$\ddot{U}_n = \frac{1}{\Delta t^2}(U_{n+1} - 2\mathrm{U}_n + U_{n-1}) \qquad [6.47]$$

The next displacement can be written as:

$$U_{n+1} = 2U_n - U_{n-1} + \Delta t^2 \ddot{U}_n \qquad [6.48]$$

Since the accelerations depend on the forces

$$\ddot{U}_n = M^{-1}(F_n - K \cdot U_n) \qquad [6.49]$$

the new point can be calculated as

$$U_{n+1} = 2U_n - U_{n-1} + \Delta t^2 \cdot M^{-1} \cdot (F_n - K \cdot U_n) \qquad [6.50]$$

Only the information about displacements and the forces (here, the space derivatives of the partial differential equations are hidden) from the previous time steps is used. In contrast, during implicit integration, the displacement at the new time point is unknown but it has to be found so that the forces during the new time step satisfy the equilibrium equation as well [6.44]

$$M \cdot \ddot{U}_{n+1} + K \cdot U_{n+1} = F_{n+1} \qquad [6.51]$$

where the acceleration $\ddot{U}_{n+1}$ is calculated analogously to equation [6.48].

After transformations and some assumptions (linear acceleration) the equation can be written in the form

$$(a_0 M + K) \cdot U_{n+1} = F_{n+1} + M \cdot (a_3\dot{U}_n + a_2\ddot{U}_n + a_0 U_n) \qquad [6.52]$$

simplified as

$$K_{\mathrm{eff}} \cdot U_{n+1} = F_{\mathrm{eff},n+1} \qquad [6.53]$$

Here the coefficients a_i depends on the time step and the chosen integration method (central differences, Newmark, Fox-Goodwin) and the effective stiffness and force matrices are named K_{eff} and $F_{\mathrm{eff},n+1}$, respectively. As equation [6.53] shows during the implicit integration a special system of linear equations has to be solved at each time step. For this, the inverse matrix of K_{eff} has to be found, which is a computational-intensive operation and not in all cases solvable. For large deformations, this matrix depends as well from the displacements, and then system [6.53] becomes nonlinear. From other point of view, if the numerical method is chosen properly, for instance Newmark, it is unconditionally stable, independent of the time step and can work with large but cost intensive time steps.

For the explicit integration only the inverse matrix of the mass matrix

$$M = \begin{bmatrix} m_{11} & 0 & & 0 \\ 0 & m_{22} & & 0 \\ & & \dots & 0 \\ 0 & 0 & & m_{nn} \end{bmatrix}$$

has to be calculated, and this is a trivial operation as

$$M^{-1} = \begin{bmatrix} \frac{1}{m_{11}} & 0 & & \\ 0 & \frac{1}{m_{22}} & & \\ & & \dots & \\ 0 & 0 & & \frac{1}{m_{nn}} \end{bmatrix} \quad [6.54]$$

The mass normally remains constant and its inverted matrix can be calculated once at the beginning of the integration process. However, for implicit integration, the inverse of the stiffness matrix has to be calculated at every time step. For nonlinear problems, the stiffness matrix depends on the displacements and thus several iterations have to be done until the displacement vector and the stiffness at the new time step are found.

Generally, because of the very small time step required, the explicit integration is used for simulations where very short timescales or nonlinear processes have to be analysed. These are for instance the crash (and generally the contact) simulations. Because the calculations can be done in parallel, the amount of steps is no longer a problem for the calculations. In all other nearly linear cases, the implicit integration offers a fast solution with large time steps. These and other differences between implicit and explicit method of the integration of the equilibrium equations are summarized in Table 6.1 (ESI Group, 2011).

6.2.6 Structure of typical FEM software

FEM calculation requires several steps and for most of these, separate programs (tools) were developed (Fig. 6.14). The preparation of the FEM model is done by the preprocessor. If the problem type (static, dynamic) is specified at the beginning, some of the software packages automatically select the elements and materials which are suitable for this kind of analysis, which reduces the risk of errors. The information about the material or geometric nonlinearities or other special properties of the modelled object has to be specified as well, in order for the proper solver type to be chosen.

Table 6.1 Comparison of implicit and explicit integration

	Explicit	Implicit
Method	Central differences (in most cases)	Newmark
Idea	Equilibrium equation at the current time step t_n	Equilibrium equation at the next time step t_{n+1}
Equation	The dynamic equilibrium is 'controlled and regulated'	Static equilibrium has to be found
Problem Types	Transient dynamics only	Static, transient, natural frequencies
Iterations	No iterations (for one time step)	Iterative solution (several iterations for one time step)
Matrix calculations	Only the element matrices; inverting of the mass matrices (trivial operation)	Full stiffness matrix; inverting of the stiffness matrix–complex operation
Equation	Not connected equation terms	Solve system of equations
Required accuracy	Single is often sufficient	Double precision is required
Computing time	'Low' time per time step, depending on: number of elements/number of integrations points and element length	Large computing time per time step depending on: number of nodes/degrees of freedom and convergence
Nonlinear problems	Simple modelling – stable solutions	Convergence problems for high nonlinear problems – no solution
Frequency of the solution	High frequency response	Low frequency response
Memory/storage	Little memory requirement, CPU speed important	Large memory required, the solutions have to fit into the RAM (while at the same time writing the matrices on the hard disc drive)
Typical number of time steps	50 000	50
Time step length	Limited: the integration is stable only if the time step is smaller than the critical limit – the highest natural frequency of the system $\Delta t \leq \Delta t^{crit} = \frac{2}{\omega_{max}}$ which depends on the smallest element See as well Courant *et al.* (1928/1967) for more information	Unlimited: the integration method is stable, independent of the time step. Practically, the time steps have to be adjusted to the required results. for nonlinear problems, the time steps have to be sufficiently small in order to ensure convergence

(Continued)

Table 6.1 Continued

	Explicit	Implicit
	Reduced integration, simple element formulations	Complex elements (with middle nodes, special shape functions)
Contact	Contact can produce energy, but solution exists, control of the energy balance required	For nonlinear problems – bisection method (or other) is required, in practice no control over the results

Preprocessor
Specify the problem definition
    Analysis type: static, dynamic, frequences
    Problem specifics – linear, nonlinear, large displacements...)
Define geometry – generated or imported from CAD
Define material properties
Choose elements
    Type
    Parameters (moment of inertia for beams...)
Define meshing controls (element size on specific places)
Mesh the volume
Apply the boundary conditions
Apply the loads
Save the data

Solver
Read the input data,
Select the solution method
Solve the problem
Save the displacements

Post-processor
Read the displacements
Calculate the stresses and strains
Visualize all data

6.14 Basic steps and modules during the FEM calculation.

The geometry of the modelled system has to be built inside the preprocessor or imported from CAD. In the case of FEM calculations of machinery, which are designed in CAD software, it is suitable to use the direct import of the geometry data. In other cases, e.g. complex textile structures, it is better to build their geometry inside the preprocessor or with special programs. For each task, enough material properties have to be specified – depending on the problem only the elasticity modulus and Poisson ratio are sufficient in case of a linear static analysis of isotropic material, but to take into account also dynamic forces, the density is required too.

The selection of the element types is the most important step. Usually, several elements from similar types are available, but these have different options which in turn define different degrees of freedom.

With all these data, each body has to be discretized into small elements from the selected type: the meshing process. In some cases, the meshing procedures create nodes very close to each other. They then have to be merged and the mesh has to be checked. Finally, the boundary conditions and the loads have to be defined. For static problems, the body has to be statically determined or over-determined. All this information is usually stored as a file and sent to the solver, which solves all the equations and writes the results (unknown displacements and other parameters) into a file again. The analysis and the visualization of the result requires separate software, known as the postprocessor, which is able to read the displacements of the nodes, to recalculate the forces within the elements and to display these in the proper places.

6.3 Elements of elastic models

The objects can be presented during modelling using different families of finite elements, presented in Table 6.2. In addition to the family type of the element, its behaviour is determined by its dimensionality, number of nodes and number and type of the degrees of freedom. The beams and trusses can exist as 2D or 3D elements. The 3D beams and trusses can move and deform only in one plane, for instance x–y, which reduces the size of the matrices and the calculations in comparison to the use of the 3D elements. The number of the nodes is at least equal to the number of the element corners, but the elements from higher order have nodes also between the corners. Elements that have modes only at their corners use linear interpolation in each direction. They are called linear elements or first order elements (Fig. 6.15a). Elements with midside nodes use square (or higher order) interpolation and are called second (or higher) order elements (Fig. 6.15b). Every node can have different numbers and types of degrees of freedom at the node from zero up to three translations and from zero up to three rotations. For instance the node of one-dimensional truss element has only one degree of freedom – the translational displacement in the x-direction.

6.3.1 Trusses and beams

Truss elements are used for structures, which can transfer loads only in one direction – the truss axis. They can work at tension and/or pressure and are defined by two nodes – both of the ends of the truss. Depending on the problem, two or three degrees of freedom per node can be available: for the plane problems only two translations per node are required, while for 3D problems all three translations have to be considered (Fig. 6.16). It is assumed that the truss element is a straight bar with a uniform cross-

Table 6.2 Commonly used element families

Family	Idealized model and simplified picture	Specifics
Truss	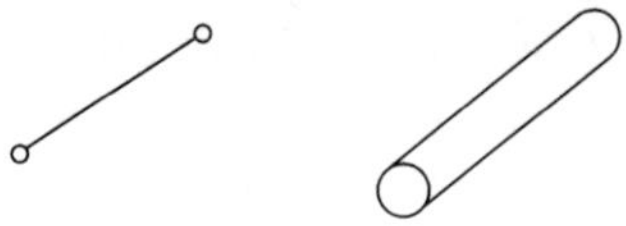	Trusses can react to deformations only by changing their length – they can transmit only tension or compression in the axis between the nodes. The trusses have their own mass (cross-section and density).
Springs and dashpots	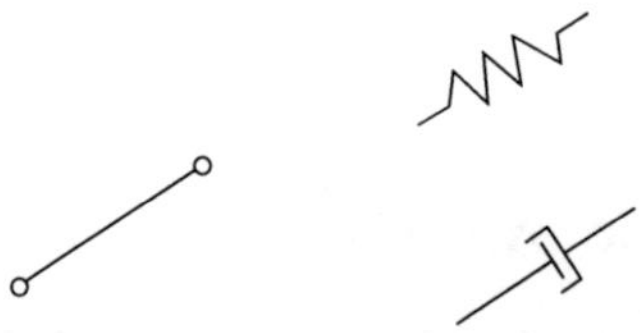	Springs and dashpots are special kind of trusses, which do not have their own mass.
Beam	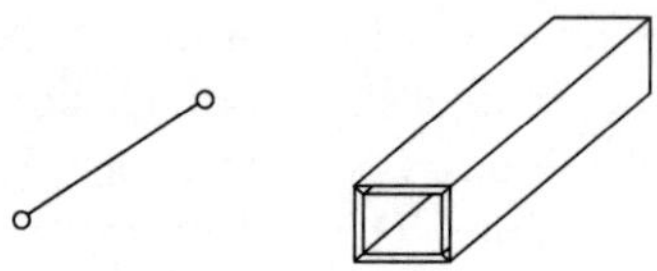	Beams are extensions of trusses, so that they can be bent. They can transmit moments (bending and twist). Thus, apart from their mass and cross-section, the moment of inertia is required.
Membrane	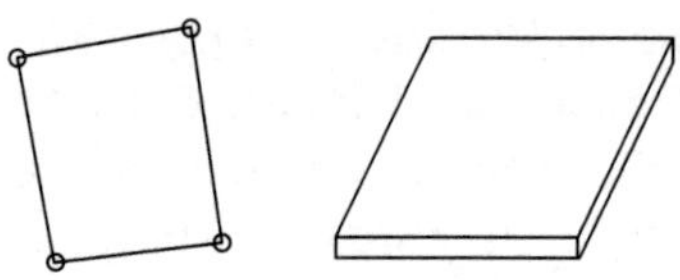	Membranes represent thin tensioned structures which carry only tension (no bending and no compression).
Plate	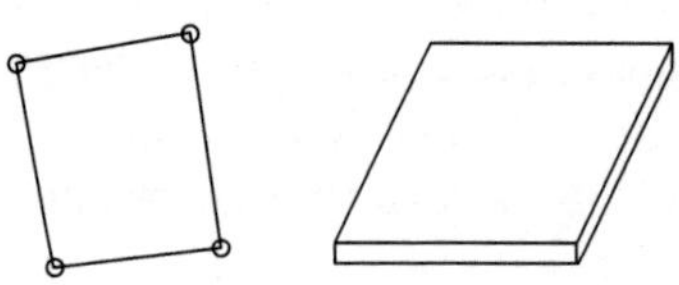	Plates are flat elements,which are able to transmit moments as well.
Shell	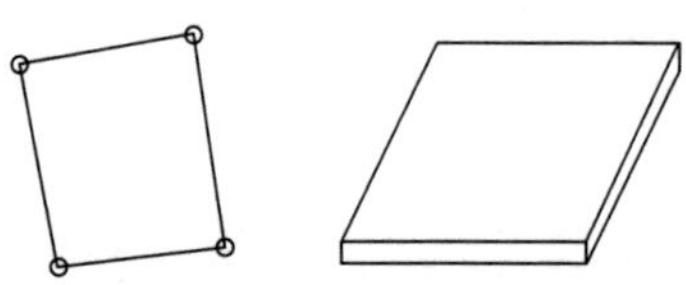	Shell elements differ from the plate elements in two features: they can be curved and they can carry membrane forces as well.
Solid (continuum) elements	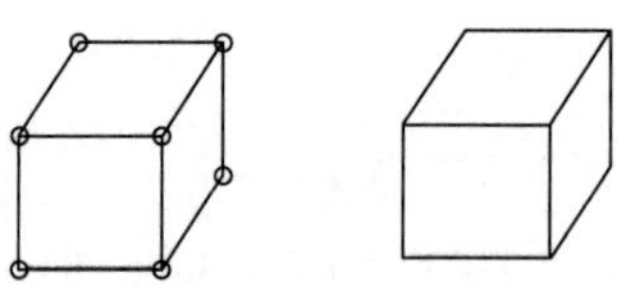	The solid elements represent a small volume of continuum.

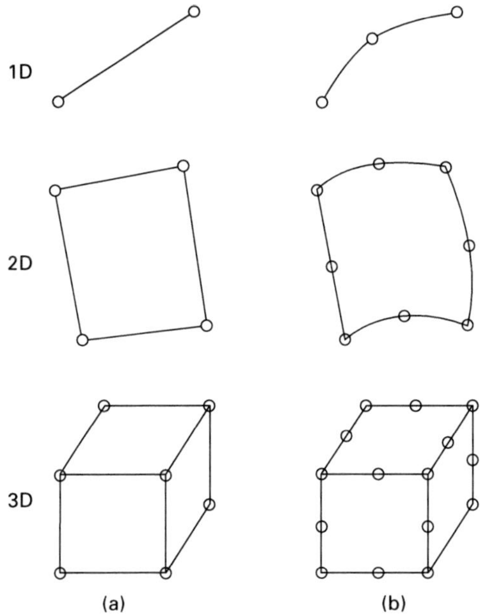

6.15 Elements with different order: (a) first order elements and (b) second order elements with midside nodes.

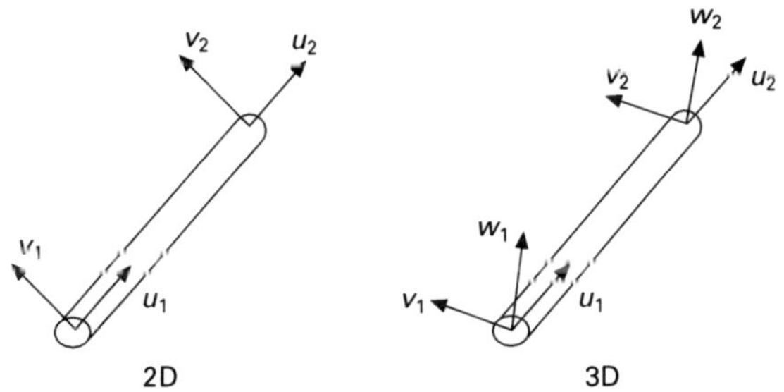

6.16 Trusses in 2D and 3D and the degrees of freedom of the nodes.

section. The displacements between the nodes are calculated using the shape functions (Fig. 6.17).

Depending on the displacements of the both ends, here called v_1 and v_2, the displacement in the one direction of any point of the truss element can be calculated using:

$$u(x) = \begin{bmatrix} 1 - \frac{x}{L} & \frac{x}{L} \end{bmatrix} \cdot \begin{bmatrix} v_1 \\ v_2 \end{bmatrix} \quad [6.55]$$

where x is the local coordinate, and L the length of the element. It is simple

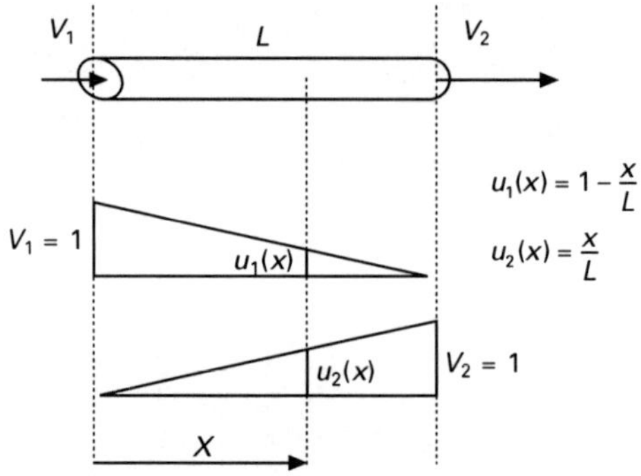

6.17 Shape functions for a truss element.

to demonstrate the strain calculation using analytical differentiation of the shape function

$$\varepsilon_x = \frac{\partial u(x)}{\partial x} = \frac{\partial}{\partial x}\begin{bmatrix} 1 - \frac{x}{L} & \frac{x}{L} \end{bmatrix} \cdot \begin{bmatrix} v_1 \\ v_2 \end{bmatrix} = \begin{bmatrix} -\frac{1}{L} & \frac{1}{L} \end{bmatrix} \cdot \begin{bmatrix} v_1 \\ v_2 \end{bmatrix} = B \cdot v \quad [6.56]$$

After that the entire element stiffness matrix K according to equation [6.38] can be calculated. The advantage of using truss elements with two nodes lies in its simplicity and shorter calculation time, but the disadvantage is in the only linear approximation of the displacements within the element. However in such an element, the loads can be applied only at both nodes at the edges. In the cases where loads have to be applied arbitrarily along the axial direction of the truss, use of higher element order trusses is more suitable. The simplest one is of second order and has a midpoint node (Fig. 6.18).

The shape function matrix for this case, depending on the coordinate ξ will be

$$N(\xi) = \begin{bmatrix} \frac{\xi}{2}(\xi - 1) & 1 - \xi^2 & \frac{\xi}{2}(\xi + 1) \end{bmatrix} \quad [6.57]$$

since x starts from zero up to the element length L, and the shape function has to have value 1 at the nodes, a change of the independent variable from ξ to x using

$$\xi = \frac{2}{L} \cdot \left(x - \frac{L}{2} \right)$$

is required and after that the differentiation about x has to be done in order the matrix B and the end the element stiffness matrix to be calculated. This operation remains hidden for the normal user of FEM software, but an understanding about the shape functions and the proper choose of element is important for the proper modelling.

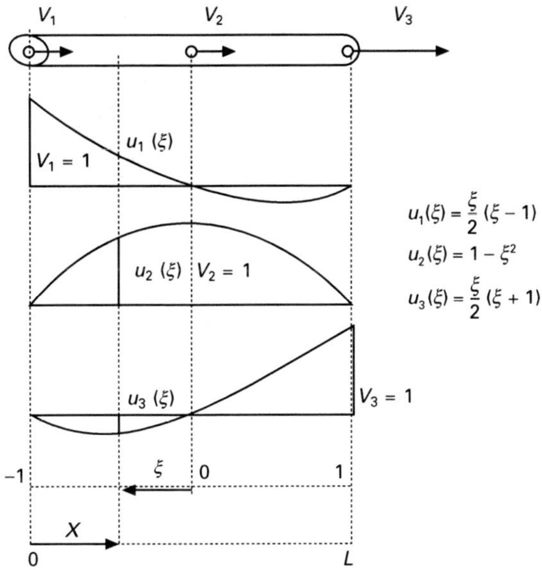

6.18 Truss element with second order shape functions and midpoint node.

Trusses are useful for the presentation of frame, yarn and rope structures, where these are working under tension without being bent. For instance, the ropes in a paraglider can be modelled by trusses (Fig. 6.19). Because of the computational and algorithmically simplicity, trusses are used very often to model different textile structures as a net shape structure. Figure 6.20 shows a model of a weft knitted structure as a hexagonal net by de Araújo *et al.* (2004) and draped structure. Figure 6.21 shows similar application for spacer warp knitted structures (Kyosev and Renkens, 2010a). Interesting application of trusses for the modelling of textile structures can be found as well in Cherif (1998).

Beams are extensions of trusses. They can transfer bending moments at the nodes as well. Thus, at the nodes, in addition to the displacements, a rotation in the beam plane for 2D beams and three rotations for 3D beams are added as degrees of freedom to each of the nodes. Thus, the 3D beam (if implemented in the software and if this option is chosen) can additionally carry twist moments due to the rotation of the cross-section (Fig. 6.22). Beam elements are used for creating models where the deformation of the beam and its bending rigidity have to be considered. The beam models differ in how their cross-section follows the deformations of the axis – and because of this in the libraries of the FEM software, different beams are available. One popular element is the Hughes–Liu beam element (Hughes, 1981), in which the rigid body rotations do not generate strains (is incrementally

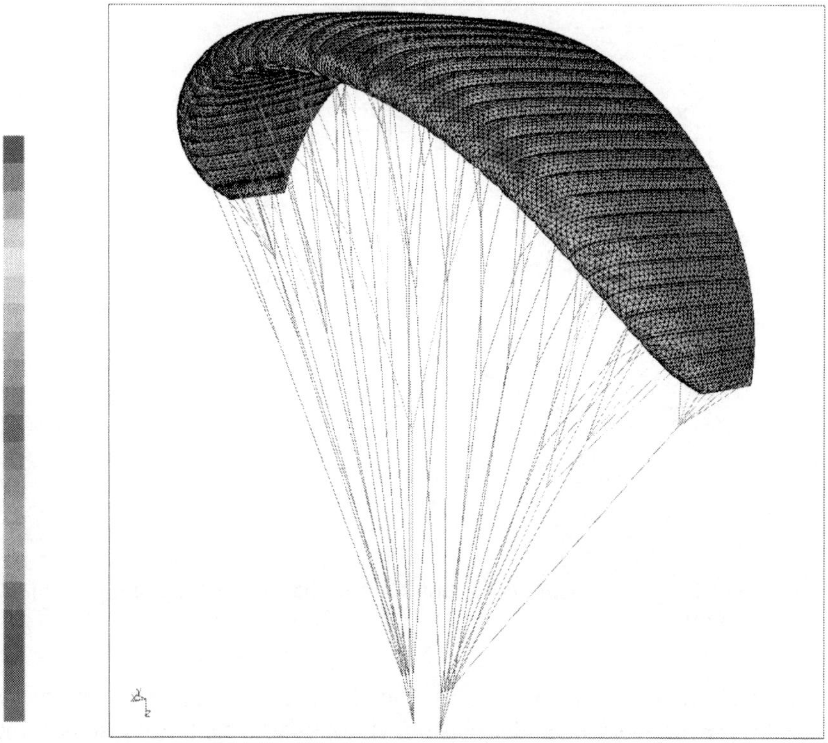

6.19 Principal stress in paragliders, simulated with CalculiX. With kind permission of Thomas Ripplinger, ADVANCE Thun AG, Thun, Switzerland.

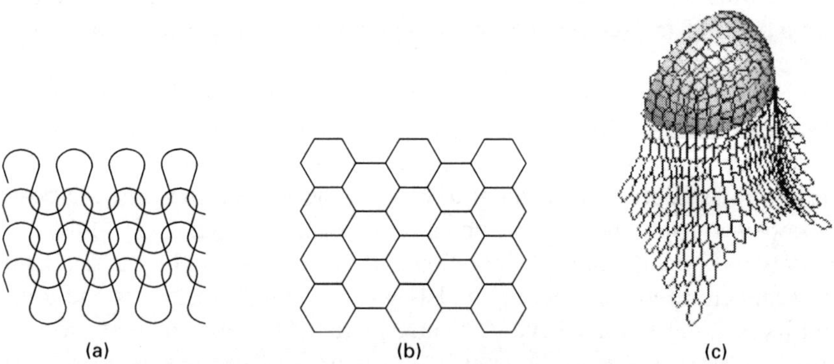

6.20 Modelling of plain weft knitted structures with trusses: (a) yarn axes of the weft knitted structures; (b) FEM model, using the yarn contact places as nodes; (c) simulated draped weft knitted structure over ball, according to de Araújo *et al.* (2004).

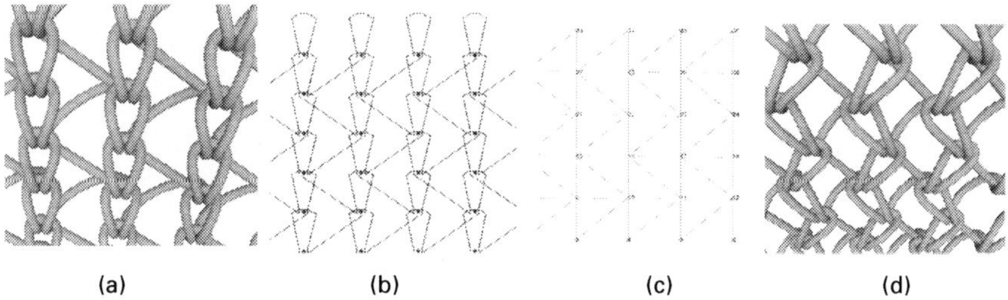

6.21 Application of truss elements for mechanical adjustment of warp knitted structure: (a) idealized tricot structure in pure geometrical model; (b) yarn axes and positions of the nodes for the modelling; (c) FEM model with trusses and directions of the nodal forces; (d) resulting structure after the FEM calculation, where the yarn forces are in equilibrium. Courtesy of Renkens Consulting and TexMind.

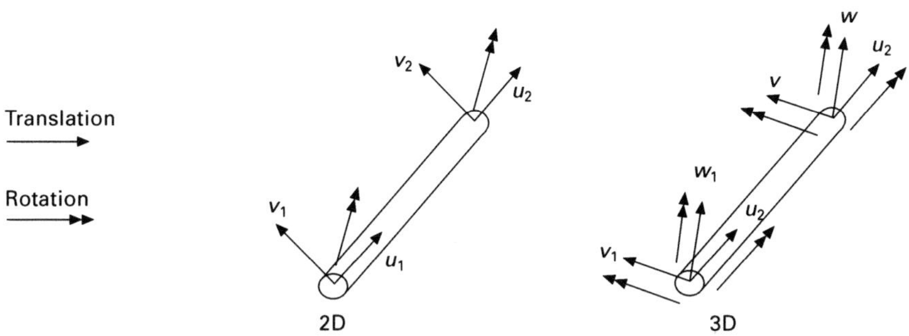

6.22 Degrees of freedom of 2D beam and 3D beam.

objective). This model is computationally efficient and compatible with brick elements and includes finite transversal shear strains. The Belytschko beam (Belytschko *et al.*, 1977), has the capability to allow large rotations. Figure 6.23 shows successful modelling of woven structure on a filament basis (Durville, 2010), where each filament is modelled as a beam (but presented on the picture as a solid body).

6.3.2 Membranes, plates and shells

Membranes, plates and shells are elements that have been developed for the presentation of thin flat structures, so that their behaviour is correctly represented without explicit presentation of the thickness of the structure. For this reason, they are suitable for structures where the thickness is negligible related to the other dimensions of the structure. Typical examples are textiles, foils, plates, etc. The thickness of the structure has to be defined as a parameter and it is used in the calculations.

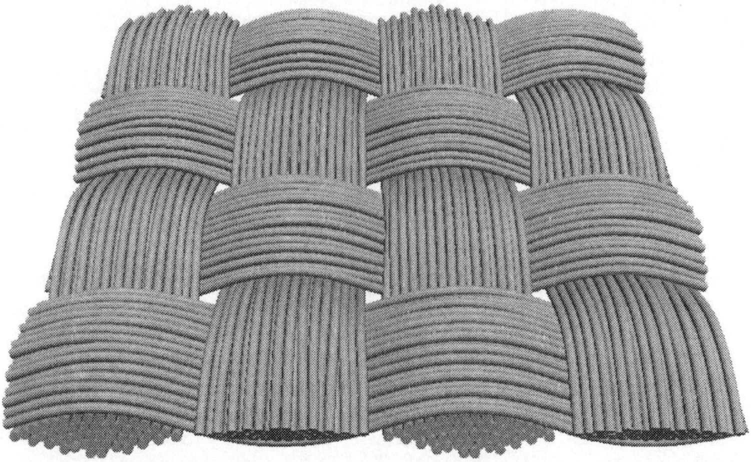

6.23 Plane weave structure, where the filaments of the yarns are modeled with beam elements. Courtesy of Damien Durvile (Durville, 2010).

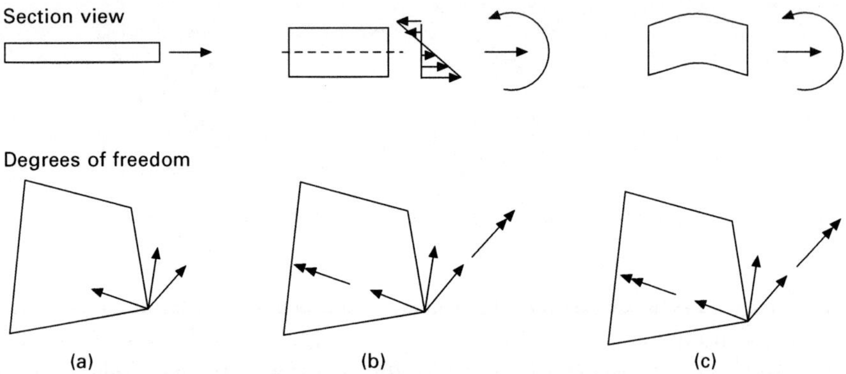

6.24 Membranes, plates and shells in a section view and degrees of freedom: (a) membrane element; (b) plate element; (c) shell element.

The membrane elements are the plane analogue of the trusses as a one-dimensional element. The nodes of the membrane elements have only translations as degrees of freedom and they can transmit only stresses which are in their plane, but no bending loads (Fig. 6.24a). These elements can be curved as well. Figure 6.19 presents a simulation of a paraglider within the software CalculiX, where the main structure is presented by membrane elements. Another typical case is the building roofs under tension, such as stadium roofs (shown in Fig. 6.25).

The plate elements can be considered as a 2D extension to the beam elements: they are able to transmit bending moments (no in-plane torsion) and

6.25 Membranes – Olympic stadium in Munich (Tobi87, 2007).

Table 6.3 Overview of membrane, plate and shell elements

Element type	Membrane	Plate	Shell
Loads	Inplane normal and shear stresses – membrane forces	Bend in two directions and twist	Combinations
Surface form	Can be curved	Flat	Can be curved
Nodal degrees of freedom	Translations	three Translations and two rotations	Three Translations and two rotations; the rotations are specially oriented
Analogue to	Truss	Beam	
When to use it	– Very small thickness relative to the length – No stress in the direction normal to thickness – No transmission of moments	– Small thickness relative to the length (1/10) – Stress distribution through the thickness is linear – Elements remain planar – No rotation about the direction normal to the element – Small displacements and rotations	Combined cases after careful checking the element properties and definitions

can model structures, which stress distribution through the thickness is linear. Important is, that the plates have to remain planar after the deformation.

For the more complex cases of combinations of the loads between membrane and plate load, the shell elements have to be used. There are several definitions of shell elements (also of plate elements, depending on their thickness) and careful studying of the programmed behaviour of the shell elements is required before choosing the proper one for a certain model. The main differences between these elements are presented in Table 6.3. All the elements can be implemented and used usually as three or four node elements from first order or as six or eight node elements of second order. The second order elements can lead to stable results even within a coarse mesh, whereas the first order elements require a finer mesh to achieve the same stability.

Elements for axisymmetric case

There are several situations of the simulation of rotational symmetric bodies like spindles, spinning rotors, carding rollers, etc. for which the calculations and the computational time can be significantly reduced if rotational symmetry is being used. In such cases, different axis-symmetric elements are used, which differ in their definitions depending on whether only the geometry is axisymmetric or the geometry and the loading are both axisymmetric.

6.3.3 Solid elements

In the general case where the solid body has to be modelled as a solid without any restrictions and simplifications, solid (3D or 'volume') elements are used. In this case, several effects can be recognized and modelled properly, which cannot be seen during the modelling with beams or 2D elements because of their simplified definition. In several cases, especially with textiles, it is harder to prepare the geometry and the mesh for solid elements. It is harder to check for errors, because several places remain hidden and only with special calculations of well-defined meshes, quality criteria can be recognized. In addition, solid elements require much greater computer resources, e.g. a much greater computing time. Despite these disadvantages, the use of solid elements allows a more complex and full instigation of the deformations and they are therefore widely used.

Figure 6.26 shows four basic types of solid elements – the four node tetrahedron element and the eight node brick element ((a) and (c)), as well as their second order versions with nodes at the middles of the element edges. An example of a ring spinning spindle, its mesh and the calculated

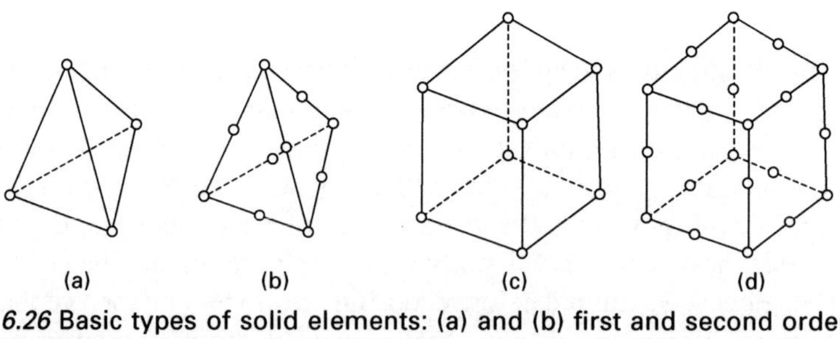

6.26 Basic types of solid elements: (a) and (b) first and second order tetrahedron element, (c) and (d) first and second order (8 and 20 node-) brick elements.

von Mises stresses using solid elements are presented in Fig. 6.27. Figure 6.28 presents a deformed monofilament under bending, modelled with solid elements, after implicit FEM calculations and Fig. 6.29 depicts an example of a deformed plain weft knitted structure, where the yarns are meshed as well with solid elements.

The modelling of the yarns with solid elements, as presented in Figs 6.28 and 6.29 is performed under the assumption of linear isotropic material behaviour. In fact, the yarns, especially staple yarns and multifilaments with less twist, have completely different behaviour in lateral directions than in

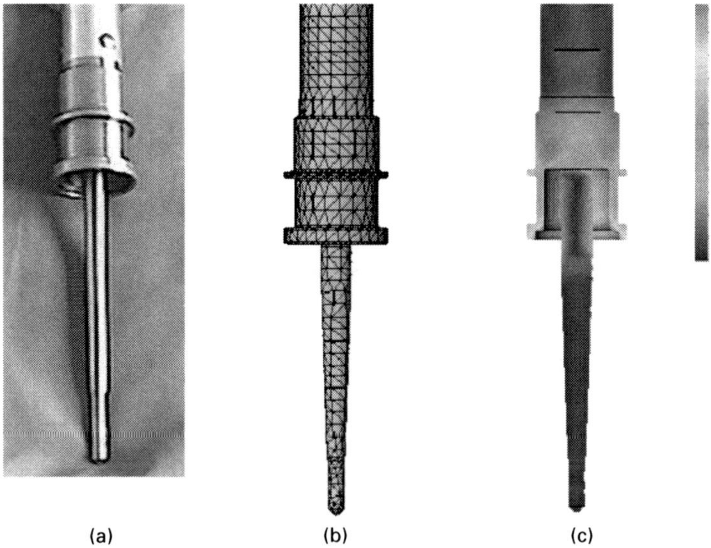

6.27 Ring spinning spindle (a); meshed (b) and its stress distribution (c) (von Mises stresses) (Kyosev, 2002).

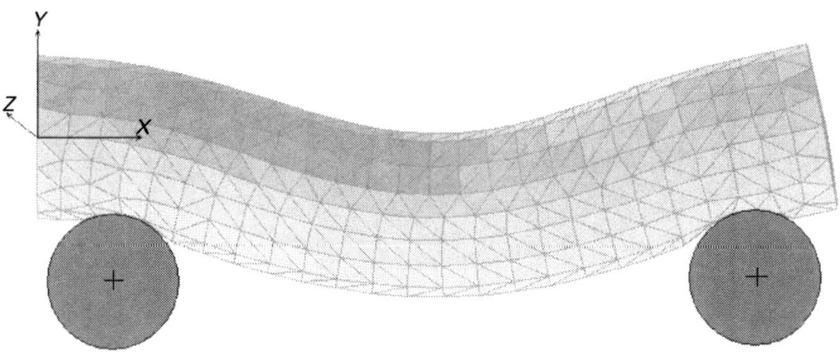

6.28 Deformed monofilament during bending, meshed with solid elements.

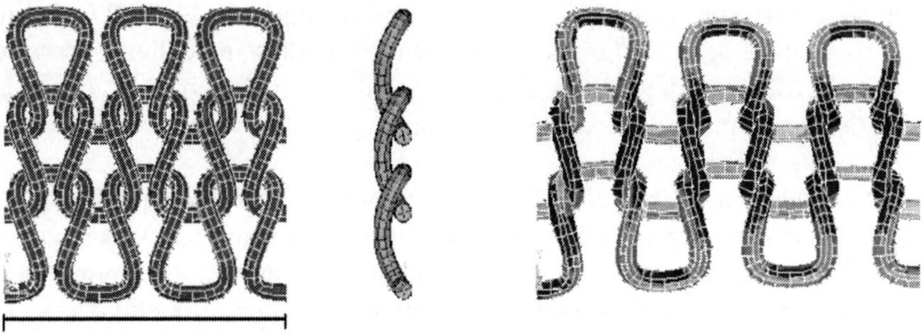

(a) (b) (c)

6.29 FEM model of a plain weft knitted structure: (a) front view; (b) side view; and (c) deformed state after explicit deformation with Ansys/LS-Dyna (Kyosev, 2006).

the axial direction and to ensure proper modelling an anisotropic material model, which also considers shear stresses in the cross-section, must be used. The other challenge during the preprocessing of yarn structures is the setting of the proper material axis orientation – at each element, the orientation of the material has to be defined properly as well (Boisse *et al.*, 2005).

6.3.4 Special elements for multilayer structures and composites

With the development and extended application of textile reinforced structures, FEM researchers and software developers started including special elements for such materials. In these elements, multiple layers of fibres with arbitrary directions can be defined. Examples of special shell elements for the draping simulation of woven composites are reported, for example, in the works of Philippe Boisse and his group, e.g. Hamila and Boisse (2008) and Hamila *et al.* (n.d.).

6.3.5 Contact elements

During the simulation of structures, where some parts are in contact with other parts or with themselves, special contact elements are required. These elements define:

- which elements or groups of elements have to be analysed, if contact occurs;
- what kind of reaction happens, if the contact is active, for instance if a penalty function has to be defined;
- how the contact time is calculated;
- whether the contact with or without friction has to be computed.

According to the first question, the contact elements can be classified into some general groups: contact between two solids, between two surfaces, between two beams or trusses, between solid/surface/beam and node.

The first group is used mainly in the CAD programs, where the solid bodies are well defined. The contact between solid bodies can be detected with very efficient computational procedures, but finally it has to be converted to contact between two surfaces. If the surfaces, which could start contacting, are known prior to the simulation, it is effective to define surface to surface (or beam to beam) contact elements only between these surfaces.

In other cases, large contact search procedures have to be started, which extend the computational time. Once detected and activated, the contact element has to generate contact forces at its nodes. The direction of these forces is usually in the contact normal and the value of the forces depends on the kind of the contact – elastic or rigid. The loops in Fig. 6.29 are calculated with surface-to-surface contact. All simulations of the textile processes like weaving and braiding require well-defined contact elements between the yarns and in the case of knitting additionally contact of the yarn with itself has to be checked. In cloth simulation, in addition to the calculation of the contact between body and the cloth, self-contact has to be considered. Examples of such simulations can be found in Schneider (2000) or Pickett *et al.* (2009) for the braiding process, and also in the several other works for weaving, knitting or braiding using LS-Dyna or Pam-Crash software.

6.3.6 Other elements

In several cases, efficient models can be built if some of the parts of the structures or the systems are not completely modelled, but only their behaviour is represented. For instance, during the modelling of the deformation of a warp yarn during weaving, it is not necessary to model the entire backrest system. If only the motion of the backrest is important, it can be represented in a model as a rigid body with a mass, connected with a spring and damper. For such cases, mass elements with one node and weight can be used for the modelling and the additional mass, or spring and damper elements with only two nodes and parameters (spring or damping constant) can be used for the calculation of the elastic forces.

6.4 Error estimation and refinement

6.4.1 Error estimation

After one model is created and calculated, it has to be proved. The most simple check of the numerical errors during the calculations is to refine the FE mesh by reducing the element size by half and to compare both solutions. The relative error for the deformations can be calculated for one node as:

$$\delta_\varepsilon = \frac{\varepsilon_L - \varepsilon_{L/2}}{\varepsilon_{L/2}} \cdot 100\% \qquad [6.58]$$

and for the stresses as

$$\delta_\sigma = \frac{\sigma_L - \sigma_{L/2}}{\sigma_{L/2}} \cdot 100\% \qquad [6.59]$$

where ε_L and σ_L are the displacement or the stress, calculated at one node, if the element size was L and $\sigma_{L/2}$ and $\sigma_{L/2}$ are displacement and stress, calculated using elements with the element size $L/2$. Usual errors reported in Reul (2010) assess the deformations to be between 5 and 20%, and for the stresses to be between 10 and 50%. If the accuracy of the calculations is not sufficient, usually the mesh has to be changed. In some cases, some nodes have to be moved, in order to avoid some singularities. The general cases for the refinements of the FE mesh are described in the following section, since most of them can be performed automatically with the software.

6.4.2 Refinement

Once a insufficient accuracy of the FE calculation is detected, the mesh has to be refined. Some of the FE packages perform this automatically, adapting the mesh in order to get a better solution. Using the *h-method*, the mesh is refined, so that the elements of the entire model are re-meshed with elements half their original size. In contrast, by using the *p-method*, the element size remains unchanged, but the order of the elements is increased. The combinations of these two strategies is known as the *hp-method.*

6.4.3 Some modelling errors

A lot of errors can occur during FEM modelling, some of which were described in Section 6.2.1. Here, only some errors are mentioned, which are connected to the meshing of the structures and the type of the solvers.

Proper problem/solver type

A major disappointment for textile researchers when they started using FEM for the simulation of textile structures was that as beginners they often started with some easy-to-use linear solver, so that the program 'crashed', giving no answer or politely explaining 'Calculations interrupted. The displacements are too large'. Since the textiles are very flexible structures, the deformations are large and the linear solvers for small displacements cannot perform the calculations stably. So, special attention has to be given of the choice of the *type of the problem* and the *type of the solver*.

Proper element type

The next common error is the *improper selection of the element type*. Modern software has built in several hundreds of element types but only a small number of these can represent the behaviour of the current structure properly.

Proper mesh size

More fine meshes in the areas with nonlinear geometry and with higher changes of the loads and more course meshes in the areas where less happens can limit the errors without significantly increasing computation time.

Avoid concentrated loads

The concentrated loads are used in the simplified mechanical calculations for beams and truss structures, but are not suitable for FEM analysis with plain and solid elements. One concentrated force is actually never applied on a single point; it is applied on a small surface. The same effect happens if beam or truss elements are connected with shell or solid elements – since the reaction of the 1D element acts as a concentrated load.

6.5 Nonlinear problems

As mentioned in Section 6.2, there are two places where nonlinearities can appear – if the displacements are not small, then the kinematic relations [6.11] and [6.16] will be no more valid and have to be replaced by other, nonlinear relations. In this case, the nonlinearity is geometric, because it is caused by pure geometric/kinematic relation. If the material law, expressed with the stress–stain equation is no longer linear, then the nonlinearity is caused by the material.

6.5.1 Geometric nonlinearities

The kinematic relations between the displacements of the points of one body (nodes) and its strains (equations [6.11] and [6.16]) are linear only in the case of small displacements. If the displacements become larger, the proper relations become nonlinear and the problem is classified to be 'with geometric nonlinearities'. For such models, different numerical methods have to be applied. The geometric nonlinearities can be classified into several groups, as presented in Fig. 6.30.

The large displacements can cause small deformations in the body if the body can rotate (Fig. 6.31). This is the case for *large rotations* (Fig. 6.31c).

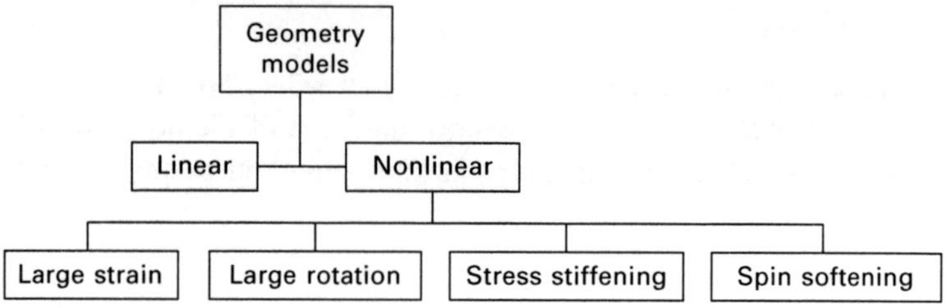

6.30 Geometry nonlinearities.

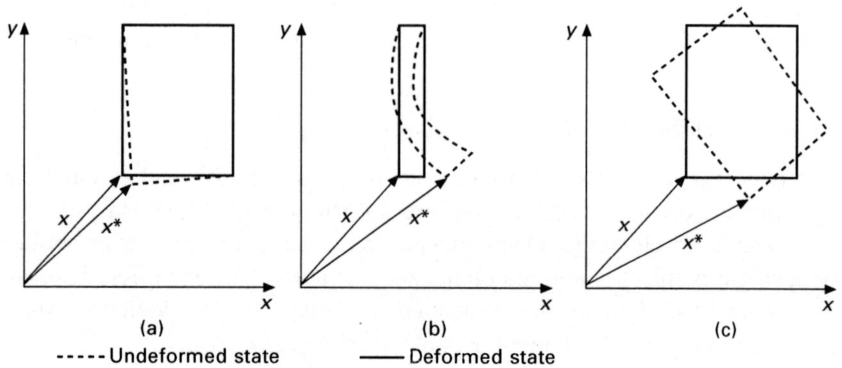

6.31 Examples for small displacements and for some geometric nonlinearities: (a) linear case with small displacements and small deformations; (b) large deformations; (c) small deformations but large rotations.

If a part of the body is fixed, then the large displacements will cause large strain as well (Fig. 6.31b). The *stress stiffening* (or weakening) is another case of a geometric nonlinearity, where the structure is stiffening due to its stress state. Such effects are common for thin structures when their bending stiffness is very small compared with their axial stiffness. Typical examples are cables, thin beams and shells, where the stiffening couples the in-plane displacements with the transverse one. In the case of pure stress stiffening, the strain and rotations are small. Another geometric nonlinearity with small strain and small rotation is the spin softening, which occurs for a spinning (rotating) body. The vibration of this body can cause relative circumferential motions, which change the direction of the centrifugal load. The centrifugal load then tends to destabilize the structure (ANSYS Inc., 2011).

For all these geometric nonlinearities, different mathematical relations between displacements and strains have to be chosen. For the users of FEM software, this means that the available options have to be carefully selected, otherwise, the mathematical model will not represent the real situation and the resulting errors can become significant. Figure 6.32 presents the Screenshot

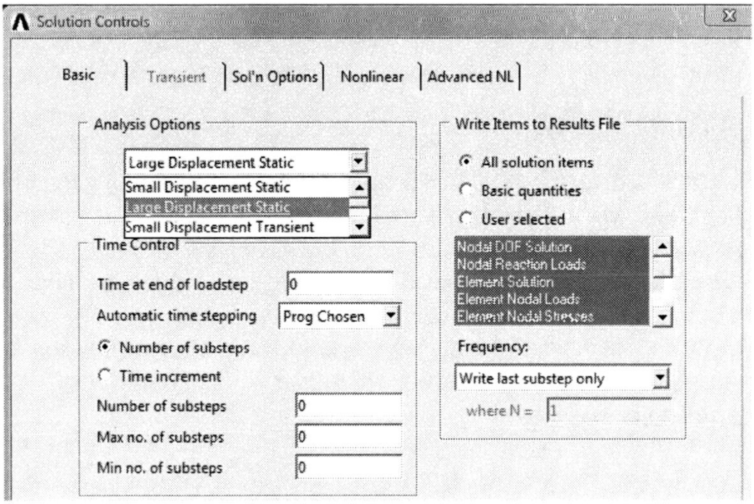

6.32 Option large displacements in Ansys®.

of the Ansys® menu, where the analysis option with large displacements in Ansys® can be chosen. All these effects can be considered almost independent of the used material, because they are connected with the kinematic relations between the displacements of the points and the body deformations, which are in turn normally not connected to the material properties.

6.5.2 Material laws and nonlinearities

Models for elastic materials, where Hooke's law between the strains and stresses is valid (equation [6.17]), are linear if there is no geometrical nonlinearity. For these models, also different material models are available, depending on the material constants from the material directions (Lekhnitskiĭ, 1963). For isotropic materials, the properties of the materials are equal in all directions and only the elasticity modulus and Poisson ratio are required as described in Section 6.2. The Poisson ratio ν describes the ratio between transversal and axial strain, if the material is stretched or compressed according to

$$\nu = -\frac{d\varepsilon_{\text{transveral}}}{d\varepsilon_{\text{axial}}} = -\frac{d\varepsilon_2}{d\varepsilon_1} \qquad [6.60]$$

The Poisson ratio has to be between −1 and 0.5. For most materials, this value is between 0 and 0.5. Most metals have a Poisson ratio of 0.3. There are some materials such as cork, for which $\nu = 0$. There are some materials (mainly structures) which have a negative Poisson ratio – they increase their thickness when stretched. For isotropic materials, the shear modulus is calculated from the elasticity modulus and the Poisson ratio according to:

$$G = \frac{E}{2 \cdot (1 + \nu)} \qquad [6.61]$$

Orthotropic materials

Textiles and textile-based structures often have different properties in the directions where the main yarns are located – they are orthotropic (Fig. 6.33). The material properties in this case depend on the space directions, named here 1, 2 and 3 instead of x, y and z and there are nine constants, which have to be defined independently (E_1, E_2, E_3, G_{23}, G_{31}, G_{21}, ν_{23}, ν_{13}, ν_{12}), three elasticity moduli, three shear moduli and three Poisson ratios. The stress–strain relation for orthotropic materials is simpler in its transposed form and is given by:

$$\begin{bmatrix} \varepsilon_{11} \\ \varepsilon_{22} \\ \varepsilon_{33} \\ 2\varepsilon_{23} \\ 2\varepsilon_{31} \\ 2\varepsilon_{12} \end{bmatrix} = \begin{bmatrix} \frac{1}{E_1} & \frac{\nu_{21}}{E_2} & \frac{\nu_{31}}{E_3} & 0 & 0 & 0 \\ \frac{\nu_{12}}{E_1} & \frac{1}{E_2} & \frac{\nu_{32}}{E_3} & 0 & 0 & 0 \\ \frac{\nu_{13}}{E_1} & \frac{\nu_{23}}{E_2} & \frac{1}{E_3} & 0 & 0 & 0 \\ 0 & 0 & 0 & \frac{1}{G_{23}} & 0 & 0 \\ 0 & 0 & 0 & & \frac{1}{G_{13}} & 0 \\ 0 & 0 & 0 & 0 & 0 & \frac{1}{G_{12}} \end{bmatrix} \begin{bmatrix} \sigma_{11} \\ \sigma_{22} \\ \sigma_{33} \\ \tau_{23} \\ \tau_{31} \\ \tau_{12} \end{bmatrix} \qquad [6.62]$$

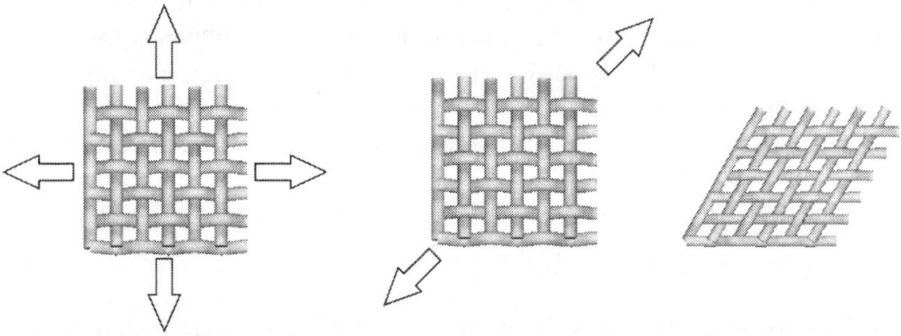

(a) (b) (c)

6.33 Woven structure as an example for an orthotropic structure: (a) the woven fabrics with main space directions (loading in these directions does not lead to shear); (b) loading different to the main directions for shear stresses; (c) deformed structure under the load in case (b).

Here, the directions 1, 2 and 3 have to coincide with the space directions of the material. In these directions, different from the three main directions, the material is anisotropic – shear and tension are applied. From the stiffness matrix it can be recognized that in such materials shear stresses do not lead to strains – the connecting elements in the stiffness matrix are zero. The following relations between the Poisson ratio and the elasticity modules are valid, hence:

$$\frac{\nu_{yx}}{E_y} = \frac{\nu_{xy}}{E_x} \quad \frac{\nu_{yz}}{E_y} = \frac{\nu_{zx}}{E_z} \quad \frac{\nu_{zx}}{E_z} = \frac{\nu_{xz}}{E_x} \qquad [6.63]$$

According to these relations, if $E_x > E_y$ then $\nu_{xy} > \nu_{yx}$. The larger Poisson ratio ν_{xy} is called major ratio and the smaller one – ν_{yx} – minor Poisson ratio.

Transversely isotropic materials

For modelling textiles, the material model for the transversely isotropic materials is also important. Nonwovens or laminates with multiple layers have an isotropic behaviour in their main plane (x–y), but a completely different behaviour in the third (z) direction. Because of the isotropic behaviour in the x–y plane, there is only one elasticity modulus in the plane E_{p} and plane Poisson ratio ν_{p}:

$$\begin{bmatrix} \varepsilon_{xx} \\ \varepsilon_{yy} \\ \varepsilon_{zz} \\ 2\varepsilon_{yz} \\ 2\varepsilon_{zx} \\ 2\varepsilon_{xy} \end{bmatrix} = \begin{bmatrix} \frac{1}{E_{\mathrm{p}}} & -\frac{\nu_{\mathrm{p}}}{E_{\mathrm{p}}} & -\frac{\nu_{z\mathrm{p}}}{E_z} & 0 & 0 & 0 \\ -\frac{\nu_{\mathrm{p}}}{E_{\mathrm{p}}} & \frac{1}{E_{\mathrm{p}}} & -\frac{\nu_{z\mathrm{p}}}{E_z} & 0 & 0 & 0 \\ -\frac{\nu_{\mathrm{p}z}}{E_{\mathrm{p}}} & -\frac{\nu_{\mathrm{p}z}}{E_{\mathrm{p}}} & \frac{1}{E_z} & 0 & 0 & 0 \\ 0 & 0 & 0 & \frac{1}{G_{z\mathrm{p}}} & 0 & 0 \\ 0 & 0 & 0 & 0 & \frac{1}{G_{z\mathrm{p}}} & 0 \\ 0 & 0 & 0 & 0 & 0 & \frac{1+\nu_{\mathrm{p}}}{E_{\mathrm{p}}} \end{bmatrix} \begin{bmatrix} \sigma_{xx} \\ \sigma_{yy} \\ \sigma_{zz} \\ \tau_{yz} \\ \tau_{zx} \\ \tau_{xy} \end{bmatrix} \qquad [6.64]$$

where $\nu_{\mathrm{p}z}/E_{\mathrm{p}} = \nu_{z\mathrm{p}}/E_z$. In this case there are five independent constants E_{p}, E_z, ν_{p}, $\nu_{\mathrm{p}z}$, $G_{z\mathrm{p}}$ which have to be determined.

Anisotropic material

The general case of the material law is anisotropic material. In this case, all constants in the matrix have to be defined separately. Since the matrix is symmetrical only 21 of these constants are independent and have to be determined separately:

$$\begin{bmatrix} \sigma_{xx} \\ \sigma_{yy} \\ \sigma_{zz} \\ \tau_{yz} \\ \tau_{zx} \\ \tau_{xy} \end{bmatrix} = \begin{bmatrix} C_{11} & C_{12} & C_{13} & C_{14} & C_{15} & C_{16} \\ & C_{22} & C_{23} & C_{24} & C_{25} & C_{26} \\ & & C_{33} & C_{34} & C_{35} & C_{36} \\ & & & C_{44} & C_{45} & C_{46} \\ & \text{symmetric} & & & C_{55} & C_{56} \\ & & & & & C_{66} \end{bmatrix} \begin{bmatrix} \varepsilon_{xx} \\ \varepsilon_{yy} \\ \varepsilon_{zz} \\ 2\varepsilon_{yz} \\ 2\varepsilon_{zx} \\ 2\varepsilon_{xy} \end{bmatrix} \quad [6.65]$$

All these relations describe the behaviour of the material if they are within the limit of the linear elasticity and if they follow the generalized Hooke's law.

Nonlinear material laws

Materials have elastic behaviour usually only up to a certain load, which is called the 'yield point'. After that, the deformations are plastic. Rubber, for instance, has an elastic stress–strain curve, but the relation is not linear – in such cases nonlinear elastic models are required. Such models are, for instance, Blatz-Ko Rubber, Mooney-Rivlin Rubber and viscoelastic models (Hallquist, 2006; ANSYS Inc, 2001). In the area of inelasticity, the different materials react differently and there is a large number of nonlinear inelastic models used and implemented in FEM software (ANSYS Inc., 2001). Foams also have a specific stress–strain behaviour and require separate models, e.g. for closed cell foam, viscous foam, for low density, for crushable foam, honeycomb models, etc.

Several discrete elements with specific behaviour are also commonly used – usually springs and dampers. These can be linear, nonlinear elastic or/and plastic, viscosity spring and different combinations of dampers. In some cases, the deformations in bodies must be neglected and a completely rigid body behaviour must be presented, for which a well-defined material model has to be used.

Most FE packages allow the use of so-called 'user materials', where the user can program their own material model in Fortran, Python or other scripting language. Since the choice of the material model, especially for polymer materials is not a simple task, there are companies and software which identify the material law and fit the material's coefficients automatically (Bergstrom, 2011).

6.5.3 Other nonlinearities

Simulation of the contact in the structures may lead to nonlinear systems of equations and it has to be treated as a nonlinear problem. Materials which are usually linear with a constant temperature, but have to be simulated in coupled simulations with temperature changes, must be considered also as nonlinear, since, for instance, the elasticity module and the stiffness matrix change during the simulation depending on the temperature of each element.

6.6 FEM Software

Nowadays, several FEM packages are available – commercial as well as open source or/and free packages. There are too many FEM packages to give a full listing in this section. In addition, the software is changing very dynamically – some companies no longer exist and some new companies have started providing their own FEM solutions as well. Here, only a few packages will be mentioned, which, according to the opinion and experience of the author, are relevant to the calculation of textile structures.

6.6.1 Commercial FEM software

A popular commercial FEM package is ANSYS® (ANSYS, 2011), which has mainly solvers for implicit linear and nonlinear calculations. There are several ANSYS products for different applications, as for example:

- Mechanical, professional, structural – different licence options for structural linear or nonlinear and dynamics analysis.
- DesignSpace – an easy-to-use package, usually included as a submenu in other CAD systems, which allows static structural and thermal, dynamic, weight optimization, vibration mode, and safety factor simulations directly in the CAD package, without the need for advanced analysis knowledge.
- Rigid Body Dynamics – for simulation of machines kinematics and dynamics, where mainly the motion has to be investigated and not the deformation of the parts.
- Composite PrepPost – for combining two or more different materials, having different properties.

For explicit calculations, the package LS-DYNA, LSTC – Livermore Software Technology Corporation (LSTC, 2011) is very popular. It is suitable for airbag simulations, crash simulations of cars with ready-to-use dummy models (human body models) and safety belts. Examples for a textile structure simulation with LS-DYNA are given for instance in Finckh (2007) or Kyosev (2006).

Abaqus (Franco-American software from SIMULIA, owned by Dassault Systemes) (Dassault Systemes, 2011) is another very popular type of FEM software, which is very useful for textile engineers due to its capability to provide both implicit and explicit solvers and having well known experience in the solving of nonlinear problems.

Pam-Crash and the entire family of packages of the ESI Group (ESI Group, 2011) is very popular FEM software with implicit as well as explicit solvers. Its module SYSPLY allows the design of composite parts from multiple layers of high performance fibres.

The NX™ Nastran® Advanced Nonlinear software solution is another package with very good capability in the nonlinearities such as contacting parts, material nonlinearities and/or geometric nonlinearities (Siemens PLM Software, 2011).

Adina is a company founded by Professor Hans Jurgen Bathe, one of the inventors of the FEM, which provides the FEM software package Adina (Aumatic Dinamic Incremental Nonlinear Analysis) (ADINA R&D, 2011), where some of the solution options are included as well in NX Nastran.

6.6.2 Open source and free FEM packages

The open source and free packages available have until now not been as well developed. This is a result of the lower number of experts having expertise in the programming of such packages in general. The majority of such packages are built firstly for a certain application and after that extended to a larger number of applications. Because of this, they are not able to solve a very wide range of problems. Before starting to use such packages, a careful check of their options should be carried out.

A useful program is the open source project CalculiX (Dhondt and Wittig, 2011). Its solver uses a partially compatible ABAQUS file format and the pre-post-processor generates input data for many finite element analysis and CFD applications.

An interesting package for starting explicit FEM learning and teaching (GPL Licence) is Impact, having several examples (Forssell and Mikhaylovskiy, 2011). The possibilities for import of 3D geometries have to be extended if complex geometries have to be calculated. Another commonly cited package is Code Aster, written in Python and Fortran and distributed under GPL licence, where not all the information is translated into English (Code_Aster, 2011).

Gmsh is an automatic 3D finite element grid generator with built-in CAD and post-processing facilities. Its design goal is to provide a simple meshing tool for academic problems with parametric input and advanced visualization capabilities. It is built around four modules: geometry, mesh, solver and post-processing. The specification of any input to these modules is done either interactively using the graphical user interface (based on FLTK and OpenGL) or in ASCII text files using Gmsh's own scripting language (Geeknet, 2011).

The LS-Dyna Post/Pro is as well distributed as freeware preprocessor and post-processor, which allows the preparation of the geometry and visualization of the results, calculated with LS-Dyna (LSTC, 2011).

6.6.3 Textile FEM preprocessors

The geometry of the textile structures can be prepared with standard CAD, but a more intelligent way is to use specialized software for textiles or to write own scripts for the geometry generation in order to be able to change the structure properties parametrically.

The WiseTex (MTM) package can be used as a preprocessor for the large part of the standard textiles. It is able to build the unit cell geometry of the most woven structures (as well as multilayer), braided structures, unidimensional (UD) laminates and the basic weft knitted structures (WeftKnit) (Plate XIV (see colour section between pages 152 and 153)) (Lomov *et al.*, 2011). The generated models can be exported with the FETex module of the same family of products to Ansys.

MeshTex software was developed by Osaka University in collaboration with the Composite Materials Group of KU Leuven. FEM calculations can be performed for mesh woven, braided and stitched non-crimp fabrics (Lomov *et al.*, 2009).

TexGen is an open source textile preprocessor, developed at Nottingham University, where the user can build the yarns point by point and after that can export the geometry in some CAD and FEM compatible formats (Fig. 6.34) (Sherburn, 2007; University of Nottingham, 2011).

For warp knitted structures, the software KnitMaster, developed by Renkens Consulting and Professor Kyosev allows export of the structures in various formats for FEM calculations (Kyosev and Renkens, 2010a; Renkens and Kyosev, 2011). The basic ideas of this software are explained in more detail in Kyosev and Renkens (2010b).

In addition to the original geometry modelling software, some engineering companies like TexMind offer also the preprocessing of textile structures like warp and weft knitted, braided and woven structures as a service. In several cases, this is a more suitable solution than using available software, especially if the textile structures to be modelled are not regular and not included in the standard libraries (Fig. 6.35).

6.7 Future trends

The trends in the development of the FEM can be summarized as follows:

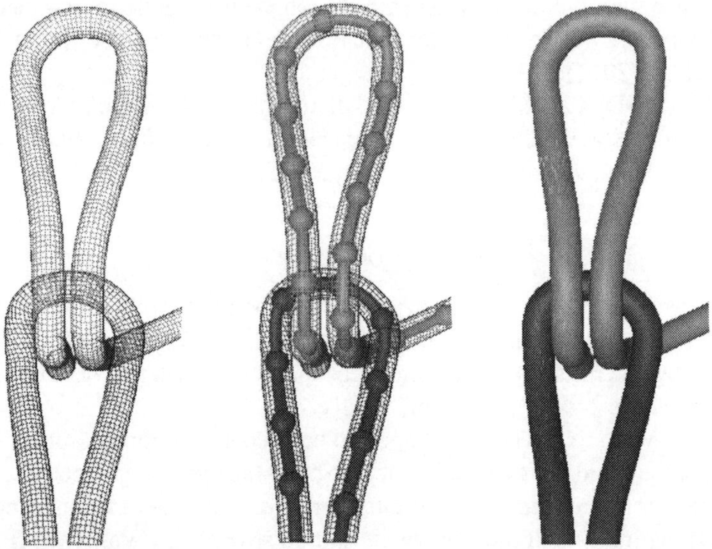

6.34 Warp knitted loops, created by KnitMaster and exported to TexGen in different views.

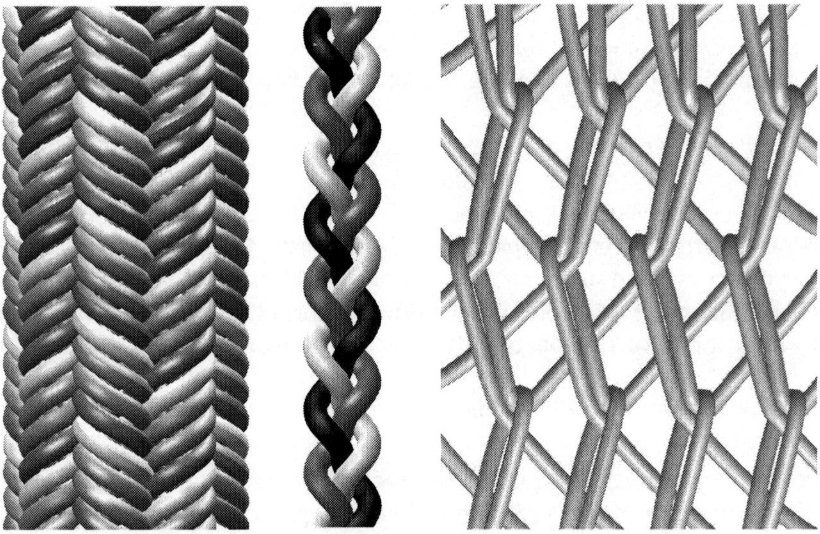

6.35 Topology of two braided structures and the geometry of warp knitted structure after relaxation. Courtesy by TexMind (2011).

- user interface;
- computational and material models;
- parallelization;

- multiscale;
- multiphysics.

The user interface includes special integration of FEM core software into the engineering CAD. This allows the constructors to check the strength and deformation of the parts during development. Such tools use well-proven families of elements and include only the main functionalities of the FEM method, usually only linear analysis. The majority of FEM programs today have an improved user interface and are not so hard to use in the preprocessing as they were many years ago. The new, better models with high order elements and special materials are also a current research area and trend.

With the huge increase in computer power in recent years, the FEM developer tries to use the power of the several CPUs and use libraries for parallel processing of the data. This allows more complex simulations to be performed within a shorter time, e.g. multiscale computations. The multiscale approach requires more computer power and is based on creating models at different levels of the structure. For instance, micro- or meso-models are used during the calculation of the properties of the unit cells, and these properties are then used to perform the simulation of the macro-structure. Most of the FEM packages provide also the possibilities for coupled simulations or multiphysics, where the interaction between fluids and structure, electric, magnetic or heat fields can be performed at the same time.

X-FEM (extended finite element method) is an extension of the FEM for structures with discontinuities. A typical example are composites, where the properties of the material in a unit cell are derived by homogenization techniques by the properties of both interfaces and their geometry (Fries, 2011). Particle FEM is another new method. It is able to deal with particles of fluids and solid bodies, which is a very powerful tool for fluid–structure interaction simulations (Idelsohn *et al.*, 2003). The idea of the method is that each node transfers the properties of the material (and not the element, as in classical FEM) and the FEM mesh is recreated at every time step.

6.8 Sources of further information

There are many different books about the finite element method and it is important that the reader finds and selects the proper books for first reading. Depending on the author and the targeted public, they can be divided into:

- books from and for mathematicians, where the theory of the method, the details about the properties and methods for solutions of the partial differential equations are explained;
- books for software developers, where usually the basic theory is provided and after that several algorithms about coding of the elements, numerical methods for solutions of sparse systems, etc. are explained;

- books for engineers, where the mechanical aspects have more attention;
- books for users of some software, where the graphical user interface and several examples are given, but little is described about the theory behind it.

Here, only a selection of a titles in the areas of FEM and connected with it are given.

The basics of the continuum mechanics are published in many university courses of mechanics for engineering schools. Popular books include Spencer (2004) and Malvern (1969).

The basics of finite element analysis for engineers are described in Bathe (1996), Reddy (1993), Zienkiewicz and Cheung (1967), Zienkiewicz (2000) and Hughes (2000). The mathematical theory of the method can be found in Brenner and Scott (2008) and the programming of the method in Bathe (1996).

More detailed works in nonlinear finite element analysis are for instance Reddy (2004), Simo and Hughes (1998) and Belytschko *et al.* (1977, 2000), while an accent on the nonlinear beams can be found in Ibrahimbegovic (1995) and for shells in Büchter *et al.* (1994) and Rouainia and Perić (1998). The popular patch test, used to assess the convergence of the calculations, is described in Taylor *et al.* (1986).

6.9 References

ADINA R&D, I., 2011. *ADINA* [online]. Available from: http://www.adina.com/index.shtml.

ANSYS Inc, 2001. *ANSYS Theory Manual: Release 5.7.*

ANSYS Inc, 2011. *Ansys(R): Theory Manual* [online]. Available from: http://www.ansys.com/.

Araújo, M. de, Fangueiro, R., and Hong, H., 2004. Modelling and simulation of the mechanical behaviour of weft-knitted fabrics for technical applications: Part III: 2D hexagonal FEA model with non-linear truss elements. *Autex Research Journal* [online], *4* (1), 24–32. Available from: http://www.autexrj.com/cms/zalaczone_pliki/5-04-1.pdf [accessed July 2011].

Bathe, K., 1996. *Finite element procedures.* Prentice Hall.

Belytschko, T., Schwer, L., and Klein, M.J., 1977. Large displacement, transient analysis of space frames. *International Journal for Numerical Methods in Engineering*, *11* (1), 65–84.

Belytschko, T., Liu, W.K., and Moran, B., 2000. *Nonlinear finite elements for continua and structures.* Wiley.

Bergstrom, J., 2011. *PolymerFEM.com* [online]. Available from: http://polymerfem.com/content.php?17-about.

Boisse, P., *et al.*, 2005. Analysis of the mechanical behavior of woven fibrous material using virtual tests at the unit cell level. *Journal of Materials Science*, *40* (22), 5955–5962.

Brenner, S. and Scott, L.R., 2008. *The mathematical theory of finite element methods*. 3rd ed. Springer.
Büchter, N., Ramm, E., and Roehl, D., 1994. Three-dimensional extension of non-linear shell formulation based on the enhanced assumed strain concept. *International Journal for Numerical Methods in Engineering*, *37* (15), 2551–2568.
Cherif, C., 1998. *Drapierbarkeitssimulation von Verstärkungstextilien für den Einsatz in Faserverbundkunststoffen mit der Finite-Elemente-Methode*. Shaker.
Code_Aster, 2011. *Code_Aster* [online]. Available from: http://www.code-aster.org/V2/spip.php?rubrique1.
Courant, R., Friedrichs, K., and Lewy, H., 1928/1967. On the partial difference equations of mathematical physics. *IBM Journal of Research and Development*, *11*(2), 215–234. Available from: http://www.stanford.edu class/cme324/classics/courant-friedrichs-Lewy-pdf [accessed February 2012].
Dassault Systemes, 2011. *Abaqus* [online]. Available from: http://www.simulia.com.
Dhondt, G. and Wittig, K., 2011. *CalculiX: A Free Software Three-Dimensional Structural Finite Element Program* [online]. Available from: http://www.calculix.de/.
Durville, D., 2010. Simulation of the mechanical behaviour of woven fabrics at the scale of fibers. *International Journal of Material Forming*, *3* (S2), 1241–1251.
ESI Group, 2011. *PAM Crash 2G* [online]. Available from: http://www.esi-group.com.
Finckh, H., 2007. Textile micromodels as a result of idealized simulation of production processes. *In: Finite Element Modeling of Textiles and Textile Composites*. Katholieke Universiteit Leuven & St-Petersburg State University of Technology and Design.
Forssell, J. and Mikhaylovskiy, Y., 2011. *Impact: Dynamic Finite Element Program Suite* [online]. Available from: http://sourceforge.net/projects/impact/.
Fries, T.-P., 2011. *The eXtended Finite Element Method* [online], RWTH Aachen University. Available from: http://www.xfem.rwth-aachen.de/.
Geeknet, I., 2011. *GMSH* [online]. Available from: http://freshmeat.net/projects/gmsh/?branch_id=51601&release_id=272840.
Hallquist, J.O., 2006. *LS-DYNA(R) – Theory manual* [online], www.lstc.com.
Hamila, N. and Boisse, P., 2008. Simulations of textile composite reinforcement draping using a new semi-discrete three node finite element. *Composites Part B: Engineering*, *39* (6), 999–1010.
Hamila, N., Boisse, P., and Chatel, S. n.d. Semi-discrete shell finite elements for textile composite forming simulation. *Proceedings ESAFORM* [online]. Available from: http://www.worldcat.org/oclc/711641946.
Hughes, T., 1981. Nonlinear finite element analysis of shells: Part I. Three-dimensional shells. *Computer Methods in Applied Mechanics and Engineering*, *26* (3), 331–362.
Hughes, T.J.R., 2000. *The Finite Element Method: Linear static and dynamic finite element analysis*. Dover Publications.
Ibrahimbegovic, A., 1995. On finite element implementation of geometrically nonlinear Reissner's beam theory: three-dimensional curved beam elements. *Computer Methods in Applied Mechanics and Engineering*, *122* (1–2), 11–26.
Idelsohn, S., *et al.*, 2003. The meshless finite element method. *International Journal for Numerical Methods in Engineering* [online]. Available from: http://www.cimne.com/pfem/docs/mfem.pdf.
Kyosev, Y., 2002. Investigation about the winding of textile yarns on the internal side of the formers. PhD. Technical University of Sofia.
Kyosev, Y., 2006. Computational model of loops of a weft knitted fabric. *In:* Z. Dragčević, ed. *Magic world of textiles: Book of proceedings ; ITC&DC, 3rd International Textile*

Clothing & Design Conference, October 8th to October 11th, 2006, Dubrovnik, Croatia. Zagreb.

Kyosev, Y. and Renkens, W., 2010a. Numercial simulation of the tension and compression behaviour of warp kintted reinforcements. *In:* C. Binetruy and F. Boussu, eds. *Recent Advances in Textile Composites*: October 26–28, 2010, Lille Grand Palais, Lille, France. Lancaster, Pa: DEStech Publications, 397–404.

Kyosev, Y. and Renkens, W., 2010b. Modelling and visualization of knitted fabrics. *In:* X. Chen, ed. *Modelling and Predicting Textile Behaviour*. Woodhead Publishing, in association with the Textile Institute; 225–262.

Lekhnit͡skiĭ, S.G., 1963. *Theory of Elasticity of an Anisotropic Elastic Body*. Holden-Day.

Lomov, S. *et al.*, 2009. Finite element modelling of progressive damage in non-crimp 3D orthogonal weave and plain weave e-glass composites. *In:* X.Chen, ed. *Second World Conference on 3D Fabrics and Their Applications.*

Lomov, S. *et al.*, 2011. *WiseTex* [online]. Available from: http://www.mtm.kuleuven.be/Onderzoek/Composites/Research/meso-macro/textile_composites_map/textile_modelling/textile_modelling_fe [Accessed 2011].

LSTC, 2011. *LS-DYNA(R)* [online]. Available from: http://www.ls-dyna.com/.

Malvern, L.E., 1969. *Introduction to the Mechanics of a Continuous Medium.* Prentice-Hall.

Pickett, A.K., Sirtautas, J., and Erber, A., 2009. Braiding simulation and prediction of mechanical properties. *Applied Composite Materials*, *16* (6), 345–364.

Reddy, J.N., 1993. *An Introduction to the Finite Element Method.* 2nd ed. McGraw-Hill.

Reddy, J.N., 2004. *An Introduction to Nonlinear Finite Element Analysis.* Oxford Univ. Press.

Renkens, W. and Kyosev, Y., 2011. Geometry modelling of warp knitted fabrics with 3D form. *Textile Research Journal*, *81* (4), 437–443.

Reul, S., 2010. *Numerische Qualität von FEM-Analysen – Vergleich der h- und p-Methode* [online], CAE-Forum.de. Available from: http://cae-forum.de/sites/default/files/20100527_SReul_sim_h_p_methode_0.pdf.

Rieg, F., and Hackenschmidt, R., 2003. *FE Analyse Fur Ingenicure*. Hanser.

Rouainia, M. and Perić, D., 1998. A computational model for elasto-viscoplastic solids at finite strain with reference to thin shell applications. *International Journal for Numerical Methods in Engineering*, *42* (2), 289–311.

Schneider, M., 2000. *Konstruktion von dreidimensional geflochtenen Verstärkungstextilien für Faserverbundwerkstoffe*. Shaker.

Sherburn, M., 2007. Geometric and mechanical modelling of textiles. PhD. University of Nottingham.

Siemens PLM Software, 2011. *NX Nastran Advanced Nonlinear – Solution 601/701* [online]. Available from: http://www.plm.automation.siemens.com/de_de/Images/4989_tcm73-4483.pdf.

Simo, J.C. and Hughes, T.J.R., 1998. *Computational Inelasticity.* Springer.

Spencer, A.J.M., 2004. *Continuum Mechanics.* Dover Publications.

Taylor, R.L., *et al.*, 1986. The patch test – a condition for assessing FEM convergence. *International Journal for Numerical Methods in Engineering*, *22* (1), 39–62.

TexMind, 2011. *TexMind* [online]. Available from: www.texmind.com.

Tobi87, 2007. *Olympic stadium of Munich (1972) during the European Cup 2007* [online]. Available from: http://en.wikipedia.org/wiki/File:Olympiastadion_Muenchen.jpg.

University of Nottingham, 2011. *TexGen* [online]. Available from: http://texgen.sourceforge.net/index.php/Main_Page.

Zienkiewics, O.C. and cheung, UY.K., 1967. The Sinite Element method *in Structural and Continuum Mechanics: Numerical solution of problems in structural and continuum mechanics.* McGraw-Hill.

Zienkiewicz, O.C., 2000. *The Finite Element Method.* 5th ed. Butterworth-Heinemann.

7

Simulation of fibrous structures and yarns

B. NECKÁŘ and M. VYŠANSKÁ, Technical University of Liberec, Czech Republic

Abstract: This chapter discusses several models which describe the mechanical properties of staple fibre yarns. Starting with definitions of yarns and their structure, a simple model of a yarn is developed. This is then used to describe the packing density of a yarn and the maximum twist it can bear without coiling. The model is then applied to determine the tensile force utilization coefficient. Finally, a theoretical model based on Koechlin's concept is used to determine a suitable yarn twist. Applying Hamburger's theory, a model is described that allows calculation of the mechanical properties (tenacity, breaking strain, etc.) of staple fibre yarns.

Key words: fibrous assemblies, mechanical properties, helical model, Koechlin.

7.1 Introduction

Fibre to yarn conversion is based on two basic technology steps: spinning preparation and the actual yarn formation process. The properties of the yarns spun by various technologies depend on the yarn forming process and the level of the fibre arrangement. Twisted yarns are a special type of fibrous assembly. Whole yarns can be intuitively divided into two parts – core and cover. The core is the sphere near the yarn axis, where the fibres are well arranged and the packing density is relatively high. The cover (characterized by fibre hairiness) comprises ends, loops and other fibre segments with a lower packing density. Strict boundary lines between these two spheres do not exist. The yarn diameter is placed in this intermediate layer, making its accurate determination difficult.

A helical model (see Section 7.6) is usually used to describe an idealized fibre assembly in the yarn structure. During twisting of the yarn, some fibres are displaced from their central position to the yarn surface (migration effect). Structural and mechanical parameters depend on the fibre arrangement. Mechanical parameters are related to the fibre arrangement in the yarn core. The degree of hairiness is determined by the fibres' configuration on the outer layer of the yarn. Yarn unevenness is connected to the distribution of the fibres in the whole yarn.

7.2 Fibres and yarns: definitions and factors affecting quality

Yarn quality depends on the types of fibre (fineness, diameter, shape factor, length, flexural rigidity, torsion rigidity, tenacity, extension to break, friction, crimp and compression resistance), the geometrical parameters of the yarn (twist, count, blending ratio – migration effect), the technology of production (classical or innovative, production rate, cost of yarn, end purpose of yarn, setting of production line) and quality control (testing of randomly selected yarn samples, systematic analysis of yarn production).

Staple fibres are the fundamental units of yarns. Any yarn model must include the required parameters of fibres (Fig. 7.1) and their relationships. Some of them are introduced here. One of the most frequently used fibre parameters is fibre fineness – the ratio between fibre mass and fibre length. The main physical unit of fibre fineness is 1 tex, which is equal to 1000 m of a fibre/yarn weighing 1 g. For fibres, the unit dtex is often used, which is equivalent to 1 g/10 000 m. Typical values for micro-fibres are below 1 dtex, for cotton fibres around 1.6 dtex, for wool often around 3.5 dtex and for coarse carpet fibres above 7 dtex.

From a geometrical standpoint, fibre 'fineness' is characterized by the ratio between volume V and length l, but standard fineness is also influenced by the fibre density ρ. Therefore, it is not correct to compare the finenesses of fibres having different densities solely by their standard fineness; it is better to use the ratio t/ρ. Fineness and density define the cross-sectional area of a fibre:

$$\frac{t}{\rho} = \frac{V}{l} \qquad [7.1]$$

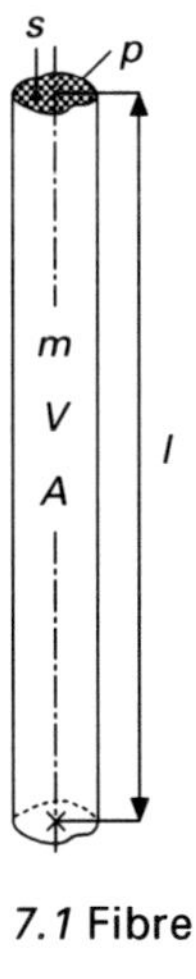

7.1 Fibre with main parameters.

The calculation of the cross-sectional area allows evaluation of the equivalent fibre diameter as shown in Fig. 7.2. For cylindrical fibres, the derivation is trivial. For non-cylindrical fibres, the same equation represents the diameter of a ring having the same cross-sectional area.

The equivalent fibre diameter d is hence given by:

$$d = \sqrt{\frac{4 \cdot t}{\pi \cdot \rho}} \qquad [7.2]$$

The perimeter of a non-cylindrical fibre is always larger than the perimeter of a ring having the same cross-sectional area. In this context, the fibre shape factor q, according to Malinowska (1975) is illustrated in Table 7.1.

$$q = \frac{p}{\pi \cdot d} - 1 \geq 0 \qquad [7.3]$$

If we know the (typical) shape factor of a fibre in advance, its perimeter p can be calculated with Equation [7.3].

Many important properties of textiles (especially transport of moisture) depend on the total area of the fibrous surface A. Therefore, it is very useful to know the surface area per unit mass of fibrous material a. The derivation of the so-called specific surface area is shown here. (Note: we can neglect the topmost and bottommost cross-sectional areas of fibres because, as the fibres are very long, these two areas are relatively small – e.g. 0.025%.) The specific area can be calculated according to

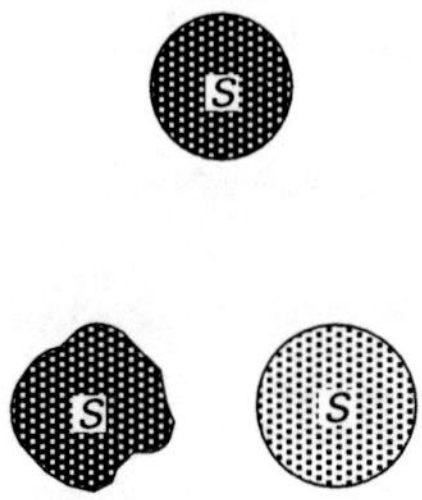

7.2 Description of equivalent diameter derivation.

Table 7.1 Values of shape factor for various fibres

Shape of fibre cross-section	q
Circle – ideal ○	0
Circle – real	0 to 0.07
Triangle – ideal △	0.29
Triangle – real	0.09 to 0.12
Mature cotton	0.20 to 0.35
Irregular saw	> 0.60

$$a = \frac{A}{m} = \frac{\pi \cdot d \cdot (1+q) \cdot l}{(\pi \cdot d^2/4) \cdot l \cdot \rho} = \frac{4 \cdot (1+q)}{\rho \cdot d} \qquad [7.4]$$

The physical definition of the mechanical (normal) stress is the ratio of force F and the corresponding area s measured in [MPa].

$$\sigma^* = \frac{F}{s} \qquad [7.5]$$

The term stress in textiles, the ratio of force and the corresponding fibre fineness is measured in [N/tex], [cN/dtex] and also called 'breaking length km':

$$\sigma = \frac{F}{t} = \frac{F}{s \cdot \rho} \qquad [7.6]$$

Hence we have

$$\sigma = \frac{\sigma^*}{\rho} \qquad [7.7]$$

It is apparent that it is not correct to compare stress or strength of fibres having different densities by using the standard textile units (e.g. N/tex, breaking length km). Typical values for cotton, PET and steel are given in Table 7.2.

7.3 Fibrous assemblies: volume, density and mass

Inside a textile fibrous assembly (Fig. 7.3) there is the fibre volume V. The total volume of this body is called V_c. The compactness of this body is characterized by the ratio between these two volumes and is known as the fibre packing density μ (equation [7.9]). Alternatively, this value is called packing ratio or volume fraction. The fibre packing density value lies in an interval from 0 to 1. With the volume of fibres V and the total volume V_c including air, we have:

$$V \le V_C \qquad [7.8]$$

Table 7.2 Tensile strength of cotton, polyester and steel

Material	Tensile strength (N/tex)	Density (g/cm^3)	Mechanical strength (MPa)
Cotton	0.32	1.52	487
Polyester (PET)	0.43	1.36	585
Steel		7.85	500

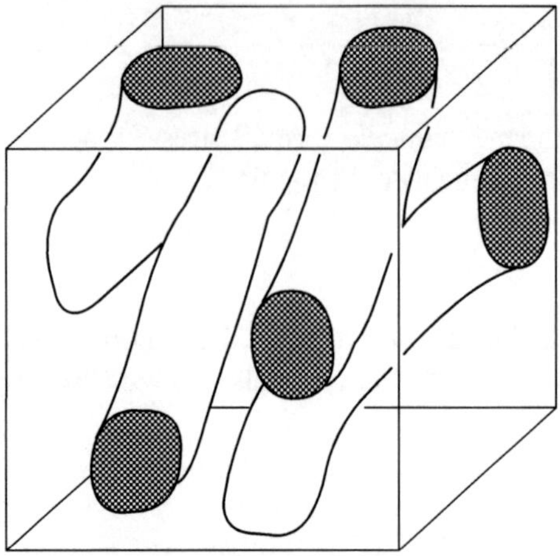

7.3 Fibrous assembly.

7.3.1 Compactness of fibrous assemblies – fibre packing density μ

The fibre packing density μ is defined by:

$$\mu = \frac{V}{V_C} \quad [7.9]$$

with $\mu \in [0;1]$.

Table 7.3 gives typical values of the fibre packing density for different materials. (Note: the packing density values around 0.4 to 0.5 are specific just for twisted staple yarns and practically no other fibrous structures have this value. Therefore, the yarns represent very specific types of fibrous assemblies.) Table 7.3 lists typical values of the packing density for various fibrous assemblies.

7.3.2 Areal interpretation

In a very (elementary) flat box as shown in Fig. 7.4, the lengths of the fibre sections inside this box are short; hence they can be interpreted as straight abscissas. It is now possible to derive the total volume and the volume of fibres according to:

$$\mathrm{d}V = \sum_{j=1}^{n} (s_j^* \mathrm{d}h) = \mathrm{d}h \sum_{j=1}^{n} s_j^* = S\mathrm{d}h \quad [7.10]$$

where S is the total sectional area of all elements as shown in Fig. 7.4. Hence we get

Table 7.3 Typical values of the packing density for various fibrous assemblies

Group	Fibrous assemblies	μ
Linear textiles	Monofilament	1
	Limit structure*	0.907
	Hard twisted silk	0.75 to 0.85
	Wet spun linen yarn	about 0.65
	Combed cotton yarn	0.5 to 0.6
	Carded cotton yarn	0.38 to 0.55
	Worsted yarn	0.38 to 0.50
	Woolen yarn	0.35 to 0.45
	Cotton roving	0.10 to 0.20
	Sliver	about 0.03
Other textiles	Woven fabric	0.15 to 0.30
	Knitted fabric	0.10 to 0.20
	Cotton wool**	0.02 to 0.04
	Leather (textiles)**	0.005 to 0.02
Other materials	Earthenware**	0.20 to 0.23
	Wood**	0.3 to 0.7
	Animal leather**	0.33 to 0.66

*Theoretical value, **approx. value.

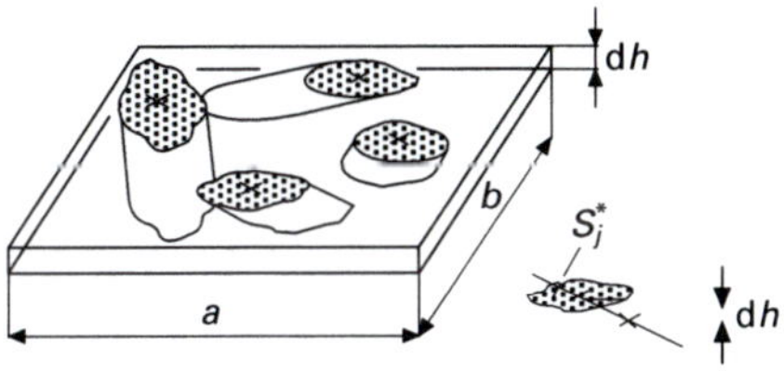

7.4 Very flat box of fibrous assembly.

$$\mu = \frac{dV}{dV_C} = \frac{Sdh}{S_C dh} = \frac{S}{S_C} \quad [7.11]$$

The fibre packing density of the flat box is then defined as the ratio between two areas. (Note: the fibre packing density in yarn is often expressed as the ratio between two cross-sectional areas.)

7.3.3 Mass interpretation

Yarn mass is based on density. The mass of a fibrous assembly is given by the mass of the fibrous material only. Using the total volume of the fibrous assembly, the density of the fibrous assembly can be defined. Using only the volume of fibres, the density of fibres can be derived. The fibre packing density is then given by the ratio between these two densities:

$$\mu = \frac{V}{V_C} = \frac{m/\rho}{m/\rho^*} = \frac{\rho}{\rho^*} \quad [7.12]$$

with ρ^* standing for the density of fibrous assembly and ρ as the density of the fibre material.

7.4 Yarn parameters: fineness, density, diameter and twist

Twisted staple yarn is a special type of fibrous assembly. For its description we use quantities as shown in Fig. 7.5. The yarn fineness ('yarn count') is one of most important standard values, but it is not sufficient to describe a yarn structure properly. The ratio between fibre volume and unit length characterizes the yarn 'size' geometrically well, but the yarn fineness is moreover influenced by the fibre mass density. Therefore, it is not possible to compare the yarns from different fibres using yarn 'count' (e.g. cotton and PET yarns). In this case, only ratios of yarn count to fibre mass density can be compared.

7.4.1 Yarn fineness

Fibre fineness in [tex] is defined as:

$$T = \frac{m}{l} = \frac{V \cdot \rho}{l} \quad [7.13]$$

Figure 7.6 shows an elementary short part of a yarn. The sum of the areas of all fibre sections in yarn cross-section is called the substance cross-sectional area of the yarn. The substance cross-sectional area of yarn is equal to the ratio yarn count to fibre mass density, or fibre volume per unit length of yarn.

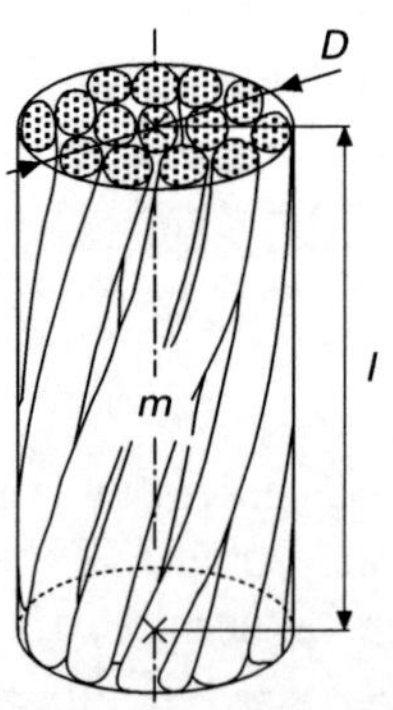

7.5 Yarn and its parameters

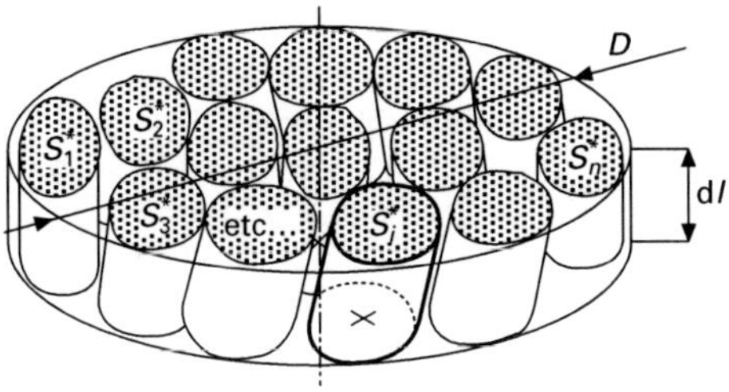

7.6 Cross-sectional area of yarn.

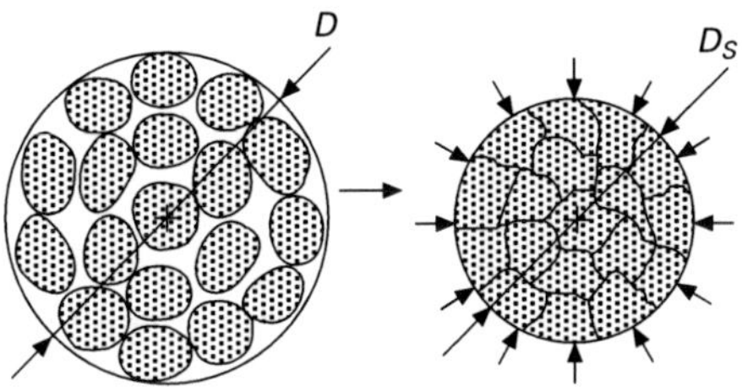

7.7 Yarn real diameter D and substance diameter D_s.

Definition of substance cross-section area is:

$$S = \sum_{i=1}^{n} s_i^* \tag{7.14}$$

with

$$dV = \sum_{i=1}^{n} s_i^* dl - dl \cdot \sum_{i=1}^{n} s_i^* - dl \cdot S \tag{7.15}$$

and

$$T = \frac{dm}{dl} = dl \cdot S \cdot \rho / dl = S \cdot \rho \tag{7.16}$$

$$dm = dV \cdot \rho = dl \cdot S \cdot \rho \tag{7.17}$$

Hence we have

$$S = \frac{T}{\rho} = \frac{V}{l} \tag{7.18}$$

If the yarn is compressed, all air is removed from the inside of the yarn. Then, the yarn cross-sectional area is equal to the substance cross-sectional area and hence the compressed yarn will have the substance diameter according to Fig. 7.7.

7.4.2 Substance diameter

The substance diameter can then be calculated according to:

$$S = \frac{\pi \cdot D_s^2}{4} \qquad [7.19]$$

Hence we get

$$D_s = \sqrt{\frac{4S}{\pi}} = \sqrt{\frac{4T}{\pi\rho}} \qquad [7.20]$$

(Note: the diameter of real yarn is always bigger than the substance diameter.)

7.4.3 Relative fineness

The ratio of yarn fineness (count) to fibre fineness is another important quantity and is called the relative fineness of yarn. It is given by

$$\tau = \frac{T}{t} = \frac{(\pi \cdot D_s^2/4)\cdot\rho}{\pi \cdot d^2/4 \cdot \rho} = \frac{S\rho}{s\rho} = \frac{S}{s} = \frac{D^2}{d^2} \qquad [7.21]$$

(Note: this value is not identical to the number of fibres in the yarn cross-section.)

7.4.4 Coefficient k_n

The fibre sectional area is a function of the fibre slope, as shown in Fig. 7.8. The coefficient k_n is the ratio between the fibre cross-sectional area and the mean fibre sectional area. Taking into account the fibre geometry according to

$$\cos \vartheta_i = \frac{dl}{dx_i} \qquad [7.22]$$

and

$$s_i^* \cdot dl = s \cdot dx_i \qquad [7.23]$$

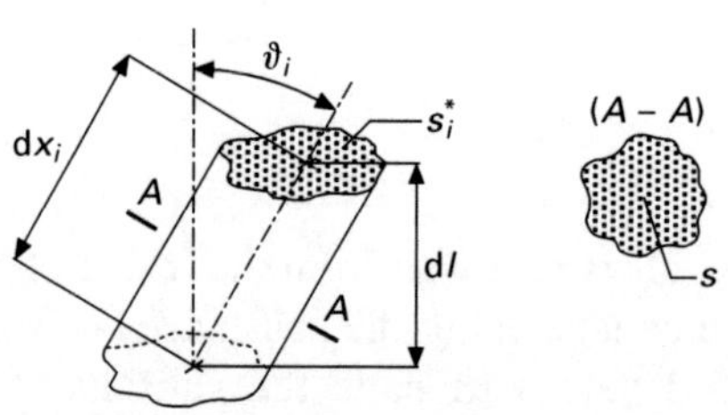

7.8 Fibre element in yarn.

we get

$$s_i^* = s \cdot \frac{dx_i}{dl} \qquad [7.24]$$

and

$$S = \sum_{i=1}^{n} s_i^* = \sum_{i=1}^{n} \frac{s}{\cos \vartheta_i} = s \cdot \sum_{i=1}^{n} \frac{1}{\cos \vartheta_i} \qquad [7.25]$$

hence

$$\bar{s}^* = \frac{S}{n} \qquad [7.26]$$

and

$$k_n = \frac{s}{\bar{s}^*} = \frac{s}{\frac{s}{n} \cdot \sum_{i=1}^{n} \frac{1}{\cos \vartheta_i}} = \frac{1}{\frac{1}{n} \cdot \sum_{i=1}^{n} \frac{1}{\cos \vartheta_i}} \qquad [7.27]$$

The value k_n is the weighted harmonic mean value of cosines of the slope angles. If each angle is equal to 0, we have $k_n = 1$. This configuration corresponds to the parallel fibre bundle.

7.4.5 Number of fibres in the yarn

The number of fibres in the yarn can be determined by

$$n = \frac{S}{\bar{s}*} = \overbrace{\left(\frac{S}{s}\right)}^{=\tau} \cdot \overbrace{\left(\frac{s}{\bar{s}*}\right)}^{=k_n} \qquad [7.28]$$

and

$$n = \tau \cdot k_n \qquad [7.29]$$

The coefficient k_n is related to the relative fineness and the number of fibres according to equation [7.29]. (Note: beyond the scope of a parallel fibre bundle, the number of fibres in the yarn cross-section is always smaller than the relative yarn fineness.) Equation [7.29] also enables us to determine the coefficient k_n experimentally.

7.4.6 Packing density of a yarn

According to equation [7.9], the packing density of the yarn is given by

$$\mu = \frac{V}{V_c} = \frac{S \cdot l}{\pi \cdot D^2 \cdot l/4} \qquad [7.30]$$

with

$$V_c = \pi D^2 l/4 \qquad [7.31]$$

we get

$$\mu = \frac{4S}{\pi D^2} = \frac{4T}{\pi D^2 \rho} \qquad [7.32]$$

Inserting the substance diameter from equation [7.19], we have

$$\mu = \left(\frac{D}{D_S}\right)^2 \qquad [7.33]$$

If the packing density is known, the yarn diameter can also be determined using equation [7.33].

7.4.7 Yarn diameter

The yarn diameter can be derived from equation [7.32]

$$D = \sqrt{\frac{4S}{\pi \rho}} = \sqrt{\frac{4T}{\pi \mu \rho}} \qquad [7.34]$$

7.4.8 Twist intensity

Some yarn characteristics are based on the yarn twist. Because the dimension of the twist Z is [m^{-1}] and the dimension of the yarn diameter is [m], the product of DZ is dimensionless. The twist intensity is then given by

$$\kappa = \pi \cdot D \cdot Z \qquad [7.35]$$

If the peripheral fibre in Fig. 7.9 follows a helical path, the twist intensity is also the tangent of angle β_D of this fibre.

7.4.9 Twist coefficients

The so-called 'twist coefficients' (twist multipliers) create another group of twist characteristics. The first type is related to the yarn substance cross-sectional area S. This coefficient is called 'areal' and only of theoretical value. The second type is related to the yarn fineness (count). This coefficient is commonly used in practice. General coefficients use a general value of twist exponent q. Koechlin's coefficients use the twist exponent 0.5. These can be also written in an alternative way, equations [7.38a, b] and [7.39a, b].

General coefficients are:

$$\alpha_s^* = Z \cdot S^q \text{ (areal)} \qquad [7.36]$$

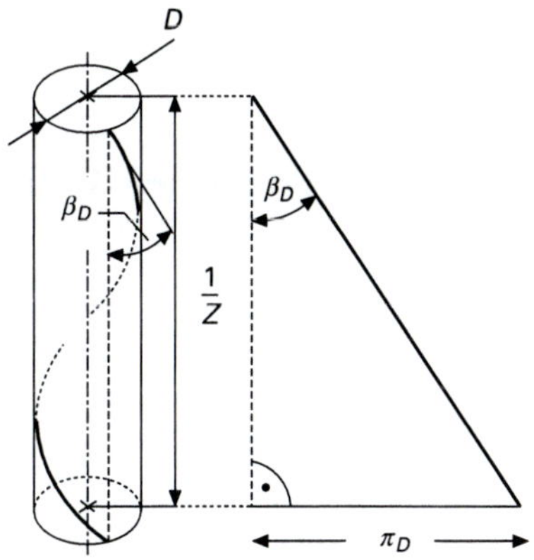

7.9 Surface fibre on yarn diameter – one coil of its helix.

$$\alpha^* = Z \cdot T^q \text{ (common)} \qquad [7.37]$$

Koechlin's type is:

$$\alpha_s = Z \cdot \sqrt{S} \text{ (areal)} \qquad [7.38a]$$

Hence:

$$\alpha_s = Z \cdot \sqrt{S} = \frac{\kappa}{\pi D} \cdot \sqrt{\frac{\pi D_s^2}{4}} = \frac{\kappa \cdot \sqrt{\mu}}{2 \cdot \sqrt{\pi}} \qquad [7.38b]$$

$$\alpha = Z \cdot \sqrt{T} \text{ (common)} \qquad [7.39a]$$

Hence:

$$\alpha = Z \cdot \sqrt{T} = \frac{\kappa}{\pi D} \cdot \sqrt{S\rho} = \frac{\kappa}{\pi D} \cdot \sqrt{(\pi D_s^2/4) \cdot \rho} = \frac{\kappa \cdot \sqrt{\mu\rho}}{2 \cdot \sqrt{\pi}} \qquad [7.39b]$$

(Note: the common version of Koechlin's coefficient is usually used in spinning mills.)

7.4.10 Dimensionless quantities

The dimensionless quantities are always very interesting, because they characterize the internal relationships in material body, processes, etc.:

- relative fineness τ;
- coefficient k_n;
- fibre packing density μ;
- areal diameter multiplier K_s;
- twist intensity κ;
- Koechlin's areal twist coefficient α_s.

In place of the yarn parameters T, Z, D and n it is possible to use four of the dimensionless parameters, e.g. τ, κ, μ and k_n.

7.5 Modelling of internal yarn geometry: helical yarn model

Practical yarn properties are the result of fibre properties, the mutual fibre interactions inside the yarn, and the interactions between yarn and outer influences. The internal structure of a yarn is very important especially for the geometrical and mechanical properties of the yarn. From Fig. 7.10 it is apparent that the specific regulations of the internal yarn geometry are relatively complicated due to the complex nature of deterministic and random influences. The general fibre trajectories have complicated shapes, a random character and show deterministic trends.

7.5.1 Fibre elements

The position of fibre elements is determined by their coordinates r, φ, ζ, their length by dl, and their direction by the elementary increments of the coordinates, i.e. dr, dφ, dζ as shown in Fig. 7.11. It is important to note:

- the increment dζ is defined always as positive; and
- the positive value of the increment dζ corresponds to the sense of direction of yarn twist.

Hence:

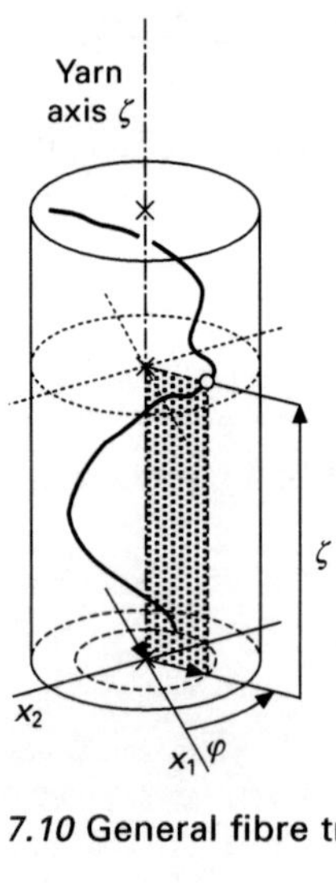

7.10 General fibre trajectory in yarn.

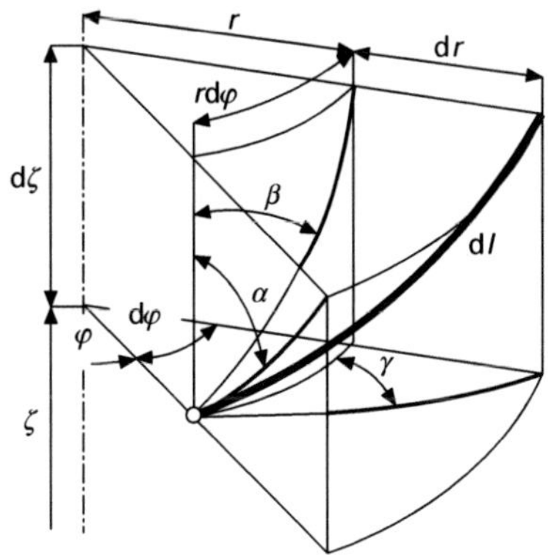

7.11 The position of fibre elements in yarn.

$$\tan\,\alpha = \frac{\mathrm{d}r}{\mathrm{d}\zeta},\ \tan\,\beta = \frac{r\mathrm{d}\varphi}{\mathrm{d}\zeta},\ \tan\,\gamma = \frac{\mathrm{d}r}{r\mathrm{d}\varphi} \qquad [7.40a–c]$$

and

$$\tan\,\alpha = \tan\,\beta \cdot \tan\,\gamma \qquad [7.41]$$

(Note: evidently, the angle β relates to the yarn twist intensity.) From Fig. 7.11, we get:

$$\cos\,\vartheta_r = \frac{\mathrm{d}r}{\mathrm{d}l},\ \cos\vartheta_t = \frac{r\mathrm{d}\varphi}{\mathrm{d}l},\ \cos\vartheta_a = \frac{\mathrm{d}\zeta}{\mathrm{d}l} \qquad [7.42a–c]$$

where

$$\mathrm{d}l = \sqrt{(\mathrm{d}r)^2 + (r\mathrm{d}\varphi)^2 + (\mathrm{d}\zeta)^2} \qquad [7.43]$$

The following relations are evident from earlier equations:

$$\tan\,\alpha = \frac{\cos\vartheta_r}{\cos\vartheta_a},\ \tan\,\beta = \frac{\cos\vartheta_t}{\cos\vartheta_a},\ \tan\,\gamma = \frac{\cos\vartheta_r}{\cos\vartheta_t} \qquad [7.44a–c]$$

Hence we have

$$\cos\vartheta_r = \frac{\tan\alpha}{\sqrt{\tan^2\alpha + \tan^2\beta + 1}} \qquad [7.45a]$$

$$\cos\vartheta_t = \frac{\tan\beta}{\sqrt{\tan^2\alpha + \tan^2\beta + 1}} \qquad [7.45b]$$

$$\cos\vartheta_a = \frac{\tan\gamma}{\sqrt{\tan^2\alpha + \tan^2\beta + 1}} \qquad [7.45c]$$

Evidently, two independent values define the direction of fibre element. Those are the two ratios of increments of coordinates: $d\varphi/d\zeta$ and $dr/d\zeta$:

$$\frac{d\varphi}{d\zeta} = 2\pi \cdot z \qquad (\text{usually: } d\zeta > 0) \qquad [7.46]$$

where z stands for the twist of the element (which is not necessarily identical to the yarn twist Z).

$$z = \frac{\tan\beta}{2\pi \cdot r} \qquad [7.47]$$

7.5.2 Helical yarn model

In a helical model, two differential equations characterize the path of each fibre:

$$d\varphi = 2\pi \cdot z_i(\zeta)\, d\zeta, \text{ with } z_i(\zeta) = Z \text{ constant} \qquad [7.48]$$

$$dr = m_i(\zeta) d\zeta \qquad [7.49]$$

Assuming that the fibre passes through the 'starting point' with coordinates $(r_0, \varphi_0, \zeta_0)$, then we obtain by integration:

$$\int_{\varphi_0}^{\varphi} d\varphi = \int_{\zeta_0}^{\zeta} 2\pi \cdot z_i(\zeta) d\zeta = \varphi - \varphi_0 = 2\pi \cdot Z(\zeta - \zeta_0)$$

and hence

$$\frac{(\varphi - \varphi_0)/2\pi}{(\zeta - \zeta_0)} = Z \qquad [7.50]$$

The yarn twist has a sense of constant Z. The height of one fibre coil is $1/Z$ and the same for all fibres. By rearrangement:

$$d\varphi = 2\pi \cdot Z d\zeta : \frac{r d\varphi}{d\zeta} = 2\pi \cdot r \cdot Z$$

and hence

$$\tan\beta = 2\pi \cdot r \cdot Z \qquad [7.51]$$

The angle β is constant at a given radius r. By integration:

$$\int_0^r dr = \int_{\zeta_0}^{\zeta} m_i(\zeta) d\zeta \qquad [7.52]$$

with

$$r - r_0 = 0, \text{ hence } r = r_0 = \text{constant}. \qquad [7.53]$$

The fibre lies on the same radius of the cylindrical surface. (Note: the same conclusions can be obtained from Fig. 7.12.) Assumptions of a helical model can be formulated also in the following way:

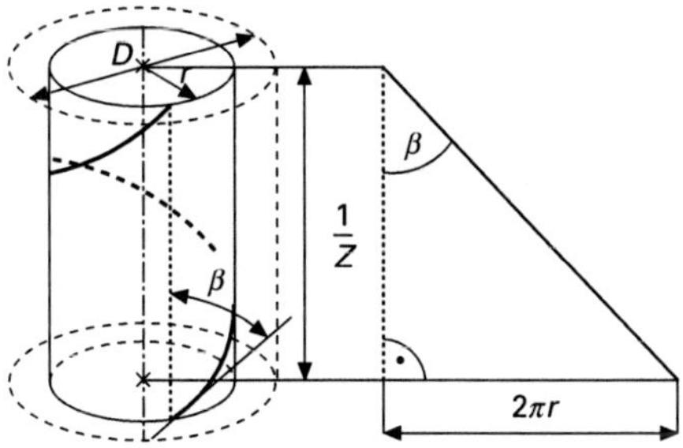

7.12 One coil of general fibre helix.

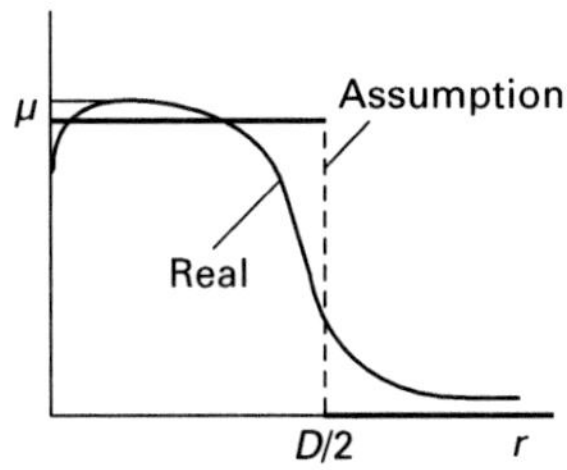

7.13 Radial packing density of real yarn and helical model of yarn.

- helical paths of fibres (same sense of rotation);
- common axis of all helixes is yarn axis;
- same coil height for all fibres;
- packing density is constant.

The fibre positions in the yarn ('starting points'), which are not determined by the assumptions stated before, can be characterized by the radial function of packing density $\mu = \mu(r)$ as shown in Fig. 7.13. Because it is relatively complicated, a simplifying assumption is used: the packing density must be constant in all places inside the yarn.

If all these four assumptions are valid, then this model is called an ideal helical model. Helical models and their applications are the oldest concepts of yarn modelling, associated with the scientists A. Köchlin (1828), E. Müller (1880), S. Marschik (1904), A.CH. Gegauff (1907), R. Schwarz (1933), E. Braschler (1935), V.I. Budnikov (1945), L.R.G.Treloar (1956) and others.

7.5.3 Number of fibres and shortening of yarn

Based on Braschler (1939), a differential annulus ('differential layer') is created at the general radius *r* in a cross-section of the general helical model as shown in Fig. 7.14. The area *A* of the differential annulus is then given by

$$A = 2\pi \cdot r\mathrm{d}r \qquad [7.54]$$

The packing density is then

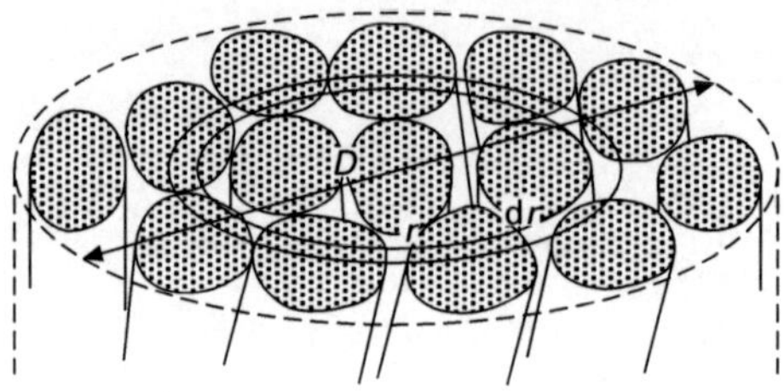

7.14 Differential annulus in a cross-section of general helical model.

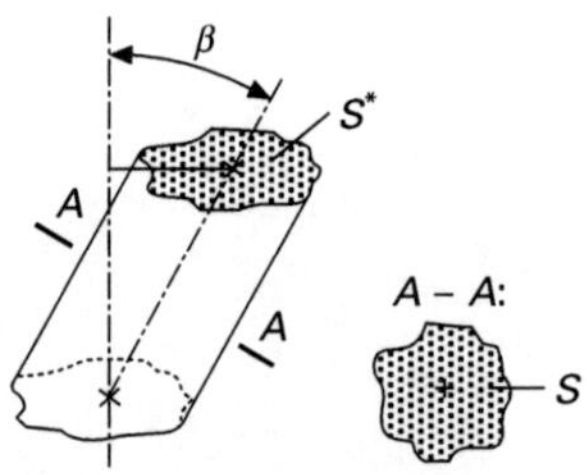

7.15 Yarn fibre element.

$$\mu = \frac{dS}{2\pi \cdot rdr} \quad [7.55]$$

where dS represents the area of the fibre sections in the differential annulus (dark grey in Fig. 7.14). The area of the fibre section can then be determined by

$$dS = 2\pi \cdot \mu rdr \quad [7.56]$$

The area of the oblique section of one fibre is then given by (according to Fig. 7.15):

$$s^* = \frac{s}{\cos\beta} \quad [7.57]$$

where s stands for the fibre cross-section.

The number of fibres in the differential annulus is hence

$$dn = \frac{dS}{s^*} = \frac{2\pi \cdot rdr \cdot \mu}{s/\cos\beta} = 2\pi \cdot \cos\beta \cdot \mu rdr/s \quad [7.58]$$

The substantial cross-sectional area of the yarn is given by

$$S = \int_{r=0}^{r=D/2} dS = 2\pi \int_{r=0}^{r=D/2} \mu rdr \quad [7.59]$$

The mean packing density of the yarn is given by

$$\bar{\mu} = \frac{S}{\pi D^2/4} = \frac{2\pi \int_{r=0}^{r=D/2} \mu r \mathrm{d}r}{\pi D^2/4} = \frac{8}{D^2} \cdot \int_0^{D/2} \mu r \mathrm{d}r \qquad [7.60]$$

The number of fibres in the yarn cross-section is then

$$n = \int_{r=0}^{r=d/2} \mathrm{d}n = \frac{2\pi}{s} \cdot \int_0^{D/2} \cos\beta \cdot \mu r \mathrm{d}r = \frac{2\pi}{s} \cdot \int_0^{D/2} \frac{\mu r \mathrm{d}r}{\sqrt{1 + \tan^2\beta}}$$

$$= \frac{2\pi}{s} \cdot \int_0^{D/2} \frac{\mu r \mathrm{d}r}{\sqrt{1 + (2\pi r Z)^2}} \qquad [7.61]$$

The coefficient k_n can then be determined by

$$k_n = \frac{2\pi}{s} \cdot \int_0^{D/2} \frac{\mu r \mathrm{d}r}{1 + (2\pi r Z)^2} \qquad [7.62]$$

In order to calculate S, μ, n and k_n in a numerical simulation, the relation $\mu = \mu(r)$ must be known. The function $\mu(r)$ can be obtained by experiments or by applying a theoretical model as described, e.g. in Hearle *et al.* (1969). A complete description of the ideal helical model can be found in Neckar and Das (2012).

7.5.4 Yarn retraction in the ideal helical model

Figure 7.16 shows the yarn retraction in the ideal helical model. The yarn retraction can hence be calculated according to

$$\delta = \frac{(\zeta_0 - \zeta)}{\zeta_0} = 1 - \frac{\zeta}{\zeta_0} \qquad [7.63]$$

The latent twist coefficient is then given by

$$\alpha_0 = Z_0 \cdot \sqrt{T_0} \qquad [7.64]$$

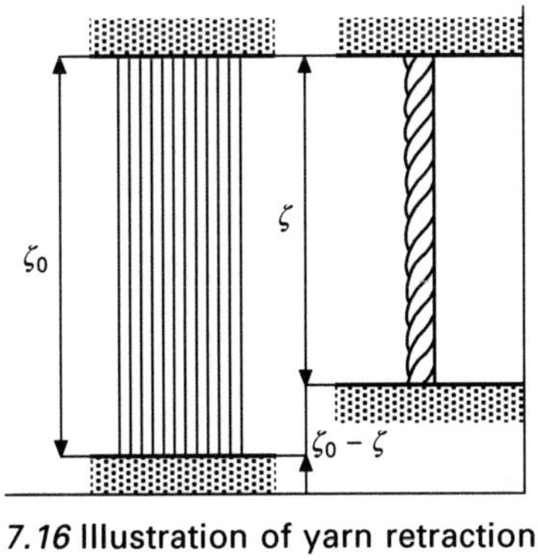

7.16 Illustration of yarn retraction.

where $Z_0 = N_c/\zeta_0$, N_c stands for the number of coils per length ζ_0 and $T_0 = m/\zeta_0$, with m representing the fibre mass. The real twist coefficient is then given by

$$\alpha = Z \cdot \sqrt{T} \qquad [7.65]$$

and hence

$$\alpha = \frac{\alpha_0}{(1-\delta)^{3/2}} \qquad [7.66]$$

The yarn retraction can also be calculated for an ideal helical model with

$$\delta = 1 - \left(\frac{\zeta}{\zeta_0}\right) = 1 - k_n = 1 - \frac{2}{(\pi DZ)^2}\left[\sqrt{1 + (\pi DZ)^2} - 1\right] \qquad [7.67]$$

Hence we get

$$\delta = \frac{\sqrt{1 + \tan^2\beta_D} - 1}{\sqrt{1 + \tan^2\beta_D} + 1} = \frac{1 - \cos\beta_D}{1 + \cos\beta_D} = \tan^2\frac{\beta_D}{2} \qquad [7.68]$$

Using equation (7.39b), we can derive

$$\delta = \frac{\sqrt{1 + 4\pi\alpha^2/\mu\rho} - 1}{\sqrt{1 + 4\pi\alpha^2/\mu\rho} + 1} \qquad [7.69]$$

Using equation (7.66), we can express δ as a function of latent twist coefficient α_0 as follows:

$$\delta = \left[\sqrt{1 + \frac{4\pi\alpha_0^2}{\mu\rho(1-\delta)^3}} - 1\right] \Bigg/ \left[\sqrt{1 + \frac{4\pi\alpha_0^2}{\mu\rho(1-\delta)^3}} + 1\right] \qquad [7.70]$$

This square equation has the solution

$$\delta = \frac{1}{2} \pm \frac{1}{2} \cdot \sqrt{1 - \frac{4\pi}{\mu}\frac{\alpha_0^2}{\rho}} \qquad [7.71]$$

Since the discriminant of the equation must not be negative, we get

$$1 - \frac{4\pi}{\mu}\frac{\alpha_0^2}{\rho} \geq 0, \; \alpha_0 \leq \frac{\mu\rho}{4\pi} \qquad [7.72]$$

Hence we have

$$\frac{\alpha_0}{\sqrt{\mu\rho}} \leq \frac{1}{\sqrt{4\pi}} \qquad [7.73]$$

which means that the latent twist coefficient is limited. The limit case is given by

$$\frac{\alpha_0}{\sqrt{\mu\rho}} = 0.281 \qquad [7.74]$$

For this case we get from equation [7.71]: $\delta = 0.5$.

With equation [7.68] we then get $\frac{1}{2} = \tan^2 \frac{\beta_D}{2}$ and hence $\beta_D = 70.5°$. From

$$\kappa_n = \pi DZ = \tan \beta_D \qquad [7.75]$$

we get $\kappa = 2\sqrt{2} = 2.828$. This means that the value of the twist intensity is also limited as is shown in Fig. 7.17. Staple yarns have smaller retractions than filament yarns since fibres slip in the region around the yarn surface during the twisting process ('false draft') and therefore the 'effective' yarn diameter D_δ is smaller than D.

The saturated twist (limit case) can be observed experimentally, but for a smaller value of the latent twist (usually for β_D from 45° to 55° in the case of filament yarns). The axial asymmetry of twisted yarn is the probable reason of this phenomenon. If the twist level exceeds the saturated twist of a yarn, this additional twist cannot be absorbed 'inside' its structure thus coils form (twist of the second order) as shown in Fig. 7.18.

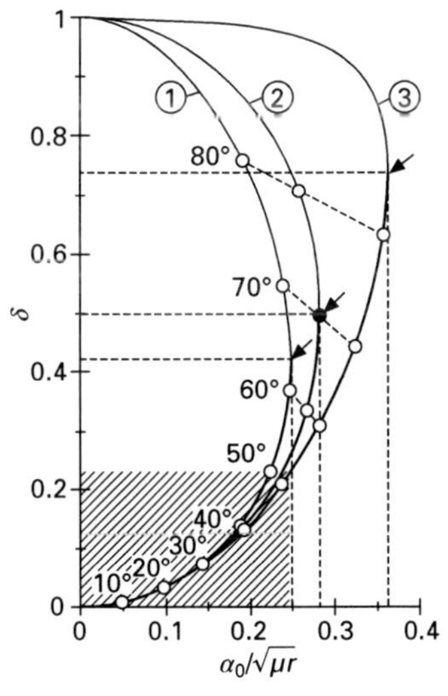

7.17 Graphical representation, curve 1, of equation [7.71]. Curves 2 and 3 are from other models of yarn retraction.

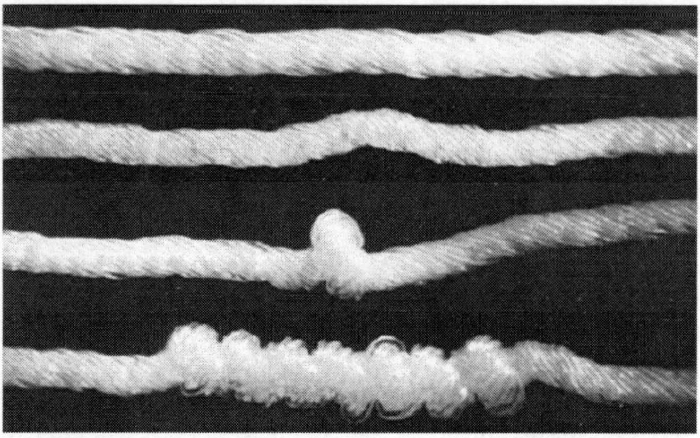

7.18 Phases of coil formation (twist of second order) in an over-saturated yarn.

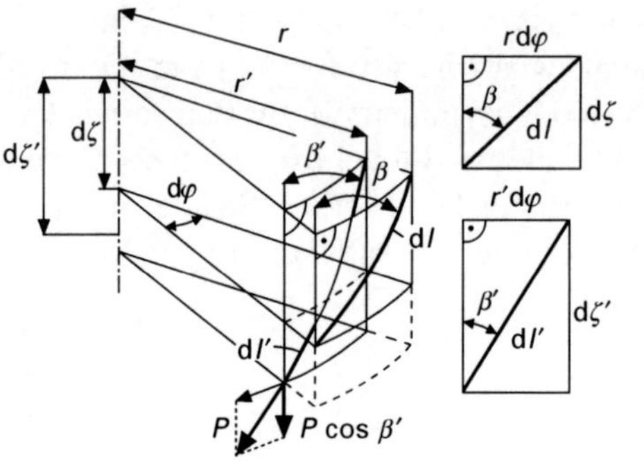

7.19 General helical fibre element before and after elongation.

7.6 Stress–strain relations

The general element of the helical fibre (length dl, angle β) as shown in Fig. 7.19, lying at the radius r determines an elementary cylindrical surface with certain dimensions (right, top). After the yarn has been subjected to elongation, the same element shifts itself to a new (right, bottom) position at a smaller radius r' with new angle β' and new dimensions. Hence we have

$$\tan\beta = \frac{r\mathrm{d}\varphi}{\mathrm{d}\zeta} \text{ and } \tan\beta' = \frac{r'\,\mathrm{d}\varphi}{\mathrm{d}\zeta'} \qquad [7.76]$$

The yarn axial strain is given by

$$\varepsilon_a = \frac{\mathrm{d}\zeta' - \mathrm{d}\zeta}{\mathrm{d}\zeta} \qquad [7.77]$$

The radial strain is given by

$$\varepsilon_r = \frac{r' - r}{r} \qquad [7.78]$$

The contraction ratio according to Poisson is defined as

$$\eta = -\frac{\varepsilon_\gamma}{\varepsilon_a} \qquad [7.79]$$

The fibre strain can be calculated using

$$\varepsilon_l = \frac{\mathrm{d}l' - \mathrm{d}l}{\mathrm{d}l} \qquad [7.80]$$

Based on Pythagoras's theorem from Fig. 7.19, we have before deformation:

$$\mathrm{d}^2 l = \mathrm{d}^2\zeta + (r\mathrm{d}\varphi)^2 \qquad [7.81]$$

after deformation:

$$\begin{aligned} \mathrm{d}^2 l' &= \mathrm{d}^2\zeta' + (r'\mathrm{d}\kappa)^2 = (1 + \varepsilon_a)^2\mathrm{d}^2\zeta + (1 + \varepsilon_\gamma)^2(r\mathrm{d}\varphi)^2 \\ &= (1 + \varepsilon_a)^2\mathrm{d}^2\zeta + (1 - \eta\varepsilon_a)^2(r\mathrm{d}\varphi)^2 \end{aligned} \qquad [7.82]$$

(Note: because of the continuity of the yarn body, $\mathrm{d}\varphi$ must be the same.) Using earlier derived equations, we obtain finally

$$\begin{aligned} (1 + \varepsilon_l)^2 = 1 &+ 2\varepsilon_a(\cos^2\beta - \eta\sin^2\beta) \\ &+ \varepsilon_a^2(\cos^2\beta + \eta^2\sin^2\beta) \end{aligned} \qquad [7.83]$$

Assuming that strains are small, we have

$$\varepsilon_l^2 \to 0, \ \varepsilon_a^2 \to 0 \qquad [7.84]$$

Then

$$\varepsilon_l = \varepsilon_a(\cos^2\beta - \eta\sin^2\beta) \qquad [7.85]$$

Note: Gegauff (1907) used $\eta = 0$ and obtained

$$\varepsilon_l = \varepsilon_a\cos^2\beta \qquad [7.86]$$

Assuming the easiest case, that the fibre tensile stress–strain relation is linear, we get

$$\sigma = E\varepsilon_l \qquad [7.87]$$

where σ represents the tensile stress and E the Young modulus. Figure 7.20 shows the load on a helical fibre element. The axial force in the fibre is then given by

$$F_1 = \sigma s = E\varepsilon_l s \qquad [7.88]$$

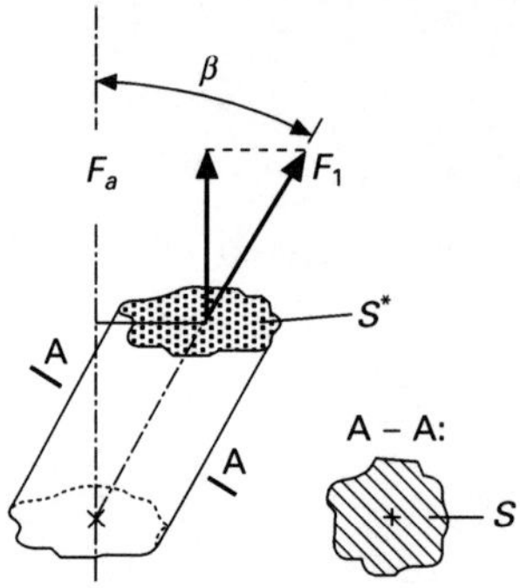

7.20 Load on a helical fibre element.

and the component force in the direction of yarn axis is

$$F_a = F_1 \cos \beta = E\varepsilon_l s \cos \beta \quad [7.89]$$

The fibre sectional area is given by

$$s^* = \frac{s}{\cos \beta} \quad [7.90]$$

Assuming small deformations, the normal stress on this area is then given by

$$\sigma_a = \frac{F_a}{s^*} = \frac{E\varepsilon_l s \cos \beta}{s/\cos \beta} = E\varepsilon_l \cos^2 \beta \quad [7.91]$$

The fibre sectional area inside the differential annulus can be determined using equation [7.56] and then the yarn axial force is

$$P = \int_{\gamma=0}^{\gamma=D/2} \sigma_a \mathrm{d}S \quad [7.92]$$

After rearrangement and integration we get the axial yarn force

$$P = \pi\mu E\varepsilon_a (D/2)^2[(1 + \eta) \cos^2\beta_D + \eta(\ln \cos^2 \beta_D)/\tan^2 \beta_D] \quad [7.93]$$

At the same strain ε_a, the untwisted fibre bundle of the same yarn count has the same substantial cross-sectional area and the axial force is then

$$P = \sigma_a S = \pi\eta E\varepsilon_a (D/2)^2 \quad [7.94]$$

The tensile force utilization coefficient in the twisted yarn is then

$$\varphi = \frac{P}{P^*} = \frac{(1 + \eta) \cos \beta_D + \eta(\ln \cos^2 \beta_D)}{\tan^2\beta_D} \quad [7.95]$$

A graphical representation of equation [7.95] is shown in Fig. 7.21. It is

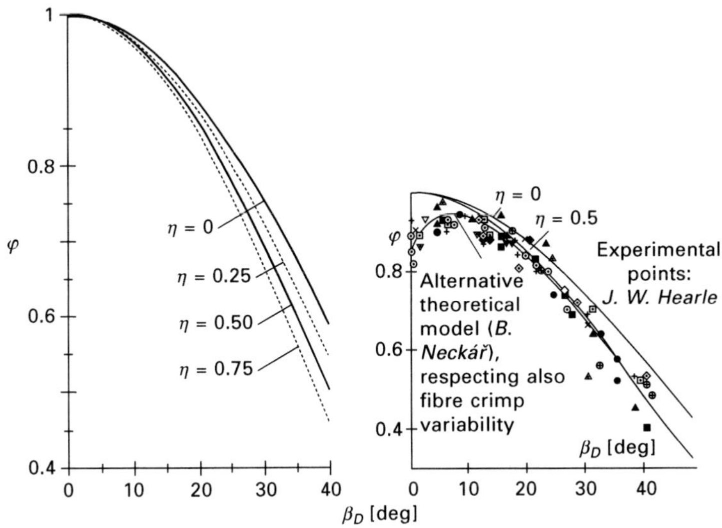

7.21 Graphical representation of equation [7.95] and comparison with experiment results.

important to note that φ is also applicable for the determination of filament yarn strength (by $\eta = 0.5$) (Fig. 7.21).

The strength of a staple fibre yarn can usually be interpreted as:

- a result of fibre path and fibre strain in the yarn; and
- a complex of frictional mechanisms.

Figure 7.22 shows the breaking tenacity versus twist. The top curve shows the model above. The frictional effect is still an open problem to be described theoretically, hence the resulting curve is only an approximation of the real case. Note: the helical model is the best known theoretical concept to describe the internal yarn geometry. Only a few basic concepts have been described here. Many other versions and their applications can be found in traditional textile literatures. A concise summary can be found in Neckar and Das (2012).

7.7 The relationship between yarn count, twist, packing density and diameter

The relations among count, twist and diameter of a yarn are related to specific geometrical and mechanical properties of the yarn. The basic quantity behind these relations is yarn packing density. The first traditional theoretical model regarding these relations was introduced by Koechlin in 1828 (Neckář, 1990). He studied the yarns produced from the same fibrous material using the same

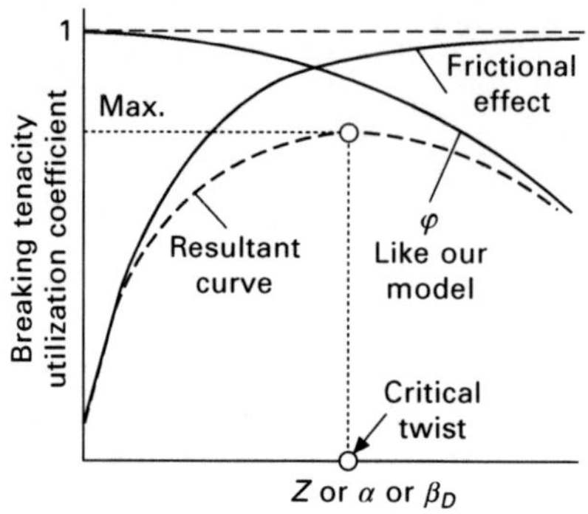

7.22 Problem of staple yarn strength components.

technology for analogous end uses. At that time, the scientific knowledge about the mechanics of fibrous assembly and yarn geometry was not sufficient to create a realistic model. Hence, it was necessary to substitute the unknown relations by the first categorical assumption as shown here.

7.7.1 Theoretical model according to Koechlin's concept

The initial assumptions (problem limitation) are:

- same fibrous material;
- same type of technology;
- same kinds of use.

Koechlin's first assumption

Koechlin's first assumption (substitution of knowledge of mechanics) was that the packing density is a function of the twist intensity only:

$$\mu = f(\kappa) \tag{7.96}$$

Hence we get for the twist coefficients:

- areal Koechlin's type (equation [7.38]):

$$\alpha_S = \frac{\kappa\sqrt{f(\kappa)}}{2\sqrt{\pi}} \tag{7.97}$$

- common Koechlin's type (equation [7.39]):

$$\alpha = \frac{\kappa\sqrt{f(\kappa)\rho}}{2\sqrt{\pi}} \tag{7.98}$$

and for the diameter multipliers, we have

- areal multiplier (equation [7.34]):

$$K_S = \frac{2}{\sqrt{\pi f(\kappa)}} \qquad [7.99]$$

- common multiplier (equation [7.34]):

$$K = \frac{2}{\sqrt{\pi f(\kappa)\rho}} \qquad [7.100]$$

The second assumption is how the yarn is twisted. If yarns of different counts are produced from the same material using the same technology, all yarn properties should be similar. But in reality, this has proven not to be possible. Therefore, we must go 'one step back': what will happen if we consider only geometrically similar yarns? (Not all values, only those that are not related to fibre fineness.) Then, the other properties may be also more or less similar. In case of geometrical similarity, the corresponding angles are the same. So, the angles β_D must be same and therefore the twist intensities κ must be the same too.

Koechlin's second assumption

Koechlin's second assumption is directed to a suitable twist. This means that the twist intensity of yarns of different counts shall be the same:

$$\kappa = \text{constant} \qquad [7.101]$$

Hence we have for the twist coefficients:

$$\alpha_S = \text{constant and } \alpha = \text{constant} \qquad [7.102]$$

For the diameter multipliers, we get

$$K_S = \text{constant and } K = \text{constant} \qquad [7.103]$$

The packing density is then

$$\mu = \text{constant} \qquad [7.104]$$

The derived equations are traditionally used in textile mills. The calculation of a suitable yarn twist is often necessary for the yarn production and the calculation of the yarn diameter is necessary for fabric manufacturing. In many cases, the resulting values are not sufficiently precise. On the one hand, the assumption about the geometrical similarity is acceptable. On the other hand, the twist intensity κ is not a function of packing density μ only as it depends also on the yarn count. Hence, the Koechlin's first assumption is not sufficiently accurate and thus an empirical correction of Koechlin's theory is necessary.

7.7.2 Empirical corrections of Koechlin's theory

Table 7.4 gives an overview about various twist coefficients that were determined by different authors. According to equation [7.36], a suitable yarn twist can be derived from

$$Z = \frac{\alpha}{T^q} \qquad [7.105]$$

where q is an empirical value of the twist exponent with normally $q > 0.5$. The yarn diameter can then be calculated with

$$D = Q_a T^{\omega} \alpha^{v} \qquad [7.106]$$

where Q_α, w, v... are empirically determined values, with $w > 0.5$ (e.g. cylindrical yarn: $w = 0.56$) and $v < 0$ (e.g. cylindrical yarn: $v = -0.22$).

The problems with models like Koechlin's concept initiated interest for a deeper knowledge of the internal yarn mechanics and motivated researchers to think about alternative ways to solve the problems. The concept of radial forces equilibrium based on the differential equation (Hearle *et al.*, 1969) is general, but very complicated and requires knowledge on some not yet sufficiently known inputs (e.g. stress–strain tensors relation). The model presented in the next section is 'something in-between'. It is not as difficult as the concept of radial forces equilibrium but is sufficient for the present knowledge we have about the mechanics of fibrous assemblies.

Table 7.4 Various twist coefficients according to various authors

Author	Year	Twist exponent q
Koechlin	1828	0.5
Strub	1900	0.6
Johansen	1902	0.644
	1905	0.785 warp
Laetch	1905	0.720 weft
	1941	0.62 – 0.75
	1937	0.565
Oeser	1940	0.47
Phrix	1942	0.666
Neckář	1971	0.577, 0.6
	1975	0.518 comb.
Salaba	1975	0.551, 0.570

7.8 A simple mechanical model of yarn properties

The model is based on the helical geometry of fibres inside the yarn (for assumptions see Section 7.6). A fibre coil lying on the imaginary cylinder of radius r is shown in Fig. 7.23. Unwinding the surface of this cylinder results in a triangle, whose geometry determines the fibre slope angle β. It is known from analytical geometry that the curvature of the helix is given by:

$$k_1 = \frac{\sin \beta}{r} \qquad [7.107]$$

Obviously, the radius of curvature (radius of the so-called oscillating circle) is the reciprocal value of the curvature:

$$r_k = \frac{1}{k_1} \text{ and } r_k = \frac{r}{\sin^2 \beta} \qquad [7.108]$$

7.8.1 Centripetal force per unit volume of a fibre

According to Fig. 7.24, the centripetal force of a fibre can be calculated for small angles φ by

$$dP = 2f \sin(d\varphi/2) = 2F d\varphi/2 = F d\varphi \qquad [7.109]$$

The volume of the unit is given by

$$dV = \frac{r_k d\varphi s}{\sin^2 \beta} \qquad [7.110]$$

where s stands for the fibre cross-sectional area. The centripetal force per unit volume of the fibre is then

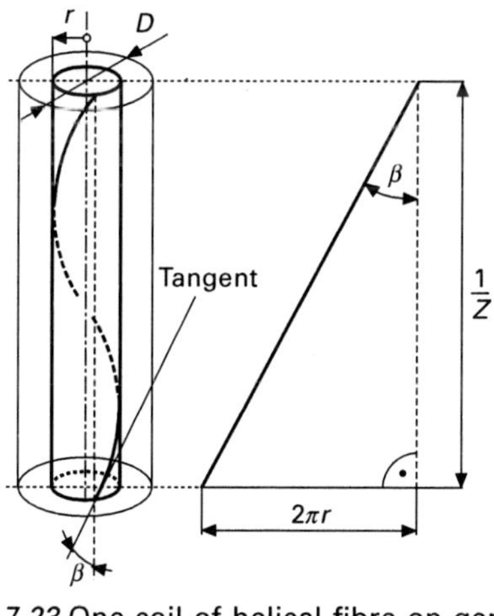

7.23 One coil of helical fibre on general radius *r*.

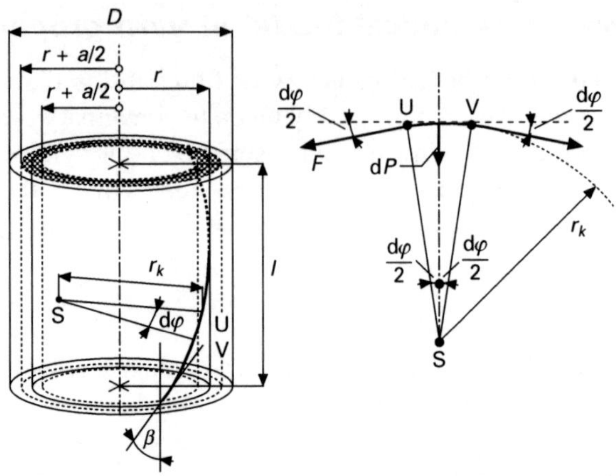

7.24 Force analysis for unit volume of fibre.

$$P_1 = \frac{dP}{dV} = \frac{F d\varphi}{(r d\varphi s/\sin^2\beta)} = F\,\frac{\sin^2\beta}{rs} \qquad [7.111]$$

7.8.2 Compressing zone

The model described here is based on the idea of a compressing zone. Around the yarn surface, the interaction of fibres is not very intensive, since no fibre in this layer is compressing from the outside. Therefore, the number of contacts, the frictional forces, the packing density and the fibre axial forces F are very small. So, the specific centripetal force P_1 must also be very small. Around the yarn axis, the slope angle β is very small. This means that the curvature of these fibres is also very small, hence the radius of curvature is very high and so the specific centripetal force P_1 must be very small too. Only the compressing zone between these layers can produce a significant centripetal force P_1. This leads to the assumption that the only significant centripetal force is present in the compressing zone between the layers, as shown in Fig. 7.24. The cross-sectional area of the compressing zone is given by

$$\pi(r + a/2)^2 - \pi(r - a/2)^2$$
$$= \pi(r^2 + ra + a^2/4 - r^2 + ra - a^2/4) = 2\pi ra \qquad [7.112]$$

The total volume of the compressing zone is hence

$$V_{c.a} = 2\pi ral \qquad [7.113]$$

The fibre volume in the compression zone is then

$$V_a = V_{c,a}\mu = 2\pi ral\mu \qquad [7.114]$$

The total centripetal force in the compressing zone is hence given by

$$P = P_1 V_a = \left[\frac{F \sin^2\beta}{rs}\right] 2\pi ral\mu = \frac{F \sin^2\beta 2\pi al\mu}{s} \quad [7.115]$$

For the following conclusions, we assume that the centripetal force *P* acts on the cylinder at radius *r*, hence all partial forces at radius *r* are centralized (Fig. 7.25). The surface area of cylinder with radius *r* is given by

$$A = 2\pi rl \quad [7.116]$$

The pressure developed in the compressing zone is then

$$p = \frac{P}{A} = \frac{F \sin^2\beta 2\pi al\mu/s}{2\pi rl} = \frac{F \sin^2\beta}{rs} a\mu \quad [7.117]$$

The normal stress is hence

$$\sigma = \frac{F_a}{s^*} = \frac{F \cos\beta}{s/\cos\beta} = \frac{F}{s} \cos^2 \beta \quad [7.118]$$

thus

$$F = \frac{\sigma s}{\cos^2\beta} \quad [7.119]$$

From equation [7.116] we get then after rearrangement

$$p = \frac{2\sigma a\mu}{D_{S/\sqrt{\mu}}} \kappa^2\left(\frac{2r}{D}\right) = 8\pi\sigma\left(\frac{a}{d}\right)\left(\frac{2r}{D}\right) \sqrt{\mu}\, \frac{\alpha_s^2}{\sqrt{\tau}} \quad [7.120]$$

where $\tau = D_S/d$.

Summarizing the constant parameters with

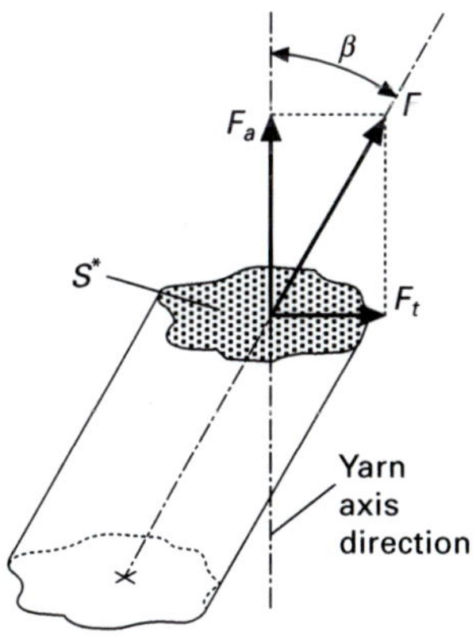

7.25 Forces resolution of fibre element in yarn.

$$C = 8\pi\sigma\left(\frac{a}{d}\right)\left(\frac{2r}{D}\right) \qquad [7.121]$$

we get

$$p = C\sqrt{\mu}\,\frac{\alpha_s^2}{\sqrt{\tau}} \qquad [7.122]$$

The value C depends on the three quantities: $2r/D$, σ and a/d. The position of the compressing zones may be relatively the same with regard to yarn surface cylinder and so $2r/D$ is perhaps constant. The axial stress σ, acting on the yarn substance cross-section, is determined by the centrifugal forces and the tensioning of the fibres during twisting. This stress is perhaps constant too. It is very difficult to comment on the relative thickness of the compressing zone a/d. There is not yet a clear enough physical imagination about it, but from practical experiences this value may be considered as a constant too. (This is an open problem for the future.)

It can be assumed that the quantity C is a parameter that depends on the fibrous material and spinning technology used. The derived equation hence expresses pressure p, as a result of the yarn geometry. Simultaneously, the pressure p is a function of the packing density (compression of fibrous assembly) according to

$$p = k_p b\,\frac{\mu^3}{[1-(\mu/\mu_m)^{2+a}]^3} \qquad [7.123]$$

From equations [7.122] and [7.123], we get the following equivalence:

$$k_p b\,\frac{\mu^3}{[1-(\mu/\mu_m)^{2+a}]^3} = C\sqrt{\mu}\,\frac{\alpha_s^2}{\sqrt{\tau}} \qquad [7.124]$$

The packing density can hence be expressed as a specific function of the yarn geometry;

$$\frac{\mu^{2.5}}{[1-(\mu/\mu_m)^{2+a}]^3} = \left(C\,\frac{d\sqrt{\pi}}{2k_p b\sqrt{\rho}}\right)(ZT^{1/4})^2 \qquad [7.125]$$

The parameter

$$Q = C\,\frac{d\sqrt{\pi}}{2k_p b\sqrt{\rho}} \qquad [7.126]$$

depends on the material and the technology used. Hence, we get the final equation for the packing density

$$\frac{\mu^{2.5}}{[1-(\mu/\mu_m)^{2+a}]^3} = Q(ZT^{1/4})^2 \qquad [7.127]$$

Note: μ here is a function of $ZT^{1/4}$, whereas Koechlin's exponent is ½.

The parameter a in equation [7.127] is usually equal to 1 and μ_m is usually equal to 0.8 (influence of yarn surface). Typical values for Q are summarized in Table 7.5. Example: for carded cotton yarn of $T = 29.5$ tex, $Z = 719.43\,\text{m}^{-1}$ ($\alpha = 123.57\,\text{m}^{-1}\,\text{ktex}^{-1/2}$), the right-hand side of the derived equation is 0.27016 and this corresponds to $\mu = 0.460$ on the left-hand side (calculated by numerical method). Then, the yarn diameter is $D = 0.232$ mm.

A problem in the practical application of this equation is the determination of the root μ. This can be solved by applying a numerical method (e.g. interval splitting method) or by approximating the value if the packing density of the yarns is around μ^* as follows:
With

$$b = 3\,\frac{1+2(\mu^*/0.8)^3}{1-(\mu^*/0.8)^3} \text{ and } c = \frac{1}{[1-(\mu^*/0.8)^3]^3(\mu^*)^{b-3}} \qquad [7.128\text{a, b}]$$

we get the packing density to

$$\mu = (Q/c)^{\frac{1}{b-0.5}}(ZT^{1/4})^{\frac{2}{b-0.5}} \qquad [7.129]$$

The yarn diameter can then be calculated with

$$D[\text{mm}] = \frac{2\times 1000^{\overline{2(b-0.5)}^{\,1}}}{\sqrt{\pi}\sqrt{\rho}(Q/c)^{\frac{1}{2(b-0.5)}}}\,T^{\frac{b}{2(b-0.5)}}\,\alpha^{\frac{1}{(b-0.5)}} \qquad [7.130]$$

with $\rho[g/\text{cm}^3]$, $Q[\text{m}^2\,\text{tex}^{-0.5}]$, T [tex], α [$\text{m}^{-1}\text{ketx}^{1/2}$].

Example: let $Q = 9.61 \times 10^{-8}\,\text{m}^2\,\text{tex}^{-1/2}$, $\rho = 1520\,\text{kg}\,\text{m}^{-3}$ and $\mu^* = 0.46$. Then we get $b = 5.11261$, $c = 9.70918$. For an approximated μ we have

$$\mu = 0.018393 \times Z^{0.43359}\,[1/\text{m}]T^{0.10840}\,[\text{tex}]$$

Table 7.5 Values of parameter Q ($\text{m}^2\text{tex}^{-1/2}$; $\mu_m = 0.8$)

Fibrous material		Spinning technology		
Type	Density ρ ($\text{kg}\,\text{m}^{-3}$)	Combed	Carded	Open end (OE) type BD
Cotton	1520	1.46×10^{-7}	9.61×10^{-8}	6.18×10^{-8}
Viscose, C-type	1500	4.12×10^{-7}		1.76×10^{-7}
PET, C-type	1360	2.98×10^{-7}		1.29×10^{-7}
Wool	1310	2.16×10^{-7}	1.20×10^{-7}	6.49×10^{-8}

and for the approximate D we get

$$D[\text{mm}] = 0.10093 \times T^{0.55420}\ [\text{tex}]\ \alpha^{-0.21680}\ [\text{m}^{-1}\ \text{ktex}^{1/2}]$$

Experiments were undertaken on carded cotton yarns and roving (Fig. 7.26). Yarn count ranges were from 16.5 to 100 tex, twist factor ranged from 83 to 140 m^{-1} $\text{ktex}^{1/2}$. Equation [7.125] helps in deriving the relation for a suitable yarn twist.

7.8.3 Suitable yarn twist

In order to determine a suitable yarn twist, the model described above together with a more precise form of the theory according to Koechlin is applied. The slope of surface fibres is then not characterized by the angle β_D at a diameter D of the cylinder but by the slope of surface fibre axis, e.g. the angle β' at a diameter D' on the cylinder as shown in Fig. 7.27. The modified diameter D' characterizes the position of the axes of surface fibres, with $D' < D$. Following this we have

$$C_D = \frac{D'}{D} \qquad [7.131]$$

In case of the theoretical limit structure (Fig. 7.27), we have

$$D - D' = d \qquad [7.132]$$

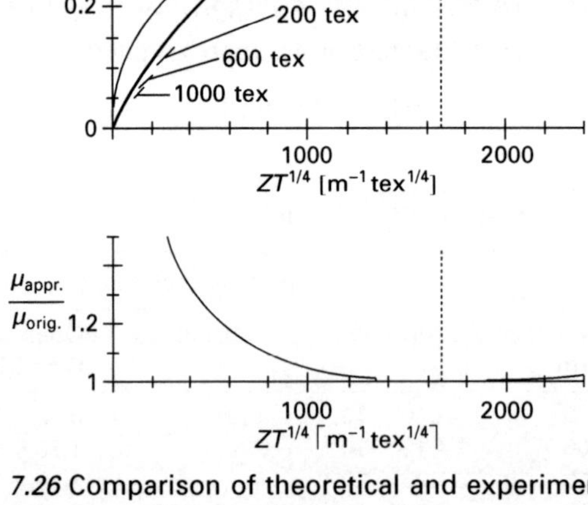

7.26 Comparison of theoretical and experimental valus for carded cotton yarns and rovings.

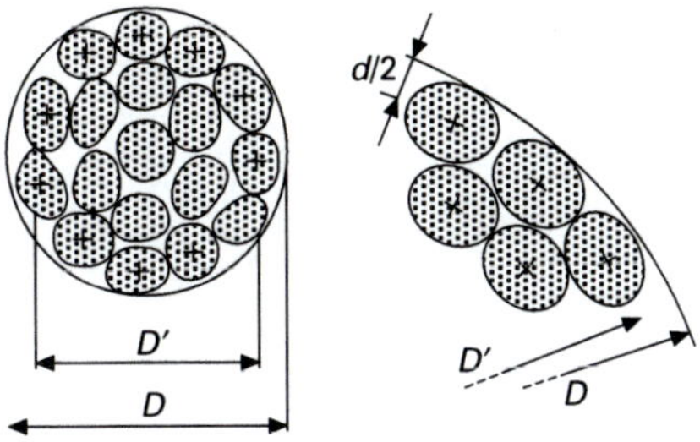

7.27 Definition of a modified yarn diameter.

Inserting [7.132] into [7.131] leads to

$$C_D = \frac{D'}{D} = \frac{D - 2(d/2)}{D} = 1 - \frac{2(d/2)}{\underbrace{D}_{= \sqrt{4S/(\pi\mu_{\text{lim}})}}} = 1 - \frac{2(d/2)}{\sqrt{4 \underbrace{S}_{= \pi D_S^2 (4)} \Big/ (\pi\mu_{\text{lim}})}}$$

$$= 1 - \sqrt{\mu_{\text{lim}}}\, d/D_S = 1 - \sqrt{\underbrace{\mu_{\text{lim}}}_{= \pi/(2\sqrt{3})} \Big/ \tau} = 1 - \frac{0.952}{\sqrt{\tau}}$$

[7.133]

Hence we get

$$C_D \cong 1 - \sqrt{\frac{t}{T}} \quad [7.134]$$

The real yarn has a diameter D and a proportionally longer modified diameter D' and therefore the value of the Schwarz constant is the same. The modified angle β' (slope of fibre axes on yarn surface) is then given by

$$\tan \beta' = 2\pi(D'/2)Z = \pi D'Z = \pi \overbrace{DZ}^{=\kappa}\, \overbrace{D'/D}^{=C_D} = \kappa\, C_D \quad [7.135]$$

Rearrangement of equation [7.127] leads to

$$\frac{\mu^{2.5}}{[1 - (\mu/\mu_{\text{lim}})^{2+a}]^3} = Q(ZT^{1/4})^2 = Q\left(\frac{\overbrace{\pi DZ}^{=\kappa}}{\pi \underbrace{D}_{= \sqrt{4T/(\pi\mu\rho)}}} T^{1/4} \right)^2 = Q\left(\frac{\kappa}{\pi \sqrt{\dfrac{4T}{\pi\mu\rho}}} T^{1/4} \right)^2$$

[7.136]

$$\frac{\mu^{1.5}}{[1-(\mu/\mu_{\lim})^{2+a}]^3} = Q\,\frac{\overbrace{\kappa^2}^{=\tan\beta'/C_D}\ \mu\rho}{4\pi\sqrt{T}} = Q\,\frac{\tan^2\beta'\rho}{4\pi\sqrt{T}\ \underbrace{(C_D)^2}_{=1-\sqrt{t/T}}}$$

$$= \left(Q\,\frac{\tan^2\beta'\rho}{4\pi}\right)\frac{1}{\sqrt{T}}\,\frac{1}{\left(1-\sqrt{t/T}\right)^2} \qquad [7.137]$$

We define the quantity

$$R = Q\,\frac{\tan^2\beta'\rho}{4\pi} \qquad [7.138]$$

and therefore

$$\frac{\mu^{1.5}}{[1-(\mu/\mu_{\lim})^{2+a}]^3} = \frac{R}{\sqrt{T}\left(1-\sqrt{t/T}\right)^2} \qquad [7.139]$$

In order to determine the quantity R, there are two possibilities. First, yarns are from the same material and of the same end use. Then we can distinguish between case 1 (same technology and different yarn counts) and case 2 (different technologies and same yarn count):

- For case 1, the modified angle β' (slope of axes of peripheral fibres) should be constant, hence, R = constant.
- For case 2, the contact density υ (it is valid $\upsilon = k_\upsilon \mu^2$) should be constant, hence μ = constant and the left part of equation [7.138] is constant and therefore R = constant.

In general, we have R = constant for all yarn counts and technologies. The parameter a in equation [7.139] is usually equal to 1 and $\mu_{\lim}$ is usually equal to 0.8. Typical values for R are given in Table 7.6.

In order to illustrate the practical use of equations [7.127] and [7.139], we consider a typical example. Q and R are usually known (Tables 7.6 and 7.7), as are t, ρ as material constants and T as the required yarn count. Hence, the process is as follows:

Table 7.6 Typical values for R

Material	ρ [kg m^{-3}]	R [tex$^{1/2}$]
Cotton – long staple	1520	2.145
Cotton – medium staple	1520	2.737
Viscose, C-type	1500	4.589
PET, C-type	1360	3.563
Wool	1310	2.341

Table 7.7 Other variables and functions to describe fibres and fibre bundles

	One fibre	Fibre bundle
Number of fibres	1	n
Tensile force	S	S_Σ
Force–strain relation	$S = S(\varepsilon)$	$S_\Sigma = S_\Sigma(\varepsilon)$
Strength	P (max. of S)	P_Σ (max. of S_Σ)
Breaking strain	a, $[P = S(a)]$	a_Σ, $[P_\Sigma = S_\Sigma(a_\Sigma)]$

1. Calculate the right-hand side of equation [7.139].
2. Find the packing density μ as a root of equation [7.137] either by calculation or using previously prepared tables.
3. Calculate the left-hand side of equation [7.127].
4. Find the suitable yarn twist Z as a root of equation [7.127].
5. Calculate the yarn diameter according to equation [7.34].

Example: for carded cotton yarn $T = 29.5$ tex, we have $t = 0.16$ tex (medium staple), $\rho = 1520\,\text{kg}\,\text{m}^{-3}$, $Q = 9.61 \times 10^{-8}\,\text{m}^2\,\text{tex}^{-1/2}$, $R = 2.737\,\text{tex}^{1/2}$. The process is:

1. The right-hand side of equation [7.139] is then 0.58723.
2. The root of equation [7.139] is $\mu = 0.460$ (calculated by numerical method).
3. The left-hand side of equation [7.127] is 0.27013.
4. The suitable twist is $Z = 719.38\,\text{m}^{-1}$ from equation [7.127] with $\alpha = 123.56\,\text{m}^{-1}\,\text{ktex}^{1/2}$.
5. The yarn diameter is $D = 0.232$ mm.

A typical problem for a practical application is the root μ in equation [7.139] which needs to be determined. This can either be done using a pre-prepared table, a numerical method (e.g. interval splitting method) or the following approximation of equation [7.139].

In the first step, an average yarn out of a set of produced yarns of the same material (t, ρ, R) and the same technology (Q) is defined. Then the quantities Z^*, T^*, μ^* of the average yarn are determined by applying equations [7.127] and [7.139]. Then, the value b in equation [7.129] is calculated and hence

$$B = \frac{2}{\left(\sqrt{T^*_{\text{tex}}/t_{\text{tex}}}\right) - 1} \qquad [7.140]$$

$$\xi = \frac{b - 0.5}{b - 1.5} \qquad [7.141]$$

$$q = \frac{\xi(1 + B) + 1}{4} \quad [7.142]$$

$$\alpha_{q[\mathrm{m}^{-1}\mathrm{tex}^q]} = Z^*_{[\mathrm{m}^{-1}]}(T^*_{[\mathrm{tex}]})^q \quad [7.143]$$

The suitable twist can then approximated by

$$Z_{[\mathrm{m}^{-1}]} = \frac{\alpha_{q[\mathrm{m}^{-1}\mathrm{tex}^q]}}{T^q_{[\mathrm{tex}]}} \quad [7.144]$$

Example: let $t = 0.16$ tex, $T^* = 29.5$ tex, $Z^* = 719.38\,\mathrm{m}^{-1}$, $\mu^* = 0.460$ (values from previous examples). Then we get $b = 5.11237$, $B = 0.15900$, $\xi = 1.27683$ and $q = 0.61996$. This leads to

$$\alpha_q = 5864\ [\mathrm{m}^{-1}/\mathrm{tex}^{0.61996}],\ Z_{[\mathrm{m}^{-1}]} = 5864/T^{0.61996}_{[\mathrm{tex}]} \text{ and } \left(\alpha_{[\mathrm{m}^{-1}\mathrm{ktex}^{1/2}]} = 5864\left[\frac{T^{0.11996}_{[\mathrm{tex}]}}{\sqrt{1000}}\right]\right)$$

The results are illustrated in Fig. 7.28.

7.9 Mechanics of parallel fibre bundles

Bundles of parallel and more or less independent fibres create an idealized yarn. This arrangement applies to all sorts of staple and filament yarns, fibre and yarn bundles, e.g. ropes, but also warp yarns for weaving etc. The mathematical relationships underpinning the mechanical behaviour of such bundles help to determine the principal properties of these yarn types. Knowledge of these relationships is a prerequisite to solving a range of

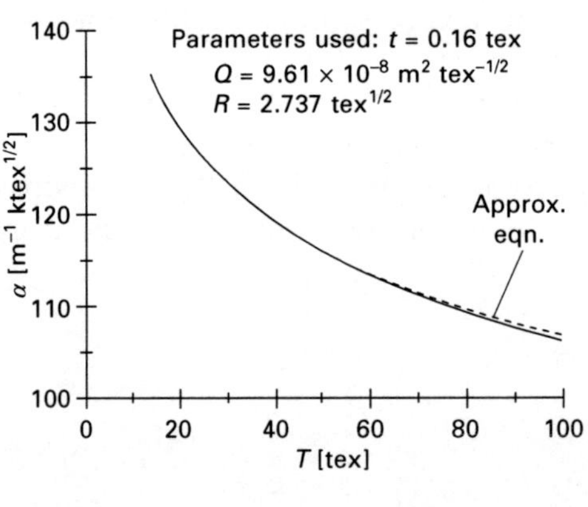

7.28 Results of example (α vs. *T*).

special mechanical textile-related problems. The following assumptions are normally made:

- the number of fibres is large;
- the fibres are straight (linear);
- each fibre is gripped by both jaws;
- fibres are mutually parallel;
- fibres are mechanically independent of each other.

The 'fibre strength' is the maximum tensile force it can bear and the 'breaking strain' stands for the strain at that point.

Common variables for one fibre and fibre bundle are shown in Fig. 7.29, where h stands for gauge length and ε represents the strain (relative elongation). In the case where all fibres have the same stress–strain curves $S = S(\varepsilon)$, the same strength P and the same breaking strain a, we have

$$S_\Sigma(\varepsilon) = nS(\varepsilon),\ P_\Sigma = nP,\ a_\Sigma = a \qquad [7.145]$$

Applying the blending theory according to Hamburger (1949), the fibre bundle is regarded as a blend of two types of fibre (Fig. 7.30) and all fibres

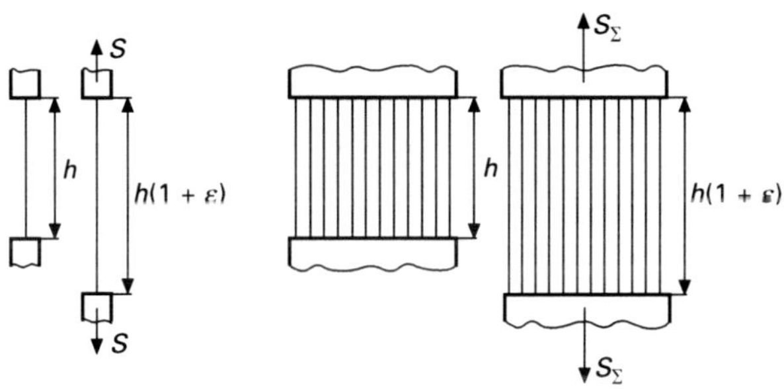

7.29 Common variables to describe fibre bundles.

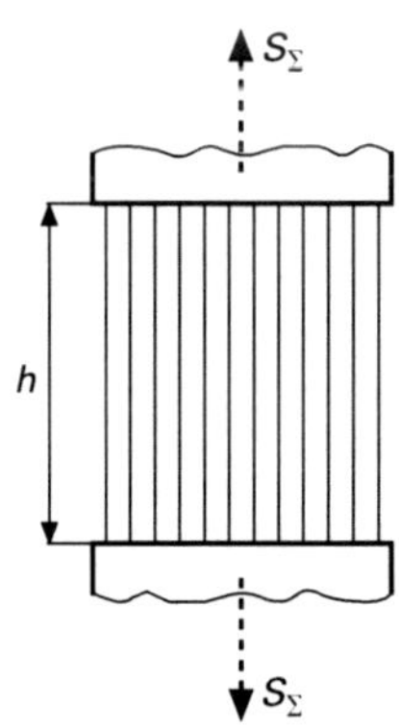

7.30 Fibre bundle consisting of two components.

of one type have the same force–strain curve $S = S(\varepsilon)$, the same strength P and the same breaking strain a.

In the following discussion, the fibre of one type having a smaller value of the breaking strain is denoted as No. 1, the other type of fibres is denoted as No. 2. (These numbers are used as subscripts, see Table 7.8.) For fibre No. 1, we have

$$m_1 = g_1 m,\ t_1 = \frac{m_1}{n_1 h},\ n_1 = \frac{m_1}{t_1 h} = \left(\frac{g_1}{t_1}\right)\overbrace{\left(\frac{m}{h}\right)}^{=T} \qquad [7.146]$$

hence

$$n_1 = g_1\left(\frac{T}{t_1}\right) \qquad [7.147]$$

For fibre No. 2, we get in a similar way

$$n_2 = g_2\left(\frac{T}{t_2}\right) \qquad [7.148]$$

The maximum forces in a bundle are shown in the stress–strain curves in Fig. 7.31. For the interval $\varepsilon \le a_1 \Rightarrow$ max. at $\varepsilon = a_1$, which results in

$$S_\Sigma(a_1) = n_1 P_1 + n_2 S_2(a_1) \qquad [7.149]$$

and hence

$$S_\Sigma(a_1) = T\left[\frac{g_1 P_1}{t_1} + \frac{g_2 S_2(a_1)}{t_2}\right] \qquad [7.150]$$

For the interval $\varepsilon \in (a_1, a_2\rangle \Rightarrow$ max. at $\varepsilon = a_2$, which results in

$$S_\Sigma(a_2) = n_1 \cdot 0 + n_2 P_2 \qquad [7.151]$$

Table 7.8 Correct marking of two components parameters

	Fibre material	
	No. 1	No. 2
Fibre fineness	t_1	t_2
Force–strain relation	$S_1(\varepsilon)$	$S_2(\varepsilon)$
Breaking strain of fibre	$a_1 \le a_2$	
Fibre strength	$P_1 = S_1(a_1)$	$P_2 = S_2(a_2)$
Number of fibres	n_1	n_2
Total number of fibres	$n = n_1 + n_2$	
Mass of fibres	m_1	m_2
Total mass of fibres	$m = m_1 + m_2$	
Bundle fineness (count)	$T = m/h$	
Mass portion	$g_1 = m_1/m$	$g_1 = m_1/m$
Sum of mass portions	$g_1 + g_2 = 1$	

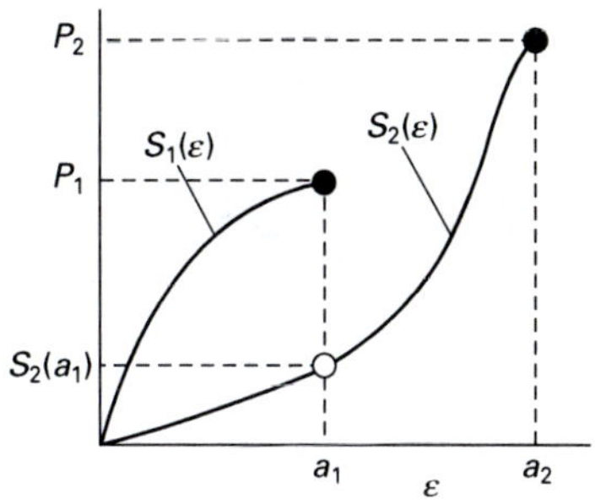

7.31 Force–strain curves of two types of fibre.

and hence

$$S_\Sigma(a_2) = \frac{Tg_2P_2}{t_2} \qquad [7.152]$$

In the interval $\varepsilon > a_2 \Rightarrow$ all fibres are broken:

$$S_\Sigma(\varepsilon > a_2) = 0 \qquad [7.153]$$

The strength of the fibre bundle can then be calculated with

$$\begin{aligned} P_\Sigma &= \max[S_\Sigma(a_1), S_\Sigma(a_2)] \\ &= T \max\left\{\left[g_1\frac{P_1}{t_1} + g_2\frac{S_2(a_1)}{t_2}\right], \left[g_2\frac{P_2}{t_2}\right]\right\} \end{aligned} \qquad [7.154]$$

where P_1/t_1 stands for the tenacity of fibre No. 1 (e.g. N/tex), P_2/t_2 stands for the tenacity of fibre No. 2 (e.g. N/tex) and $S_2(a_1)/t_2$ is the specific stress of fibre No. 2 (e.g. N/tex) at $\varepsilon = a_1$. The bundle tenacity P_Σ/T is then given by

$$\frac{P_\Sigma}{T} \max\left\{\left[g_1\frac{P_1}{t_1} + g_2\frac{S_2(a_1)}{t_2}\right], \left[g_2\frac{P_2}{t_2}\right]\right\} \qquad [7.155]$$

and the breaking strain of the bundle can be calculated with

$$a_\Sigma = a_2, \text{ if } \frac{P_\Sigma}{T} = g_1 \frac{P_1}{t_1} + g_2 \frac{S_2(a_1)}{t_2} \qquad [7.156]$$

or

$$a_\Sigma = a_2, \text{ if } \frac{P_\Sigma}{T} = g_1 \frac{P_2}{t_2} \qquad [7.157]$$

Figure 7.32 shows a graphical representation of the resulting equation [7.158]:

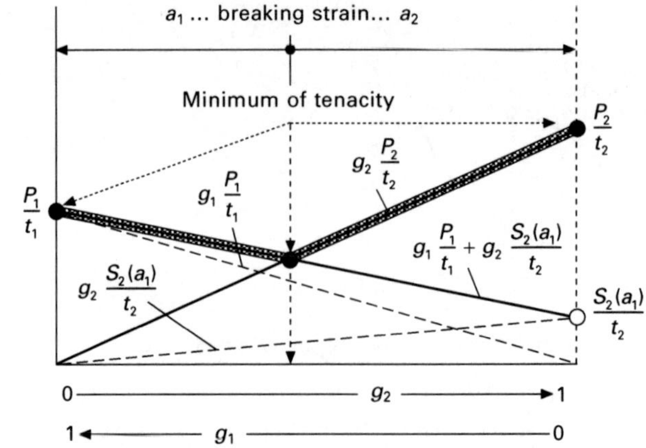

7.32 Graphical representation of equation [7.158].

$$\frac{P_\Sigma}{T} = \max\left\{\left[g_1\frac{P_1}{t_1} + g_2\frac{S_2(a_1)}{t_2}\right], \left[g_2\frac{P_2}{t_2}\right]\right\} \quad [7.158]$$

For the minimum bundle tenacity, there are two possibilities according to Fig. 7.32:

(i) $$g_2 = 0, \frac{P_\Sigma}{T} = \frac{P_1}{t_1} \quad [7.159]$$

(ii) $$\overbrace{g_1}^{=1-g_2} \frac{P_1}{t_1} + g_2\frac{S_2(a_1)}{t_2} = g_2\frac{P_2}{t_2} \text{ and } \frac{P_1}{t_1} = g_2\frac{P_1}{t_1} + g_2\frac{P_2}{t_2} - g_2\frac{S_2(a_1)}{t_2} \quad [7.160]$$

hence

$$g_2 = \frac{P_1/t_1}{P_1/t_1 + P_2/t_2 - S_2(a_1)/t_2} \quad [7.161]$$

and using this value we get

$$\frac{P_\Sigma}{T} = g_2\frac{P_2}{t_2} \quad [7.162]$$

Now, the minimum bundle tenacity is the minimum of the three calculated values P_Σ/T. It is important to note that after adding fibres with a higher tenacity, the tenacity of the resulting bundle can decrease! This theory can also be applied for a rough estimation of combination staple fibre yarns. In this case, it is necessary to use analogous values of the corresponding individual component yarns in place of the fibre parameters. In this case, P_1/t_1 then stands for the tenacity of the first single yarn (100% material No. 1) and P_2/t_2 for the tenacity of the second yarn (100% material No. 2). $S_2(a_1)/t_2$ is the specific stress of the first yarn (100% material No. 2), when the strain is equal to its breaking strain, i.e. ($\varepsilon = a_1$); all, e.g., in N/tex. A more detailed description of this model can be found in Neckar and Das (2006).

7.10 Examples of models of fibrous assemblies and yarns

In Keefe *et al.* (1992) and Keefe (1994), a mathematical model for fibrous assemblies is developed where yarns are regarded as three-dimensional structures. In the earlier model, the yarns are considered to be inextensible and incompressible whereas in the later model, this limitation is overcome. A rope-like structure can thus be described and analysed composed of a twisted bundle of individual yarns.

Ghoreishi *et al.* (2007a) describe a model for synthetic fibre ropes. They develop one model for an assembly of a large number of twisted components and another one for a structure with a straight core and six helical wires. This 1+6 model is then used to predict the axial stiffness of the rope and the correlation to measured values is good. The mechanical behaviour under tension and torsion is described in a subsequent paper (Ghoreishi *et al.*, 2007b). The tension–torsion coupling behaviour of the constituents is used to deduce the behaviour of the rope. A comparison with measurements shows the validity of the model.

Carnaby and Postle (1991) discuss different mechanisms of fibre deformation and slippage which contribute to the stored elastic strain energy and work done against friction during the deformation of staple-fibre yarns. They assume continuity in the strain and energy fields and thus suggest using a unit cell of fibres in order to describe the mechanical behaviour of a yarn.

Liu *et al.* (2007) designed a model for wool yarns. It is based on discrete-fibre modelling principles, hence the yarn is considered as an assembly of a large number of discrete fibres. The movement of fibres during deformation and their final positions after deformation are determined by their need to minimize the tensile strain. The dependency of the yarn strain on fibre tension, bending and torsion is derived. A comparison with experimental data shows an acceptable accuracy of the predicted values.

An earlier model to predict the strength of yarns based on the fibre properties is described in Iwona (1992). The mechanisms of yarn failure depending on the interfibre friction is also analysed in Broughton *et al.* (1992). They found that even slight changes in the interfibre friction can lead to large changes in the yarn strength.

Phillips *et al.* (2010) investigated the mechanical governing equations that describe the torsional behaviour of multi-ply yarns under tension. They found that the slope or gradient of the torque due to tension (Fig. 7.8), the fibre number, moduli and diameter have the greatest influence on the torsional properties. The tensile properties of textured yarns are discussed in Kawabata and Sasai (1978). They apply a model where each fibre contributes independently to the yarn's tensile property. Then, they model the tensile property of a crimped single fibre (helical coil). Applying this to determine the tensile property of a textured yarn shows a good correlation with measured values.

A comprehensive list of many more interesting papers on the subject of this chapter can be found in Neckář and Das (2012).

7.11 Acknowledgement

This chapter was supported by a Czech grant GACR No. 106/09/1916.

7.12 References

Braschler, E. (1939), *Strength of cotton yarn*, Gizlegprom, Moscow.

Broughton, R. M., El Mogahzy, Y., Hali, D.M. (1992), 'Mechanism of yarn failure', *Textile Research Journal, 62:3*, 131–134.

Carnaby, G.A., Postle, R. (1991), 'Discrete fiber versus continuum models in the mechanics of staple yarns', *Journal of Applied Polymer Science: Applied Polymer Symposium, 47*, 341–354.

Gegauff, C.M. (1907), 'Strength and elasticity of cotton threads', *Bull. Soc. Industr. Mulhouse 177*, 150.

Ghoreishi, S.R., Cartraud, P., Davies, P., Messager, T. (2007a), 'Analytical modeling of synthetic fiber ropes subjected to axial loads. Part I: A new continuum model for multilayered fibrous structures', *International Journal of Solids and Structures, 44: 9*, 2924–2942.

Ghoreishi, S.R., Cartraud, P., Davies, P., Messager, T. (2007b), 'Analytical modeling of synthetic fiber ropes subjected to axial loads. Part II: A linear elastic model for 1+6 fibrous structures', *International Journal of Solids and Structures, 44:9*, 2943–2960.

Hamburger, W. J. (1949), 'The industrial application of the stress–strain relationship', *Journal of the Textile Institute 40*, 700–718.

Hearle, J. W. S., Grosberg, P., Backer, S. (1969), *Structural Mechanics of Fibres, Yarns and Fabrics*, Cambridge.

Iwona, F. (1992), 'New approach for predicting strength properties of yarn', *Textile Research Journal, 62:6*, 340–348.

Kawabata, S., Sasai, T. (1978), 'Theoretical analysis of tensile properties of textured yarns', *Journal of the Textile Machinery Society of Japan, 24:1*, 13–18.

Keefe, M., Edwards, D.C., Yang, J. (1992), 'Solid modeling of yarn and fiber assemblies', *Journal of the Textile Institute, 83:2*, 185–196.

Keefe, M. (1994), 'Solid modeling applied to fibrous assemblies part I: twisted yarns', *Journal of the Textile Institute, 85:3*, 338–349.

Liu, T., Choi, K.F., Li, Y. (2007), 'Mechanical modeling of singles yarn', *Textile Research Journal, 77:3*, 123–130.

Malinowska, K. (1975), *Evaluation of Shape Coarseness of Fibre Cross Sections*, Przeglad Włokienniczy, 4.

Neckář, B. (1990), *Yarns: Creation, structure, properties*, SNTL Prague.

Neckář, B. , Das, D., (2006) 'Mechanics of parallel fibre bundles', *Fibres & Textiles in Eastern Europe 14:3*, 23–28.

Neckář, B., Das, D. (2012), *Theory of Structure and Mechanics of Fibrous Assemblies and Yarn*, CRC Press.

Phillips, D.G., Tran, C.-D., Fraser, W.B., van der Heijden, G.H.M. (2010), 'Torsional properties of staple fibre plied yarns', *Journal of the Textile Institute, 101: 7*, 595–612.

8
Simulation of wound packages, woven, braided and knitted structures

Y. K. KYOSEV, Hochschule Niederrhein – University of Applied Sciences, Germany

Abstract: This chapter presents the basic methods for the modelling and simulation of woven, braided and knitted fabrics and wound packages. All these structures are built by yarns and thus there are several common aspects during their modelling. Emphasis is given to the geometrical methods for the calculation of the yarn paths of these structures. Some possible extensions of the models for consideration of the mechanical interactions between the yarns are discussed and examples of the capabilities of some commercially available CAD systems for design and simulation of textile structures are presented.

Key words: woven, knitted, braided structures, wound packages, simulation, yarn-based structures.

8.1 Introduction: objectives of simulating yarn-based structures

This chapter discusses yarn-based structures, also called secondary fibrous assemblies, which consist of one or more types of primary fibrous assemblies. During the simulation of such structures, the most common and effective approach at present is to regard the yarn as a body with certain material parameters and not as a structure. This simplifies and speeds up simulations significantly. Emphasis is given to geometrical methods for simulation of woven and knitted structures. The simulation of braided structures is presented in two ways. From the structural view, they are similar to woven structures; hence, the same geometric methods can be used for their simulation. For process simulation, separate methods have to be used. Figure 8.1 gives an overview of the classification of the most common textile structures.

The simulation of textile structures aims at the prediction of their properties and behaviour prior to production of the structures themselves. Depending on the application, the important parameters (e.g. geometrical, mechanical, thermal, acoustic etc.) of the structure have to be known and selected so that the customer requirements are met in the best possible way. This is the first objective of the simulation of textile structures – to predict the structure's properties, giving some approximation for their values, e.g.

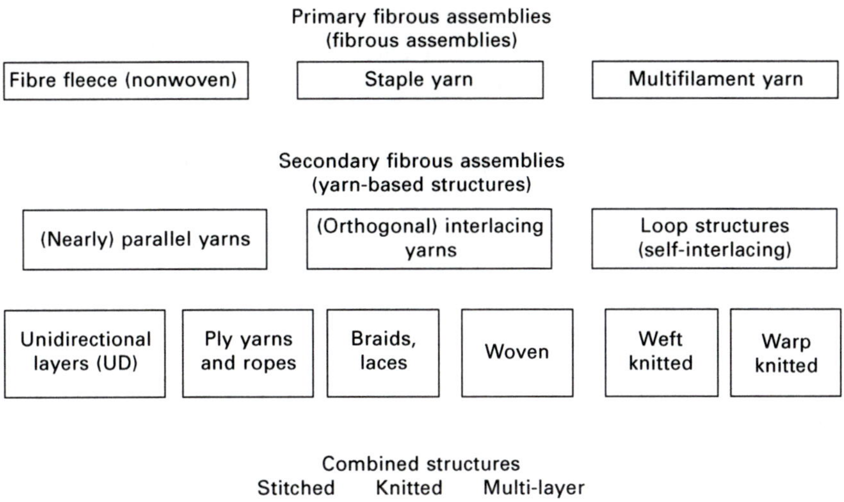

8.1 Classification of textile structures.

8.2 Example of a convertible roof: (a) as a protection against the rain only the water resistance of the structure is required; (b) to be happy with the folding, the complete folding process has to be investigated.

elasticity modulus, area density, porosity. A typical example is shown in Fig. 8.2: if the calculated water permeability of the convertible roof of the car is lower than the required value, then the roof will most likely be able to shield the passengers in the car from the rain.

In more complex cases, such as the folding of a cabrio roof, single parameters like water permeability, bending ridigity and thickness alone are not sufficient to make an accurate prediction whether the folding will work properly. In these cases, the complete folding behaviour has to be investigated, for instance using numerical simulation. This explains the second (final) objective of the simulation – to be a tool for the investigation of the complete behaviour of textile structures under certain conditions. The simulation of the complex behaviour of textiles requires information about their structure. In the case of yarn-based structures the most important aspect is yarn geometry. The interlacing between the yarns determines the behaviour, the freedom of the motion and the friction between the yarns.

Feeding the geometry of the yarns into software packages, different simulations are possible:

- with the finite element method (FEM), mechanical, thermal, acoustic and other simulations are possible;
- with computational fluid dynamics (CFD) software, the simulation of the flow of air or moisture through textiles is possible;
- applying dynamics software (mass-spring systems, particle models, etc.), drape and motion behaviour can be simulated.

Because all these simulations cannot be performed without defining the geometry of the relevant structures, the geometry of yarn structures is the main objective of this chapter.

8.2 Principles of geometrical and mechanical modelling

The simulation of textile structures is based on establishing a model to represent the most important behaviour of the structure. In the case of *experimental* simulations, the working conditions are reproduced using a real model, e.g. of a car, which will be crashed against a wall to simulate impact behaviour. Such simulations are not the objective of this chapter. Here, only computer simulations based on mathematical models are considered. The mathematical models can represent the equations of deterministic physical models (e.g. FEM, mass-point systems, beams), heuristic models, e.g. fuzzy logic, neural nets, genetic algorithms and others, or a combination of these methods.

Heuristic models, based on fuzzy logic and neural nets are described in separate chapters of this book. These can also be applied for the simulation of fabrics. However, their design for different textile structures is in most cases connected with large sets of trials by the researcher, until a suitable neural net architecture or membership functions for a fuzzy logic system are found.

This chapter will therefore concentrate on deterministic models of textile structures (Fig. 8.3). The common way to build simulation models for all yarn-based structures starts with the design of the yarn geometry. There are different ways of constructing this yarn geometry:

- reconstruction from micro-computer tomography images, microtom images, or a series of cross-sections;
- (topological) generation of the geometry, based on the information of the interconnections between the yarns;
- simulation of the production process and calculation of the resulting yarn geometry.

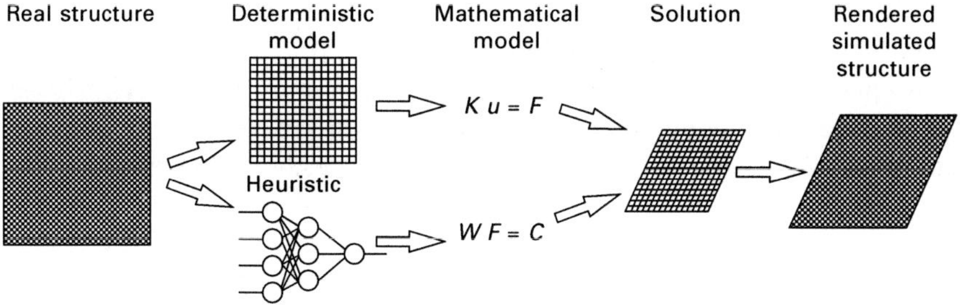

8.3 Real structure, mechanical or heuristic model, mathematical model, simulated product.

Table 8.1 Methods of geometry generation

	Advantages	Disadvantages
Image reconstruction, image analysis	Very exact path reconstruction Suitable for verification of the results from the other models	Very time and/or cost intensive, Requires special techniques Results are not applicable to other structures Very small pieces of the structure can be analysed
Production process simulation	Can produce any kind of patterns with nearly exact path	Computationally intensive Large and complex models – the entire machine and the yarns have to be modelled Requires time, especially for large numbers of repeats In several cases, the resulting accuracy is not needed
Topological generation	Suitable for building parametric models for similar structures, fast and inexpensive	Not very exact path reconstruction, requires (mechanical) corrections Limited to the implemented structures

All these approaches are widely used and applied in different cases, depending on the final goal. A brief comparison between these methods is presented in Table 8.1.

The use of image recognition is not really connected to any kind of textile structure (Wirjadi, 2009; Wirjadi *et al.*, 2009). Its algorithms work very well for a wide range of structures, if the device that makes the images is able to distinguish the fibres or yarns from the surroundings so that, during the processing of these pictures, these fibres or yarns can be described. One of the difficulties of the use of such models is the detection of the boundaries between the single yarns, especially in those cases where the tows are parallel. The methods used in these cases are extensively discussed in books on image processing and are hence not a topic in this chapter. An example of a MicroCT is of braided structure, used to verify the simulation of the geometry of a rope is shown in Fig. 8.4.

Another method for the creation of the geometry of yarn structures is based on an integrated simulation of the entire production process. This requires understanding the yarns, the motion of the main machine elements and how they interact with the yarns (needles, heald frame drives, beat up, etc.) and an accurate description of the contact pairs (yarn to element, yarn to yarn). In this case, several effects of the production process can be investigated and any structure which can be produced on the modelled machine can be generated with the same model. The disadvantage of this approach is its complexity and the large computing time required.

A third way for the simulation of the geometry of textile structures is based on the existing knowledge about the interlacing between the yarns in the structure. In each of the sub-areas, e.g. weaving, knitting and braiding, systems exist to encode these intersections into a pattern. Using this code representation of the pattern, yarn and machine parameters, the structure can be rebuilt in the computer. This method is relatively simple to implement and applicable to all textile structures. The main disadvantage is that the

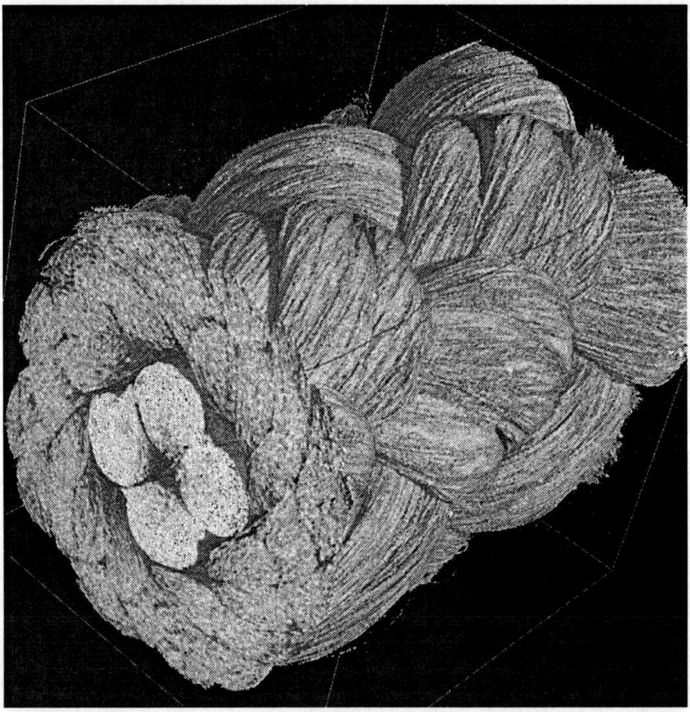

8.4 MicroCT of braided rope (courtesy of Fraunhofer Institut for Techno- and Wirtschaftsmathematik (ITWM) – Kaiserslautern).

geometry is idealized and normally requires additional 'improvement' so that the yarns are put into a position that corresponds more closely to reality. This third method is discussed in the next section.

Independent of which method is used for geometry generation, the yarns are converted into finite element meshes (see Chapter 6). These can then be used for mechanical simulation using FEM or a simulation of the fluid flow through the structure using CFD (see Chapter 5). In the case of impregnation of the structure inside a resin matrix, the homogenization method can be used as well to predict the equivalent elasticity parameters of the resulting structure.

A completely different approach is normally used for the simulation of a textile structure at the macro-scale. In this case, the generation of the yarn geometry is usually omitted and the surface is considered to be a shell or a plate (see Chapter 6) or a mass-spring system. In these cases, the main question is which kind of material properties have to be used in order for the simulated structure to react in a similar way to the original. Some references discussing such simulations are given in Section 8.7.

8.3 Simulation of wound packages

A well-prepared wound package (e.g. a bobbin) can exist as a body without any core. Wound packages are normally not classified as textile structures because they are used mainly as a 'preliminary product' for the production process. There are also several applications where wound packages are the final preform, e.g. in the composites winding of high pressure bottles. Because their properties (e.g. density, density distribution, permeability, stability, stress distribution) are very important for subsequent production processes, their modelling will be considered as well as those of woven, knitted and braided structures. Like other textiles, a bobbin can be simulated in meso-scale (winding level) or in macro-scale (layer by layer).

8.3.1 Simulation of bobbins at the winding level

The relations between the yarn guide motion and the bobbin form have been analysed by, for instance, Efremov and Efremov (1982). Several properties of bobbins in relation to the winding structure are described in detail by Simon and Hübner (1983). A method for calculating the yarn path for the filament winding of composites has been developed (Johansen *et al.*, 1998; Koussios *et al.*, 2004a; Li *et al.*, 2005). Application of this method is reported in Kyosev *et al.* (2006b).

The winding structure (Fig. 8.5b) consists of single windings (Fig. 8.5a), which are stored in the computer as a list of coordinates (r_i, φ_i, z_i) of the approximately equally distributed points of the yarns (Fig. 8.6). These

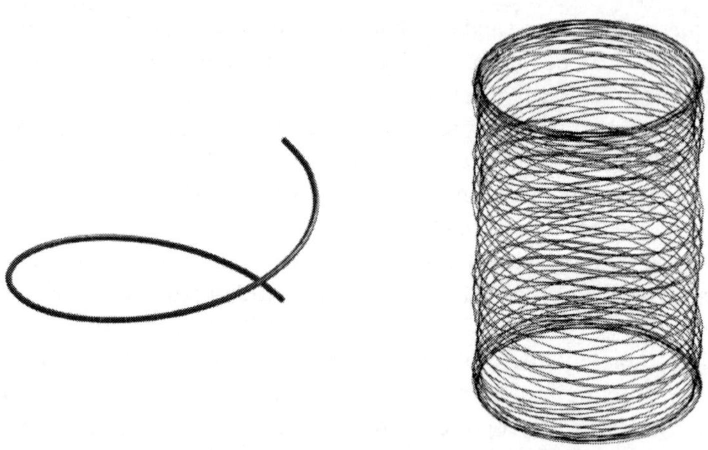

8.5 Single winding (a) and layer of a wound package (b).

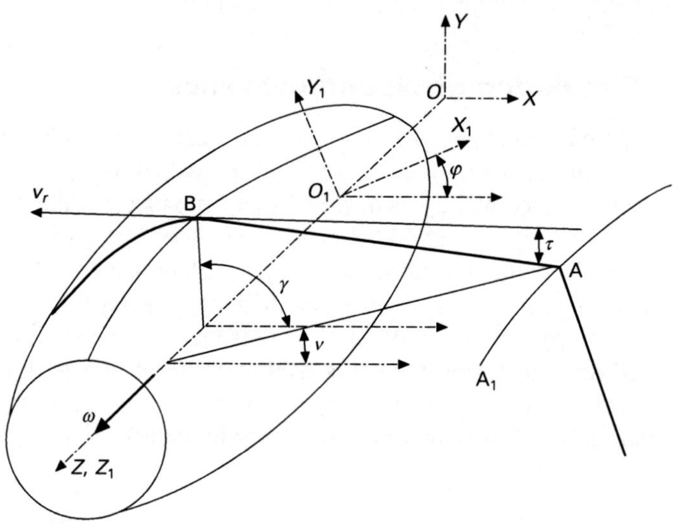

8.6 Geometry of the winding process.

coordinates can be obtained by means of experimental measurement or numerical simulation. Within the numerical approach, the coordinates of winding points are calculated from geometrical considerations about the yarn guide motion and the parameters of the winding body.

The bobbin is assumed to have a certain geometry, which can be described by the rotation of one or more (piecewise) analytically describable plane lines about the *OZ* axis. In the industrial case, this is usually a single line, but there also exist winding forms with a more complex geometry not described here.

The bobbin body rotates about its OZ axis with an angular velocity ω (Fig. 8.6). If the free length AB of the yarn between the yarn guide A and the winding point B is very small, then the winding form is almost completely controlled by the yarn guide, here represented by point A. The influence of this free length can be ignored. It can be assumed that the Z-coordinates of points A and B are the same or the yarn guide moves close to the bobbin surface (AB = 0). In those cases, where the influence of the distance AB needs to be investigated or is more significant, additional geometric relations according to Efremov and Efremov (1982) or Koussios *et al.* (2004a, 2004b) have to be used for the description of the relation between the yarn guide motion and the winding point.

The simplified case (AB = 0) is often preferred in industrial applications, because of the good control over the bobbin. Then, the following simple algorithm can be used:

0. Initialization
   (a) bobbin surface $R(\varphi, z)$ as function of the angle and z-coordinate
   (b) yarn guide motion low $Z(t)$ as function of the time t
   (c) starting winding point (r_1, φ_1, z_1) at start time t_1
1. Calculations. For each actual time $t_i = t_{i-1} + \Delta t$, if $t_i < t_{END}$ calculate:
   (a) the angular position of the current winding point $\varphi_i = \varphi_{i-1} + \omega \cdot \Delta t$, which corresponds to the rotation of the bobbin for the angle $\Delta\varphi = \omega \cdot \Delta t$
   (b) the z-position of the yarn guide eye from its motion low $z_i = Z(t_i)$
   (c) calculate the radius at the winding point $r_i = R(\varphi_i, z_i)$ + yarn diameter, as a radius, corresponds to the bobbin surface to the coordinates (φ_i, z_i)

Figure 8.7 shows the results for three simulated bobbins when using this algorithm, with very small differences in the winding ratio $n_W = n_B/n_G$. According to Simon and Hübner (1983), n_W stands for the number of windings per yarn carriage stroke, represents the bobbins revolutions per minute and n_G is the number of yarn guide cycles per minute.

The ratio $n_W = 7.9808$ is very close to a whole number and can thus lead to a defective package (Fig. 8.7b). The value $n_W = 7.9365$ can be used for controlled (precise) winding if the distance between the yarns corresponds to the yarn thickness (Fig. 8.7a). The ratio $n_W = 7.8496$ results in a layer with random winding used by drum winding (Fig. 8.7c). On the right side of those figures with windings, their cross-sections through a radial surface are shown, where each circle presents a single yarn cut. These cross-sections show the distribution of the yarns in the layers. Inside the simulated layer (this is not valid for the entire body!), the structure presented in Fig. 8.7c has the most uniform distribution of the winding, at lower density. The windings presented in Fig. 8.7a for a controlled package normally have a

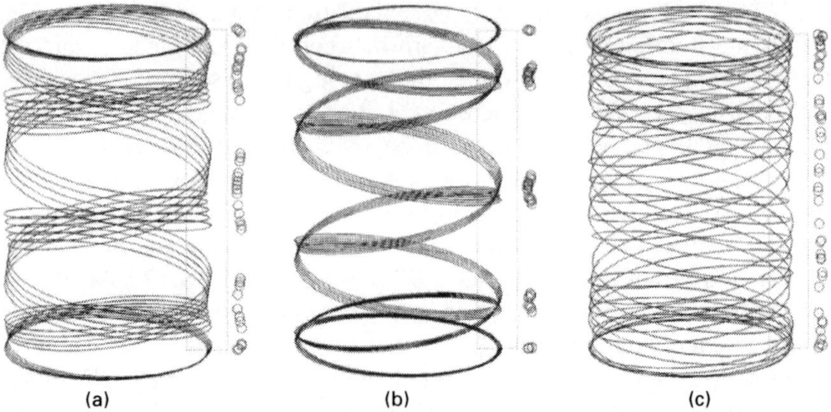

8.7 Examples of simulated winding structures with presentation of the cross-cut section of the layer: (a) precise winding; (b) bad layers – the new windings are positioned directly over the previous ones; and (c) random winding.

higher density compared to the other types. The structure in Fig. 8.7b shows a very high concentration of the windings in very narrow regions, which corresponds to a defect bobbin.

The structures depicted in Fig. 8.7 can be produced by applying very small changes of the winding ratio or the winding angle, but they have completely different densities and mechanical properties. If the simulation is continued for a longer period of time, the surface of the bobbin in Fig. 8.7b will no longer be cylindrical or smooth, but of a more complex form. This surface then has to be stored in the computer so that it can be used for the calculation of the coordinates of the point.

The algorithm shown here does not take into account the influence of the windings on the layer radius. In order to do this, the yarn thickness and the orientation of the yarns have to be considered by determining every single contact point between the windings. Such an algorithm is of course possible, but it will work fast and with sufficient accuracy only for windings with a random structure like those depicted in Fig. 8.7c.

When the windings from a half-cycle of the yarn guide intersect the windings from the previous half cycle, the contact geometry is similar to that presented in Fig. 8.8a. In all other cases, where the structure is similar to the one presented in Fig. 8.7a,b, the contact geometry can be defined as a ‘contact line’, not as ‘crossing point’, because a certain length of the yarn segments will touch each other side by side (Fig. 8.8b). This case is not critical, if a segment can be found one level below, which lies across (Fig. 8.8c), as it will determine the radius of the new yarn segment. In many cases, both the yarn segments on the previous layer will have the same orientation as the new one. The calculation of the exact position of the new yarn segment

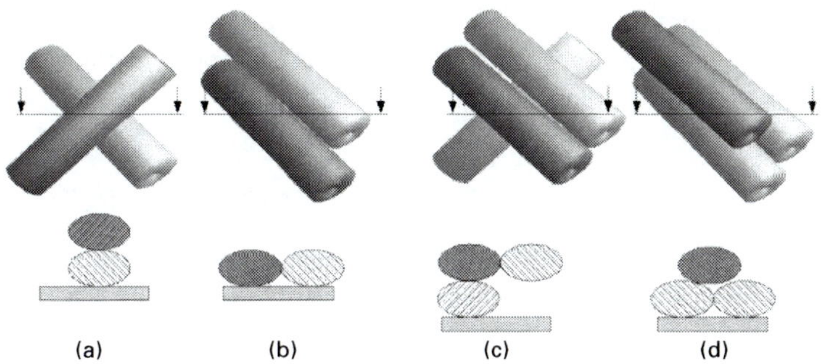

8.8 (a)–(d) Possible configurations between old winding and new yarn segments.

then leads to the contact problem of three-dimensional bodies, which takes more computational resources.

All these relationships can be implemented in a simulation algorithm, which can be used to calculate the yarn shapes with a reasonable degree of geometrical accuracy. However, up until this point, the algorithm does not take into account any mechanical interactions between the yarns. The friction between yarns, the yarn tension during winding and the deformation of the bobbin under the pressure of the windings have to be considered as well if such simulations are to be realistic. For many applications though, a geometry simulation is fully sufficient, e.g. if the yarn guide movement is to be analysed or if the stability conditions of single windings have to be checked.

8.3.2 Stability of the windings

The distribution of the yarns in a thin layer can be simulated using the algorithm described above. Besides using the coordinates of the points, the winding angle α for each yarn segment can be calculated. This angle can be used for checking the stability conditions of the yarn piece (Proshkov, 1986):

$$\tan \alpha \leq \tan \theta \leq \tan \varepsilon_{\max} = \mu \quad [8.1]$$

$$T_2 \leq T_1 \cdot e^{\int_0^{\psi} \cos\theta \sqrt{\mu_{\max}^2 - tg^2\theta} \mathrm{d}\psi} \quad [8.2]$$

where θ is the geodesic angle of the surface and $\tan \varepsilon_{\max} = \mu$ is the friction coefficient between yarn and winding surface, T_1 and T_2 are the yarn tension on the both edges of the winding between the angles 0 and ψ. Here we have to emphasize that, because of the roughness of the surface, the friction coefficient μ is not always equal to the friction coefficient measured for yarn–yarn contact. Equation [8.1] can be used to check whether the windings in the simulated structures are stable or not. In addition, the maximum winding tension T_2, ensuring that single yarn segments remain in static equilibrium (stable) over the surface, can be also calculated.

8.3.3 Stress distribution inside wound package

Knowing the stress distribution in the bobbin explains many bobbin properties. Incorrect winding tensions can cause a loss of stability of the package, making it unusable. For synthetic mono- and multi-filament yarns, a higher stress can lead to a deformation of the tube. For certain yarn types, the stress they experience in storage is significant, because different relaxation processes in the layers can cause defects and hence a poor quality of the product at the next processing step. An example of a bobbin where the inner layers are deformed is shown in Fig. 8.9.

Figure 8.10 shows the three different types of layers used in building the mathematical model of the stress equilibrium of the wound package. The difference between the layers is in their boundary conditions. The innermost layer has a fixed inner radius and the outermost layer has a known outer radius and no pressure from the outside. The mathematical model for the stress distribution is based on the static layer equilibrium. The model is identical to the one presented for the roll making operation by Ghosh *et al.* (1991b). It is based on the stress equilibrium of single elements in radial direction, with

$$\sigma_r - \sigma_\theta + \frac{\partial \sigma_r}{\partial r} r = 0 \qquad [8.3]$$

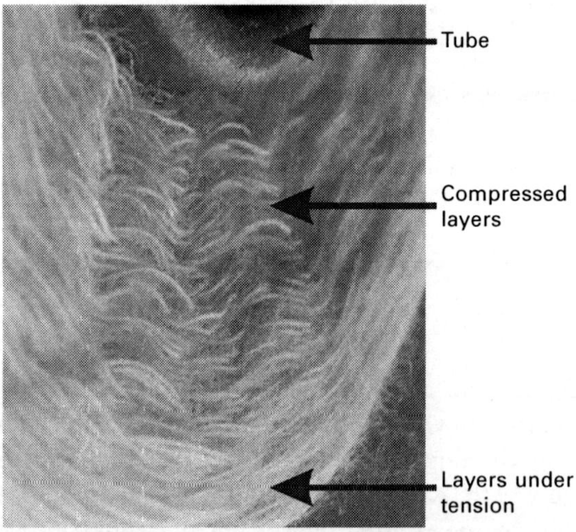

8.9 Wound package with compressed layers.

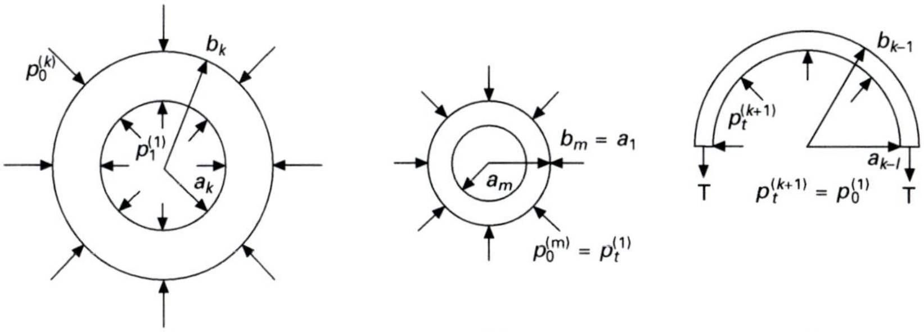

8.10 Layers of a wound package and their boundary conditions: (a) each middle layer is pressed from the inside and the outside; (b) the innermost layer has a definite inner radius and pressure from the outside; (c) the outermost layer has a known outer radius, but is pressure-free from the outside.

where σ_r and σ_θ are the radial and circumferential stresses, respectively. The relations for the strains in terms of radial displacements u can be described by

$$\varepsilon_r = \mathrm{d}u/\mathrm{d}r \qquad [8.4]$$

and

$$\varepsilon_\theta = u/r \qquad [8.5]$$

The circumferential elasticity module takes into account the initial yarn elasticity module as well as the winding angle β and the number of yarns in the layer, hence

$$E_\theta = n \cdot E_{\text{yarn}} \cdot \cos^2(\beta) \qquad [8.6]$$

The pressure–thickness curve of the textile yarns and structures is nonlinear (van Wyck, 1946; Grishanov and Lomov, 1997) (Figure 8.11). As a result, a bi-linear approximation can be used as was shown by Kyosev *et al.*, (2006a, 2006b). Here, a linear stress–strain relation in radial direction

$$\sigma_r(\varepsilon_r) = E_r(\varepsilon_r) \cdot \varepsilon_r \qquad [8.7]$$

is used. According to the current deformation ε_r of each layer, the corresponding elasticity module is then given by

$$E_r(\varepsilon) = E_r^{\text{low}} \quad \text{if} \quad \varepsilon_r^{\text{limit}} < \varepsilon < 0 \qquad [8.8]$$

$$E_r(\varepsilon) = E_r^{\text{high}} \quad \text{if} \quad \varepsilon < \varepsilon_r^{\text{limit}} \qquad [8.9]$$

This approach significantly simplifies the equation, but leads to a higher number of iterations if the calculated state is near the deformation limit $\varepsilon_r^{\text{limit}}$.

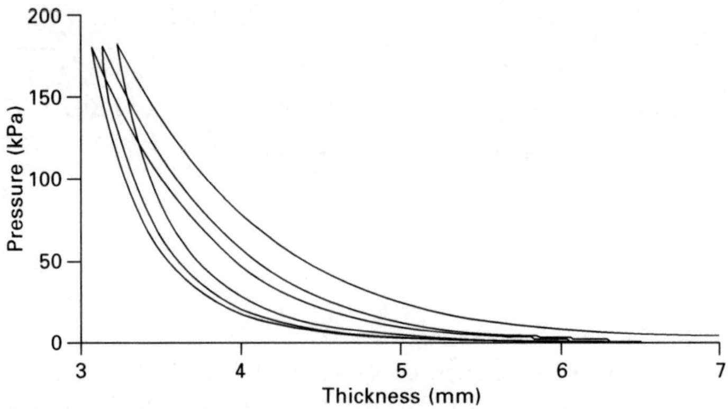

8.11 Example of the experimentally measured values of thickness–pressure curves of layers (three cycles) (Kyosev *et al.*, 2006b).

When applying $K = \sqrt{E_\theta / E_r(\varepsilon)}$, we need to solve the following ordinary differential equations for each layer

$$\frac{d^2u}{dr^2} + \frac{1}{r} \cdot \frac{du}{dr} - \frac{u}{r^2} \cdot K^2 = 0 \quad [8.10]$$

which has the analytical solution

$$u(r) = C_1 r^K + C_2 r^{-K} \quad [8.11]$$

Here, C_1 and C_2 are integration constants, which are found for each layer according to the following boundary conditions:

- inside of the first layer (tube deformation is neglected)

$$u^1(r = r_{\text{tube}}) = 0 \quad [8.12]$$

- common boundary between layers i and $i + 1$:

$$r^i_{\text{out}} + u^i(r^i_{\text{out}}) = r^{i+1}_{\text{in}} + u^{i+1}(r^i_{\text{in}}) = r^{i,i+1}_{\text{boundary}} \quad [8.13]$$

- stress equilibrium on the boundary

$$\sigma^i_r(r^{i,i+1}_{\text{boundary}}) = \sigma^{i+1}_r(r^{i,i+1}_{\text{boundary}}) \quad [8.14]$$

- outside the first layer

$$\sigma_\theta = \frac{u \cdot E_\theta}{r} = n \cdot F \quad [8.15]$$

where n is the number of yarns in layer cross-section and F is the winding tension.

Equations [8.12]–[8.15] make up a linear system for the integration constants of each layer, which can be solved in any programming environment by Gaussian elimination. The calculation of the whole body is carried out according to the following algorithm:

0. Initialize the first layer.
1. Build the next layer near the last one and calculate the new equilibrium state for all layers.
2. Calculate stresses, update the geometry of the layers by applying the displacements. Go to step 2 until the complete body is simulated.

Some examples for the calculation of radial and circumferential stresses in a bobbin with radius 0.1 m are presented in Fig. 8.12. It shows that, after winding of about 10 layers, the yarn inside these layers is no longer under tension (circumferential stress is near zero) and there is a potential risk of loss of stability or producing of a defect body, as shown in Fig. 8.9. It has to be noted that a higher anisotropy rate K can lead to an overflow problem by the calculation of the displacements, for instance for voluminous yarns.

More details about mathematical models for the simulation of the stress–strain distribution inside wound packages of filament winding and for warp beams can be found in Batra *et al.* (1976), Vicentini (1969), Neal (1967), Beddoe (1967), (Kempner and Hahn (1995), Ursiny *et al.* (1995), Ursiny (1986) and Koltze (2001). From a mathematical point of view, the stress–strain distribution of the wound package and of fabric rolls presented by Ghosh *et al.*, 1991a,1991b) and, Ghosh and Peng (1995) is identical. The differences are mainly in the mechanical properties of the layers.

8.4 Simulation of woven structures

Woven fabrics are the most thoroughly investigated textile structures. There are over 200 research papers covering different aspects of the geometrical modelling of woven structures. There are several different types of simulations of woven structures, depending on the final goal. The main simulations reported in the literature can be classified into the following groups:

- simulation of the woven fabrics as a structure from an aesthetic or artistic design/colour perspective;
- simulation of the three-dimensional woven structure for further use in mechanical, thermal and other engineering calculations or as well in artistic design;
- simulation of the drape behaviour of the structure, for the purpose of clothing simulation or for the draping of reinforcement textiles for composite parts.

Each of these simulations requires different methods and will be discussed in separate sections. There are also simulations using soft computing techniques which are not discussed here.

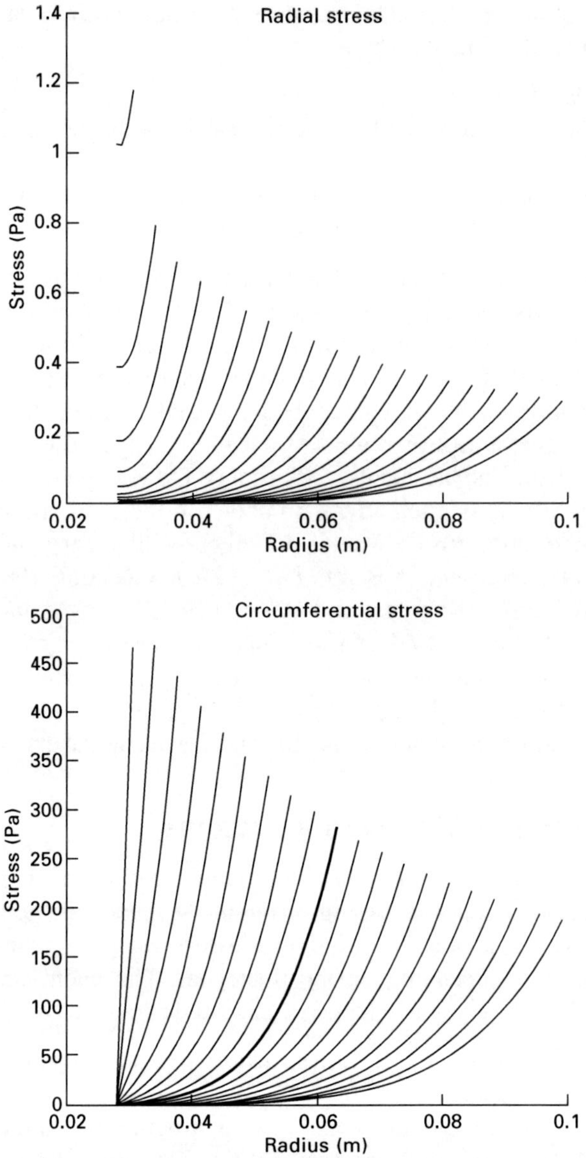

8.12 Radial and circumferential stresses in the layers, calculated at $K = 7.746$.

8.4.1 Simulation of woven fabrics for artistic design

These simulations are probably the earliest commercially available simulations of woven structures, developed to replace the sophisticated work of designers, preparing punch cards for Jacquard machines. All these simulations start from the weave pattern definition, since it contains the information about the appearance of the yarns from both sides. The usual pattern definition for single layer structures uses the marking black (or 1, if matrix form is used) of these crossing points between the warp and weft yarns, where the warp yarn lays above the weft yarn (Fig. 8.13).

This coding is used for the control of the heald frames or the Jacquard unit of the weaving machine, as well as for the reconstruction of the geometry of the woven fabric. The places marked with a '1' are those where the warp yarns are visible on the upperside. On the other places, only the weft yarn is visible, whereas the warp yarn is hidden. This principle was and is still used by most computer programs to simulate the appearance of the woven structure, without having to reconstruct the real appearance of the fabric. Using pictures of real or created yarns, appropriately placed and rotated, it is possible to present a realistic view of the structure. Figure 8.14 shows a view of the pattern editor in Arahne© software, whilst Plate XV (see colour section between pages 152 and 153) shows the simulated fabric using the same software. This example is only for demonstration. Nowadays, most programs provide a visual representation of the fabric. A list of currently known programs for weaving simulations is given at the end of this chapter.

8.4.2 Simulation of a three-dimensional woven structure

Another, more intuitive way to simulate woven structures is to generate the entire yarn geometry in three-dimensional space and to visualize this,

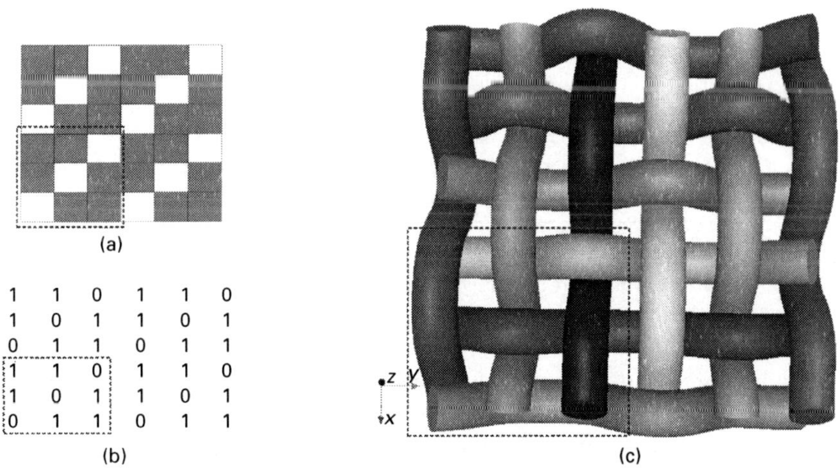

8.13 Coding of woven structures and its simulation: (a) pattern with the marked repeat; (b) the same pattern in matrix form; (c) simulated with the software TexGen (University of Nottingham, 2011) woven structure, corresponding to this pattern.

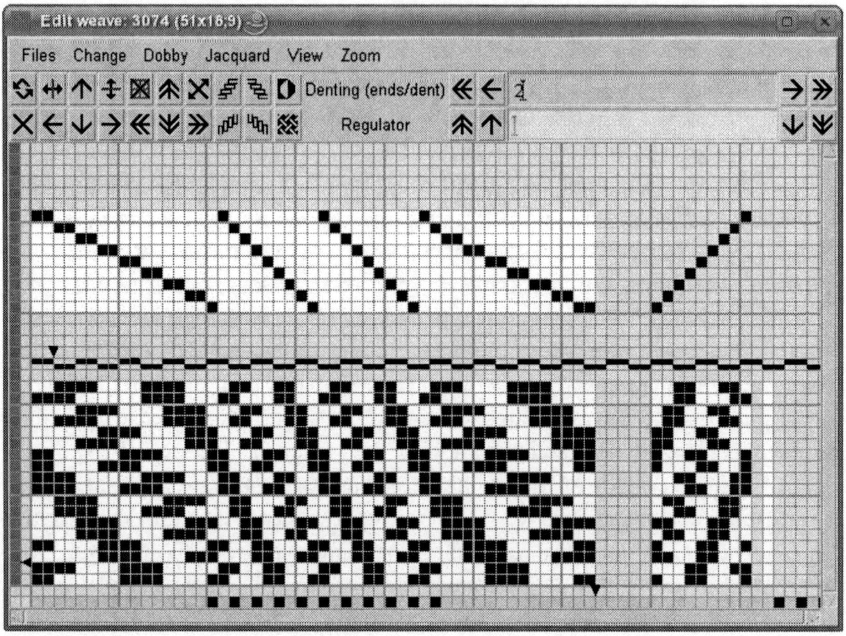

8.14 Pattern editor from Arahne (Arahne, 2011).

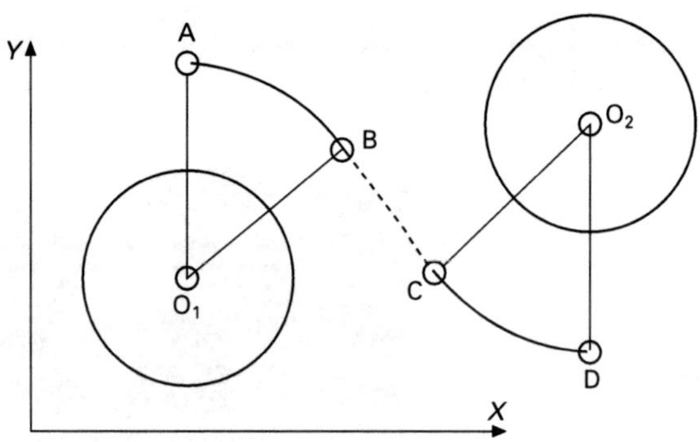

8.15 Principle of the calculation of the yarn axis coordinates in the case of yarns with circular cross-section and approximation of the yarn paths using circular arcs and lines.

including lightening, shadows, etc. This method is described in this section. There are several research papers which include a mathematical description of the yarn geometry of woven structures in three-dimensional space.

Such mathematical models are based on the description of the yarn path on its plane. In the plane of a warp yarn path, the weft yarns are shown with their cross-section form and this allows the definition of the contact points and the types of the lines between them. Figure 8.15 demonstrates one of the simplest and earliest models (Peirce, 1937) of a woven structure based on a circular cross-section of the yarns and modelling of the yarn path using circular arcs and lines. This model has been extended in several papers for different shapes of the yarn cross-section (lenticular, eliptic) and for different kinds of approximations of the yarn paths (eliptic, parabolic, cubic, natural splines, etc.). Experience shows that each of these models can be valid for certain cases but are not applicable to other cases. For plain fabrics, a typical model can be found in Behera and Hari (2010) and Postle *et al.* (1988). A very precise implementation of the different models in a software program which chooses the suitable yarn path approximation function depending on the yarn cross-section form is the program WiseTex© (Lomov *et al.*, 2011).

In this section, only the basic steps for modelling are described, based on the definition of the key points and connections. It is possible that, in some regions, the yarns are interpenetrating, whilst in other regions, they are more widely spaced. However, a refinement of the yarn path segments at some later stage is always possible. Starting from the pattern coding, and using the yarn cross-section dimensions, the axes of the yarns can be calculated. The first warp yarn on Fig. 8.16 starts with a position underneath the weft yarn. After that it lays twice above the weft yarns. For simplicity, it can be assumed that the undulation of the yarn axis is equal to the yarn diameter, so that h is equal to the yarn radius on Fig. 8.16. This leads to

$$\begin{aligned}&\text{if Matr}\ (i, j) = 1 \quad z_{i,j} = +h\\&\text{if Matr}\ (i, j) = 0 \quad z_{i,j} = = -h\end{aligned} \qquad [8.16]$$

with Matr representing the coding matrices with the pattern and i and j standing for the indices of its rows and colomns, respectively.

The x-coordinates of the warp yarns depend on their position and the space between the yarns. For the case of equally distributed warp yarns with the distance t_{wrap} between them, the ith warp yarn has the x-coordinate:

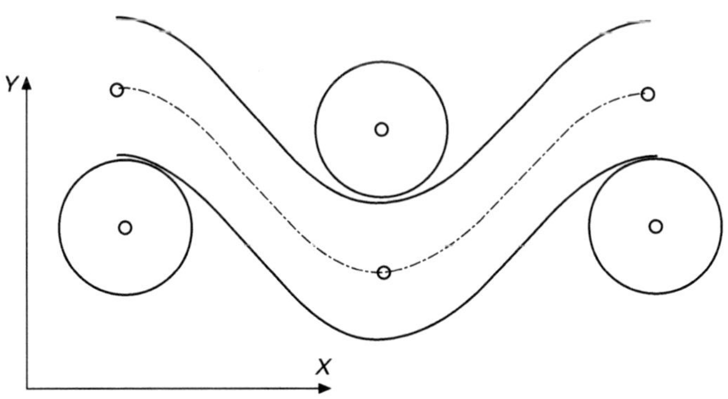

8.16 Calculation of the yarn axis position.

$$x_i = (i - 1) \cdot t_{\text{wrap}} \quad [8.17]$$

The key-point, where this warp yarn is above or underneath the weft yarn number j, has the y-coordinate:

$$y_j = (j - 1) \cdot t_{\text{weft}} \quad [8.18]$$

where the distance between the weft yarns is t_{weft}.

Similarly, the coordinates of the weft yarns can be calculated and so a list of coordinates for each yarn path can be calculated in the form

$$S_j = [(x_1, y_1, z_1), (x_2, y_2, z_2), \ldots, (x_n, y_n, z_n)] \quad [8.19]$$

Using this set of coordinates, the yarn axis can be created using a suitable kind of connecting functions between the points, as presented in Fig. 8.17a. Around this yarn axis, a yarn volume is built, and the yarn cross-section around a set of points of the yarn axis is calculated (Fig. 8.17b). Each cross-section has to lie on a plane, spanned by the normal and the binormal vector to the yarn path at that point. For a FEM calculation with volume elements, the entire yarn volume has to be meshed using small elements, as in Fig. 8.17c. More details about the algorithms for interpolation of the yarn axis, building the yarn volume and meshing can be found in Sherburn (2007) and the software package TexGen (University of Nottingham, 2011). After the geometry is created in this way and uploaded into FEM or CFD software,

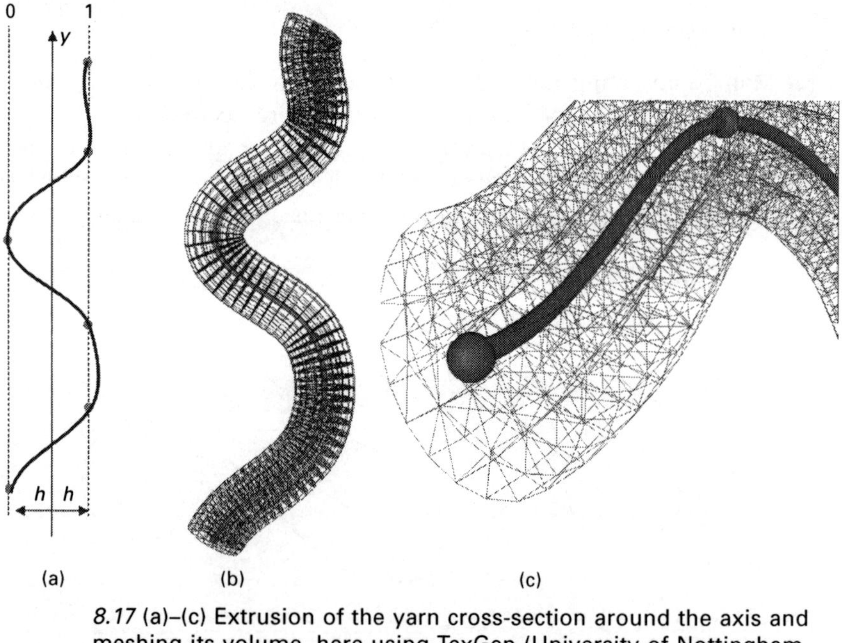

8.17 (a)–(c) Extrusion of the yarn cross-section around the axis and meshing its volume, here using TexGen (University of Nottingham, 2011).

different behaviours like mechanical, thermal, fluid dynamics, etc. can be simulated. Another use of this geometry is mapping of images from real yarns to the surface of the yarn in order to achieve a photorealistic appearance for the artistic design of the fabrics. In this way it is as possible to create realistic pictures of plush structures, as for instance by EAT GmbH 'The DesignScope Company' (EAT, 2011).

The preparation of the geometry of multilayer woven structures works in a similar way. The pattern coding for multilevel structures in this case is based on a matrix with numbers, each marking the level on which one warp yarn crosses the weft yarns (Plate XVI and Fig. 8.18). Figure 8.19 shows the topological representation of the woven structure based on the assumption that the weft yarns are straight (Fig. 8.19a) or bent at the crossing places

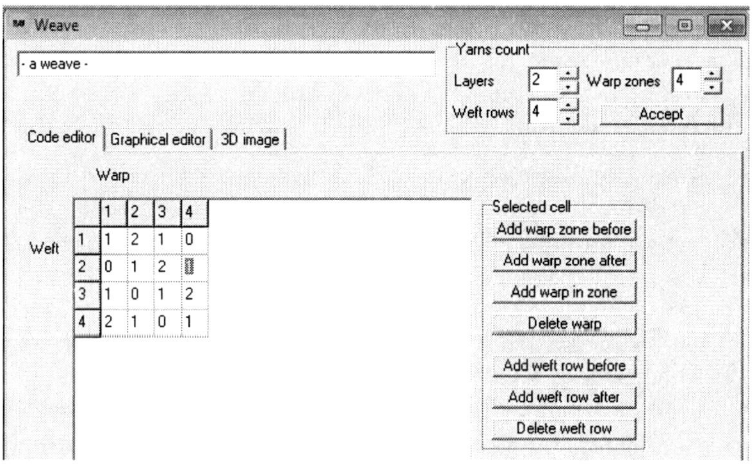

8.18 Coding of a 3D woven structure with WiseTex.

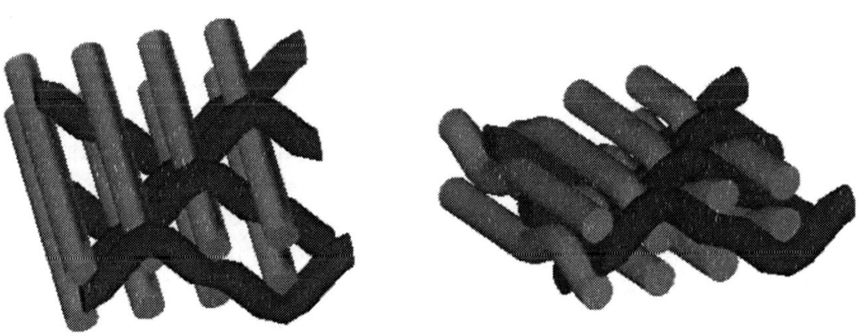

(a) (b)

8.19 Weave pattern in 3D: (a) with straight weft yarns, (b) with bend weft yarns.

with the warps (Fig. 8.19b). The crimp calculations need to be undertaken after the mechanical data of the yarns has been provided.

The method described here is suitable for the simulation of an approximate geometry of the yarn structure. The simulation of the exact geometry of the woven structure is not possible using only pure geometrical data of the pattern and the yarns. The geometry of the yarns depends significantly on their bending rigidity, the twist as well as the conditions during weaving. Because of this, only simulations which take into account all this information are able to reproduce the geometry of the woven structure correctly. Currently the only program which takes into account the mechanical parameters of the yarns is WiseTex (Lomov *et al.*, 2011). Other programs only cover geometric data.

Usually the calculations for static cases are based on the principle of the minimum of the potential energy of the structure, in this case of all yarns (Verpoest and Lomov, 2005; Lomov *et al.*, 2006). The following example demonstrates some results of the simulation of the tensile and lateral compression behaviour of a woven structure using WiseTex. A woven structure (Heuwinkel, 2011) with the following parameters is simulated:

- warp yarn fibres: polyester (PES) filaments; fineness 0.13 tex, diameter 0.022 mm;
- weft yarn fibres: PES filaments, fineness 0.39 tex, diameter 0.021 mm;
- warp yarn parameters: fineness 21 tex, modelling diameter 0.3 mm, 153 filaments in cross-section, without twist, bending stiffness 0.068 N/mm^2; strength 6 N, maximal elongation 25%, cross-section compression parameters $\eta_1 = 0.812$, $\eta_2 = 1.198$,
- weft yarn parameters: fineness 71 tex, modelling diameter 0.54 mm, 182 filaments in cross-section, without twist, bending stiffness 0.17 N/mm^2; strength 6 N, maximal elongation 26%, cross-section compression parameters $\eta_1 = 0.812$, $\eta_2 = 1.188$;
- fabric parameters: 60 warp yarns per cm; 9 weft yarns per cm; thickness 0.5 mm; width 10 mm; plain woven pattern.

The force–elongation curve from an experimental test and from the simulation of the tension behaviour with WiseTex is presented in Fig. 8.20, where it can be seen that the trend of the curves is the same with a small underestimation of the tension by WiseTex. The simulated compression curve has a similar trend as the measured one, but its values are several percent smaller (see also Fig. 8.21). These differences are a result of the differences of the behaviour of a multifilament yarn compared to a monofilament yarn. For the geometrical modelling, the yarns are presented with a constant cross-section which in reality is not constant. Figure 8.22 presents a view of a similar structure and the modelled unit cell is presented in Fig. 8.23.

If the yarn geometry has to be presented properly, the yarn cross-section

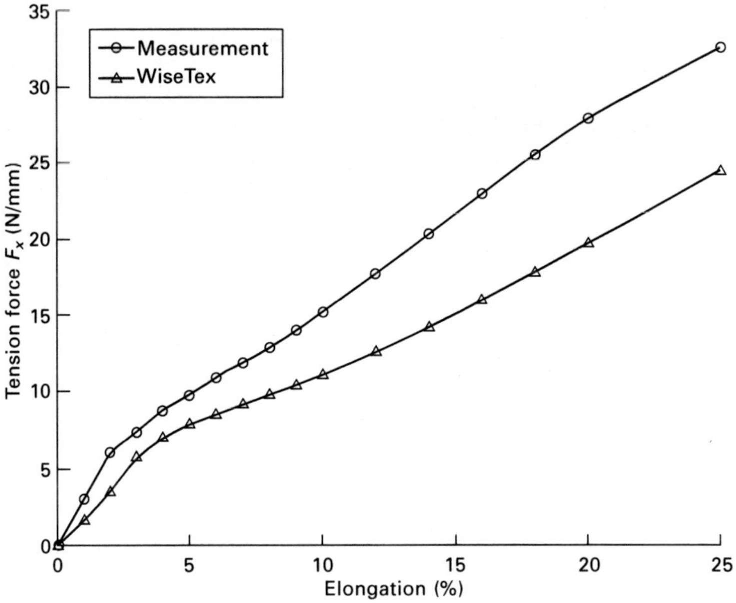

8.20 Simulated and measured load elongation curves of a woven tape (Heuwinkel, 2011).

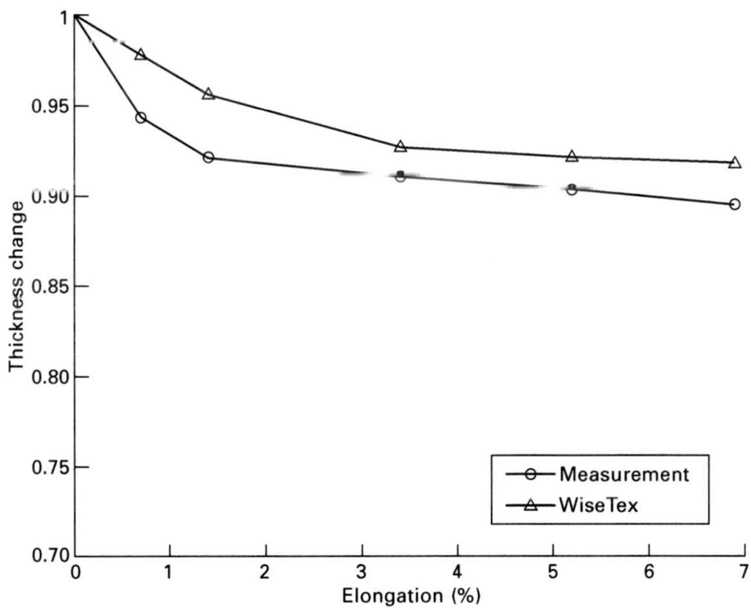

8.21 Simulated and measured compression curves of a woven tape (Heuwinkel, 2011).

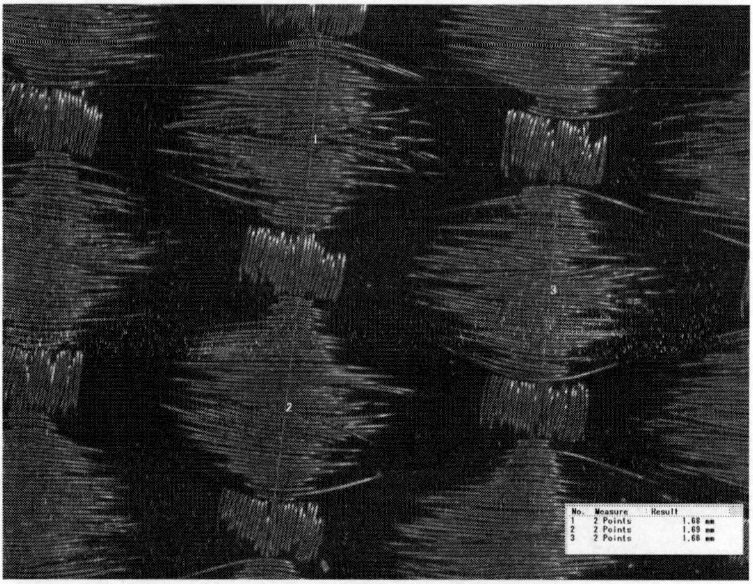

8.22 Photograph of a similar structure as the one modelled. The multifilament yarns without twist change their cross section on the places where there is no contact with the neighbouring yarns.

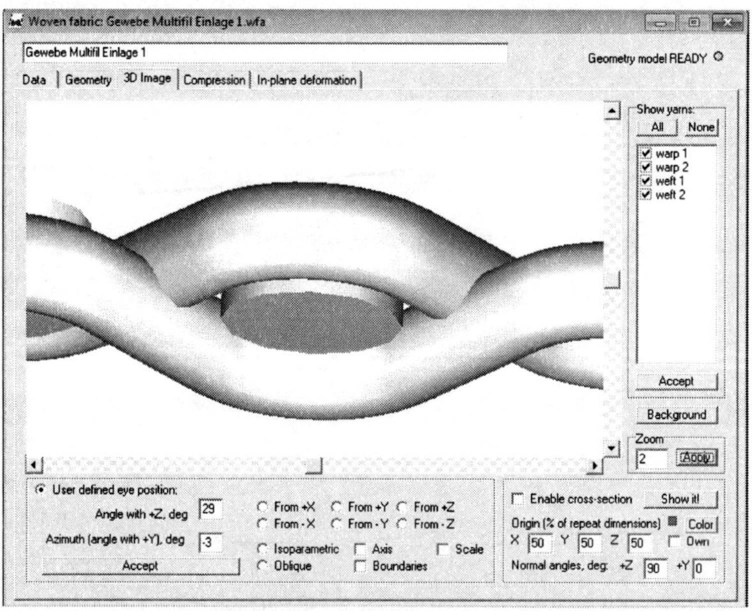

8.23 Unit cell of the modelled structure with visible interconnection between the yarns.

has to be variable. In the case of staple yarns, the yarn volume density has to be variable. Such options are currently not implemented in the available software. One solution is to model the yarns with a constant cross-section, allowing penetration between these, but keeping the proper number of single filaments and the mechanical data (Fig. 8.23) constant. For mechanical calculations based on yarn path geometry, this approach leads to a good approximation of the experimental data as in the example shown. If fluid transfer between the yarns has to be simulated, such interpenetrations between the yarns can lead to more significant differences between the simulation and experimental results.

If the yarn level modelling of the woven structures is not sufficiently precise for an application, then the yarns have to be modelled on the filament level. This approach solves the problem of the differences in the cross-section and avoids the need to investigate the in-plane behaviour of the yarn cross-section. However, this requires significantly higher computational resources and a larger preparation time for the samples (Durville, 2010) (Fig. 8.24). A key disadvantage to all the current programs for the simulation of the woven structure is that they require some kind of topological representation of the structure. This is generally not the same as the pattern for the weaving machine, especially for multilayer structures. Some special algorithms for hollow or angle interlock structures, where the simulation of the structure is based on the coding for the machine, are implemented by TexEng LTD (TexEng Software Ltd, 2011b) and Plate XVII presents an example graphical interface for a hollow structure.

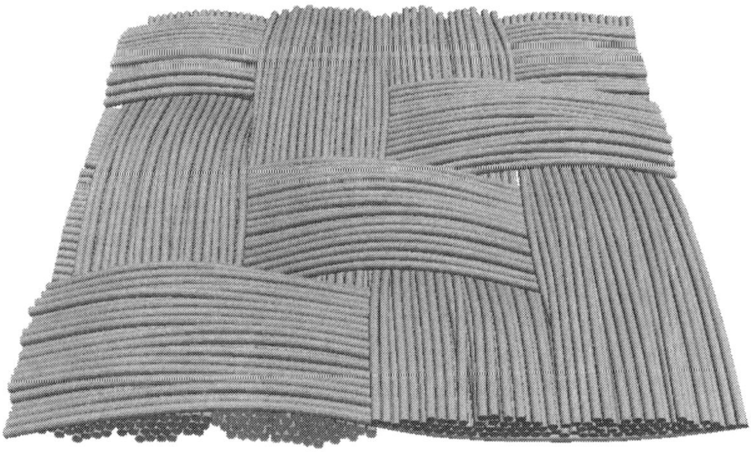

8.24 Modelled woven structure with twill pattern, where the single yarns are presented as a set of filaments, modelled completely (Durville, 2010).

8.5 Simulation of braided structures

The unit cell (smallest repeat) of braided structures is identical to those of woven structures. Because of this, the methods for the simulation of the braided structures are very similar to those used for woven structures. Due to several significant differences during the production process (e.g. beat up force during weaving which does not exist in braiding), braided structures are not as dense as woven structures and have different mechanical behaviour.

8.5.1 2D colour pattern design of braided structures

Only the areas where one yarn is visible from the upper side show the colour of the yarn. As a result, only the colours of the yarns on the top side of the interlacing are important for the colour pattern design of braided structures. The principle of pattern calculation is identical to that of woven structures, described in Section 8.4 (Fig. 8.25). Braided structures are built by one, two or more systems of yarns which interlace with each other or with other systems at a certain angle. Half of this angle is usually equal to the angle between the yarns and the product axis and is called the braiding angle (Fig. 8.26). The braiding angle is important for the realistic simulation of the fabric's appearance.

Owing to the lower number of products and customers, the use of CAD for braided products in the fashion area is uncommon. There is no software

1	1	0	0
1	0	0	1
0	0	1	1
0	1	1	0

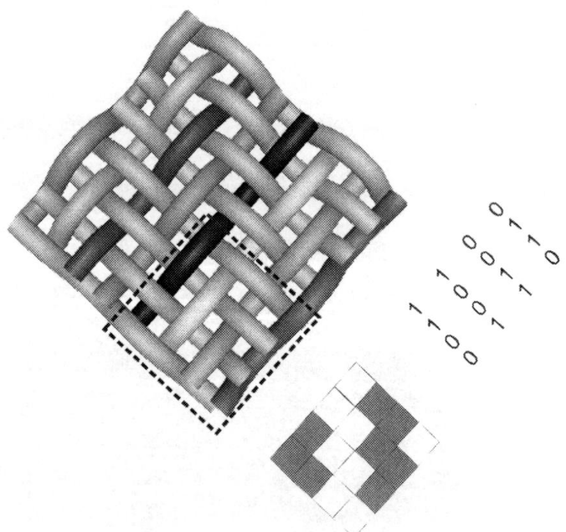

(a) (b) (c)

8.25 (a)–(c) Coding and unit cell of braided structure, similar to that of a woven structure.

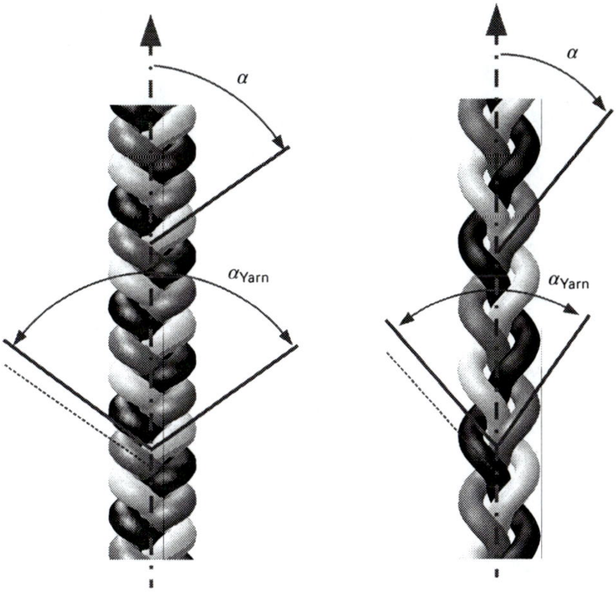

8.26 Angle between the yarns and braiding angle α.

which simulates braided structures photorealistically, unlike woven and the weft knitted structures. For most customers, the distribution of the colours is sufficient in order to check from the simulation if the product meets their requirements. Figure 8.27 shows a simulation of one sample with a diagonal colour effect with three different braiding angles, using CAB software from Herzog (Herzog Maschinenfabrik, 2011).

Since the patterns used in the braided samples are very limited, with fewer than five commonly used patterns, it is not necessary to calculate the pattern matrix for tubular and flat structures as for woven structures. After determining whether the structure is flat or tubular, and selecting the type of yarn interlacing (one over one, two over two, diamond = two over two but with two ply yarn), the algorithms for the simulation of the structure are identical. Figures 8.28 and 8.29 show the graphical user interface for patterning of tubular and flat braided fabrics respectively using CAB Design software, where the user can choose the colour for one yarn (= carrier). After picking the number of carriers, the places where this yarn will be visible are automatically coloured in the chosen colour.

It is apparent that the simulated view of the structure is not significantly different from the pattern view in Fig. 8.25c, where the yarn colours, braiding angle and positions of the yarns in the product are shown. If one yarn floats first above two other yarns and then underneath two yarns, this floating is presented as a rectangular area equal to the two squares without showing the square borders.

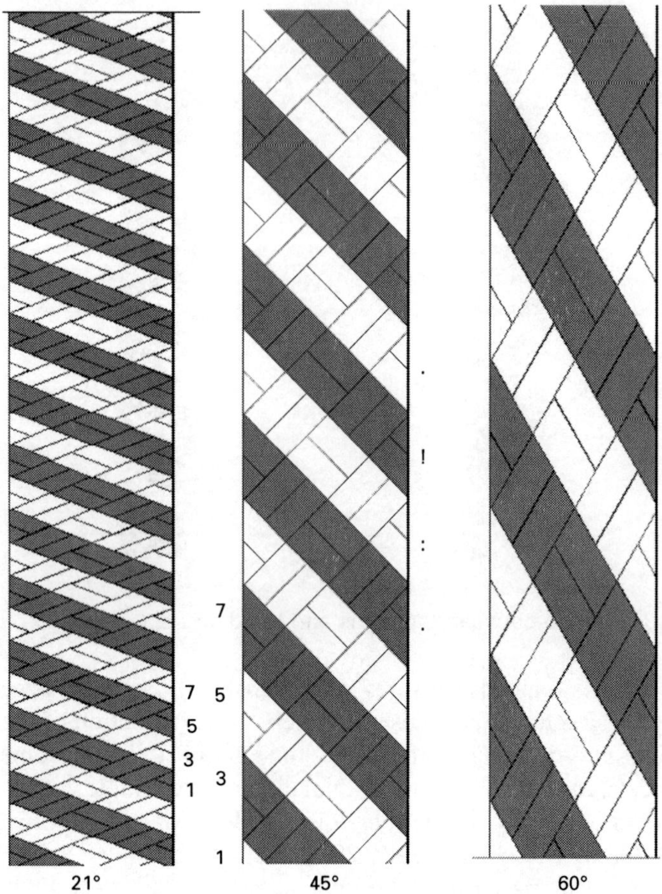

8.27 Tubular sample with diagonal colour effects, yarns in 'two over two' (diamond) arangement, simulated with different braiding angle using CAB Design from August Herzog GmbH and Co. KG (Herzog Maschinenfabrik, 2011).

8.5.2 3D design of braided structures

Reports of algorithms and software for the creation of 3D yarn geometry of braided structures can be found in several papers. These works can be divided into three groups:

- simulation of the yarn geometry for flat and tubular structures;
- simulation of the geometry and the shape of the braid during overbraiding of certain profiles;
- simulation of 3D braided products.

The geometry of flat and tubular structures can be simulated using the same principle applied to woven structures using geometric methods

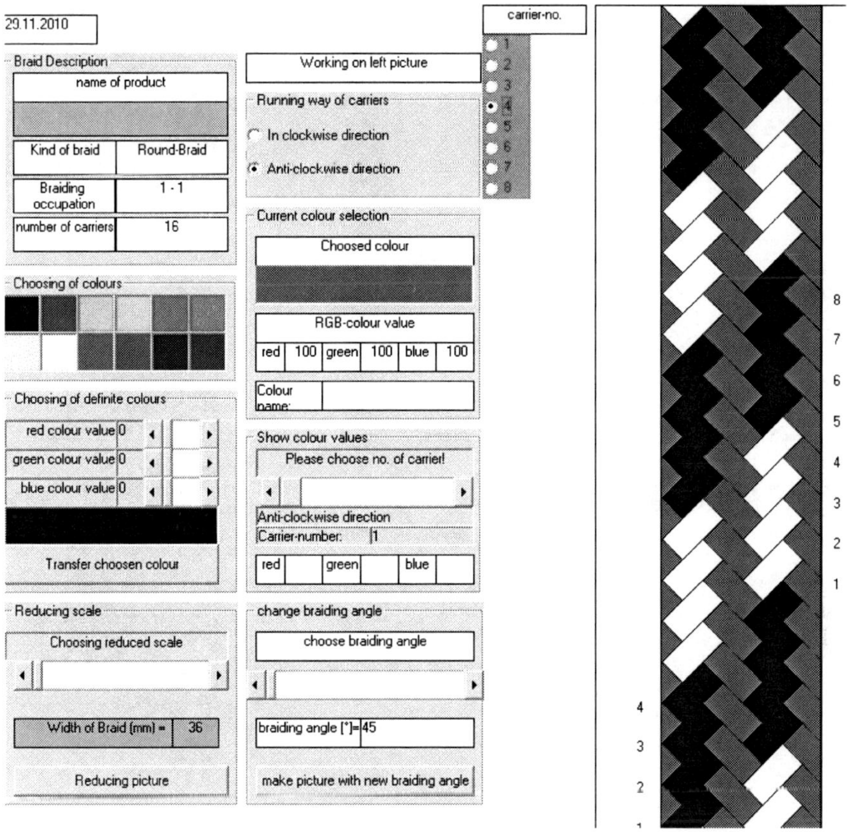

8.28 Graphical user interface for simulation of tubular braided fabrics, using the CAB Design software by August Herzog GmbH & Co. KG. (Herzog Maschinenfabrik, 2011).

(Liao and Adanur, 2000; Kyosev, 2007). This is a sufficiently accurate representation for several applications and in establishing the initial geometry for FEM computation. Figure 8.30 shows simulated flat braided structures with different numbers of carriers whilst Fig. 8.31 shows simulated tubular braided structures using TexMind software (TexMind UG, 2011).

The simulation of overbraided products is more complex then tubular and flat braids. This is due to the yarn path depending on the mandrel shape and in several cases on the dynamics of the braiding process. In these cases, purely geometric methods can be used, if the braided structure is 'open', i.e. if the neighbouring yarns work in the same direction and do not come into contact with each other. Studies by Rawal *et al.* provide a good description of the yarn paths which can be used for simulation during the overbraiding of different regular bodies, either as a pyramid (Rawal *et al.*, 2007), cone or cylinder (Rawal *et al.*, 2005) or in the case of triaxial structure (Potluri *et al.*,

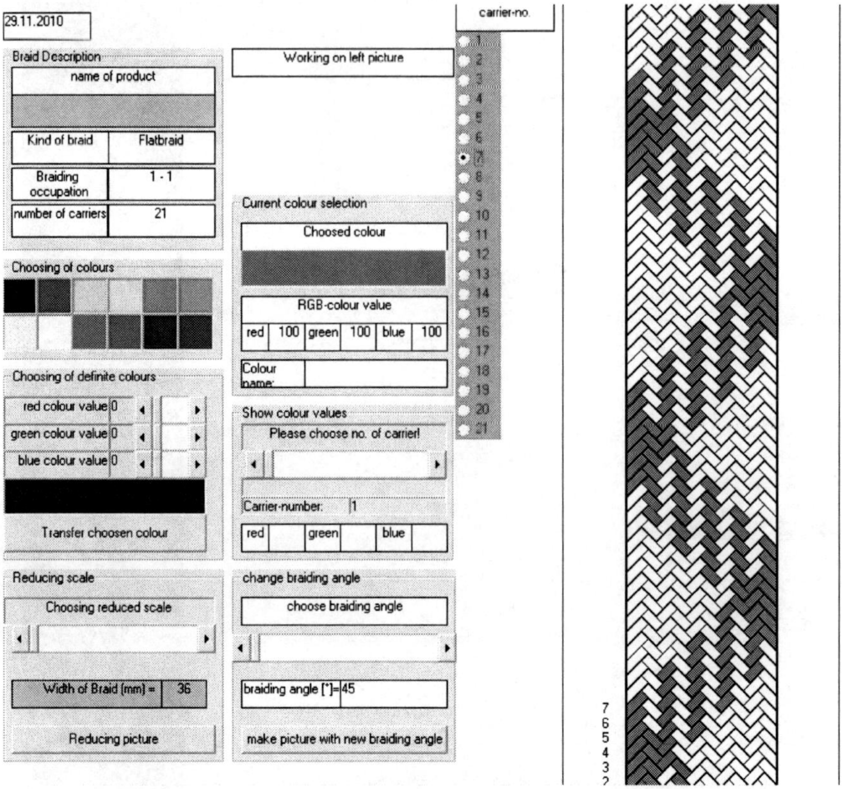

8.29 **Graphical user interface for the simulation of flat braided fabrics, using the CAB Design software by August Herzog GmbH & Co. KG (Herzog Maschinenfabrik, 2011).**

2003). If the braid has a higher density and the yarns are near to each other, then the geometric methods lead to larger differences in the path positions of the yarns and can be used only as a visualization method.

8.5.3 Simulation of braided structures using process simulation

The simulation of the braiding process and the braids themselves is based on the analytical description of the yarn paths during the braiding process. The contact between two yarns has to be detected to allow calculation of the contact points between the yarns. The differences in the approaches of different researchers are in the treatment of the crossing points between the yarns. Different models exist for cases where the yarn remains on its initial place or moves, for friction between the yarns, friction between yarn and mandrel and how the yarn cross-section is modelled.

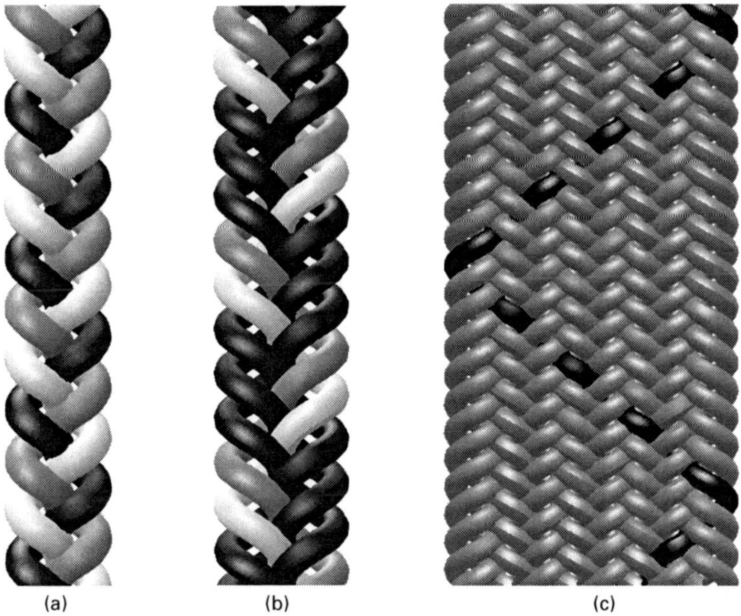

8.30 Simulated 3D geometry of flat braided structures from machines with (a) 3, (b) 5 and (c) 17 carriers (Kyosev, 2011).

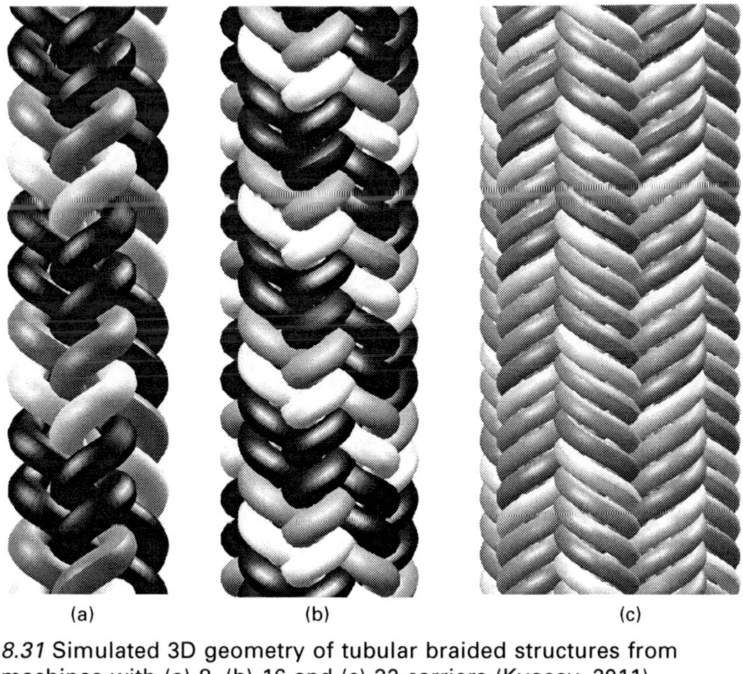

8.31 Simulated 3D geometry of tubular braided structures from machines with (a) 8, (b) 16 and (c) 32 carriers (Kyosev, 2011).

In the most common examples, the yarn axes are used for the calculations and friction is ignored. Examples of the use of this method in the case of triaxial braided structures can be found, for example, in Zhang *et al.* (2007). The analysis of local changes of the braiding angle in the structure, depending on the mandrel profile during the overbraiding process, can be found in Nishimoto *et al.* (2010). In this paper, the yarn path between the last braiding point, the current contact points over the mandrel and the carrier are described. The same principles, built into a more general algorithm and software for overbraiding simulation, are presented in Akkerman and Rodriguez (2006) and Kessels and Akkerman (2002). In these studies, the mandrel geometry is presented as a discrete (meshed) surface and the contact between the yarns and the elements of this surface is calculated in order to establish the final position of the yarn pieces. Current FEM software can also be used. As an example, the carrier arrangement and motion for a 3D braiding machine from August Herzog Maschinenfabrik GmbH is simulated prior to actual operation using special simulation software, which tests the collision control and path of the carriers (Plate XVIII). The resulting geometry from such a simulation is presented in Plate XIX (Stueve and Gries, 2009). The simulated geometry can also be used for predicting the properties of the braids using FEM or other methods with suitable boundary conditions and loads (Pickett *et al.*, 2009). In order to speed up the simulation process, the yarns are modelled with a constant cross-section. Once proper properties of the yarns in the lateral direction have been determined and defined prior to the simulation, this method can produce simulated structures which are very close to reality.

8.5.4 Simulation of the properties of braided fabrics

The geometry of the braids is not only used to visualize the braided structure, but is also a necessary step for the prediction of the mechanical and other properties of the braids. Some basic geometric calculations for tubular and flat tubular braids (expected diameter, braiding angle, etc.) can be performed by the CAB Design software mentioned above (Herzog Maschinenfabrik, 2011). Some of the calculations are also shown on the web page of the company A & P Technology (A&P Technology, 2011). The calculation of unit cell properties such as the size of the unit cell, pore size, as well as application of the homogenization method, is also possible using the WiseTex package (Lomov *et al.*, 2006). Figure 8.32 presents the graphical user interface for the preparation of the geometry and material properties of the braided structure. The results for the geometrical properties of the unit cell are presented in Fig. 8.33 and the 3D model of the fabrics in Fig. 8.34.

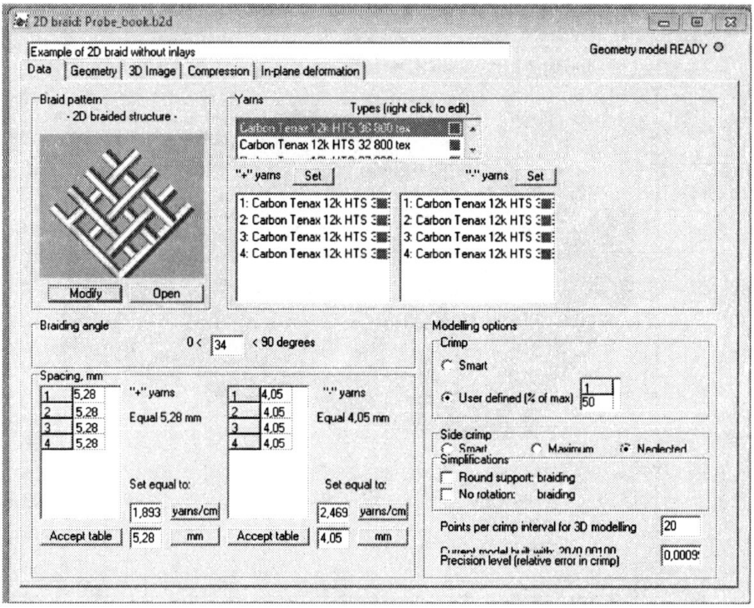

8.32 Graphical interface for the input of the pattern and yarn parameters of braids within WiseTex.

Unit cell size	**17,5 x 22,8 x 0,800 mm**
Areal density	**349,7 g/sq.m**
Porosity (inter-yarn)	**38,2 %**
Fibre volume fraction	**2,5 %**

Yarns

	d1,mm	d2,mm	Vf,%	l,mm	Crimp,%	Side crimp angle,
+:1	0,40	3,60	4,1	17,52	0,3	0
+:2	0,40	3,60	4,1	17,52	0,3	0
+:3	0,40	3,60	4,1	17,52	0,3	0
+:4	0,40	3,60	4,1	17,52	0,3	0
-:1	0,40	3,60	4,1	22,81	0,1	0
-:2	0,40	3,60	4,1	22,81	0,1	0
-:3	0,40	3,60	4,1	22,81	0,1	0
-:4	0,40	3,60	4,1	22,81	0,1	0

8.33 Results from the unit cell calculation with WiseTex.

8.6 Simulation of knitted structures

Knitted structures are built by interlooping of yarns. Depending on the type of the interloping, they can be classified as weft knitted structures

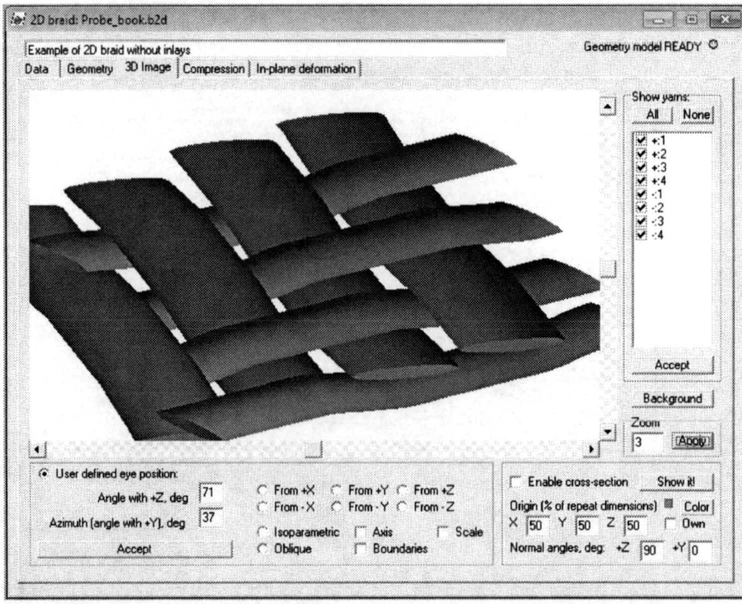

8.34 Simulated 3D image of the braided structure from WiseTex.

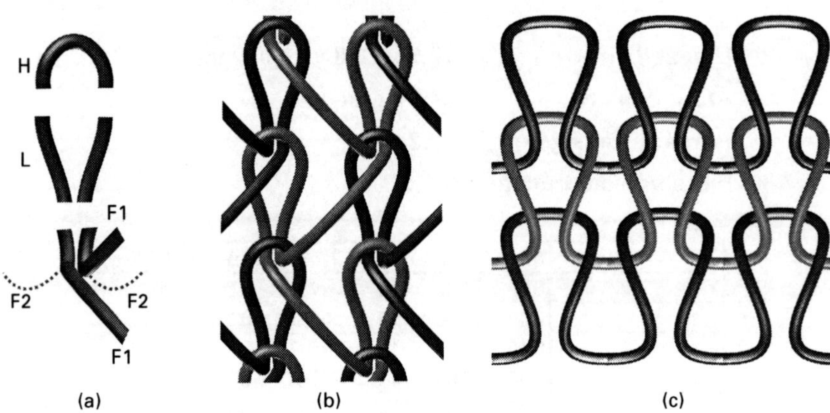

8.35 Knitted structures: (a) elements of one loop; (b) warp knitted structure; and (c) weft knitted structure (Kyosev and Renkens, 2010).

(Fig. 8.35c), where one yarn builds loops in the horizontal (weft) direction whilst, in the next row, the same (or a different) yarn closes the loops of the previous row, building new loops. In contrast, warp knitted structures are built from several (warp) yarns, where one yarn builds only one (or two in some structures) loops per row (Fig. 8.35b). In both cases, the loops consist of the same elements – head (H), legs (L) and feet (F) (Fig. 8.35a). The feet connect neighbouring loops in weft knitted structures and loops from different rows in the case of warp knitted structures.

Different mathematical descriptions of knitted loop geometries have existed for more than 50 years (Leaf, 1955), but a detailed description of the loop in the different states of the fabrics still remains an ongoing research problem. The common techniques are:

- to model an idealized geometry and refine it so that some kind of relaxed structure is built;
- to simulate the whole knitting process and create the knitted structure as result of the simulation.

The first method leads to useful results within a very short time and thus is preferred, provided the exact yarn paths are not really necessary. All geometric models for both warp and weft knitted structures in this case are based on the presentation of the loop form (Fig. 8.36) using some key points, usually the projections of the contact points above the plane surface. The exact x and y-coordinates of every key point can be derived using the loop sizes in both directions. Assuming that the yarn cross-section is circular, the z-coordinates of the key-points can also be calculated (Fig. 8.37). More details about the equations for the calculations can be found for instance in Kyosev and Renkens (2010). After calculating the coordinates of the key points, different curve types can be used for their connections. Kurbak and Ekmen (2008) and Kurbak and Soydan (2008) establish very detailed

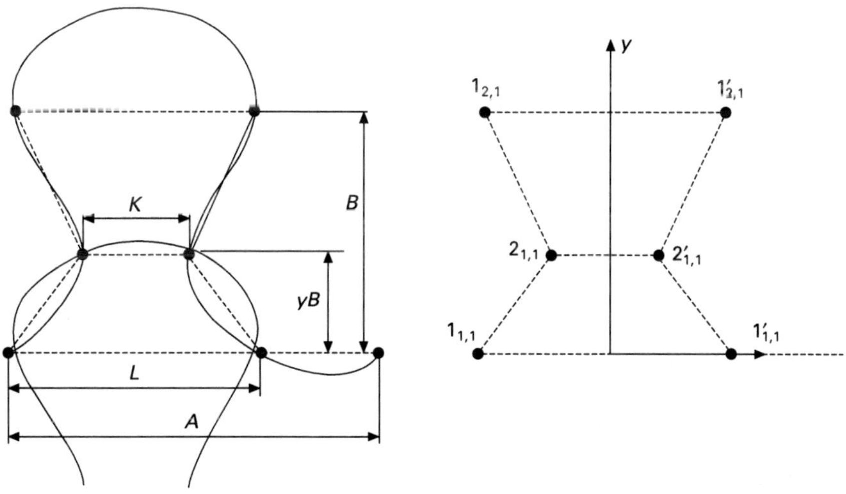

8.36 Knitted structures: (a) elements of one loop; (b) warp knitted structure; and (c) weft knitted structure (Moesen *et al.*, 2003; Kyosev and Renkens, 2010).

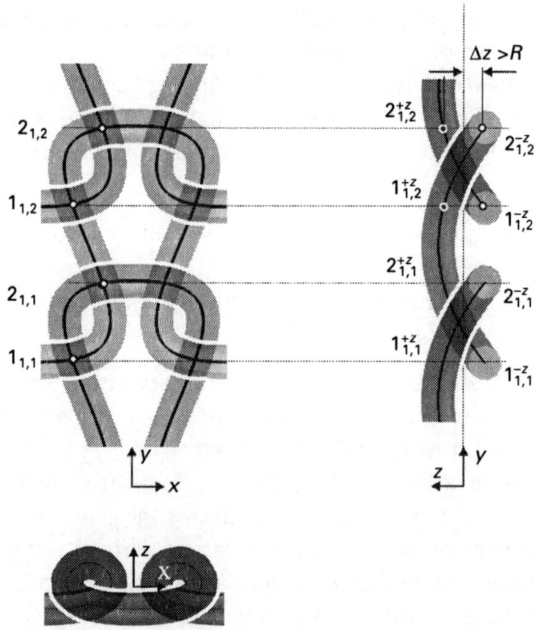

8.37 Knitted structures: (a) elements of one loop; (b) warp knitted structure; and (c) weft knitted structure (Kyosev and Renkens, 2010).

geometrical relationships, though alternative methods using cubic splinces or other approximation functions also work well.

8.6.1 Weft knitting examples

Figures 8.38 and 8.39 show 3D simulations of weft knitted structures with WeftKnit and KnitGeo Modeller software. Both programs have been developed in order to generate 3D yarn geometry to allow further calculation. Another kind of simulation of weft knitted structures uses photorealistic techniques. Such modules are based on the visibility of the yarns on the front side and include the structure of the yarns. Figures 8.40 and 8.41 show photorealistic simulations of weft knitted structures using software developed by Shima Seiki (2011). This can show the characteristic hairiness of fancy yarns. There are two ways to produce this simulation:

- recognize which elements (loops, tucks, floatings) are visible and use pictures of these elements to build a mosaic;
- compute the 3D geometry of the yarns and to map the picture of the yarn surface above the 3D yarn volume.

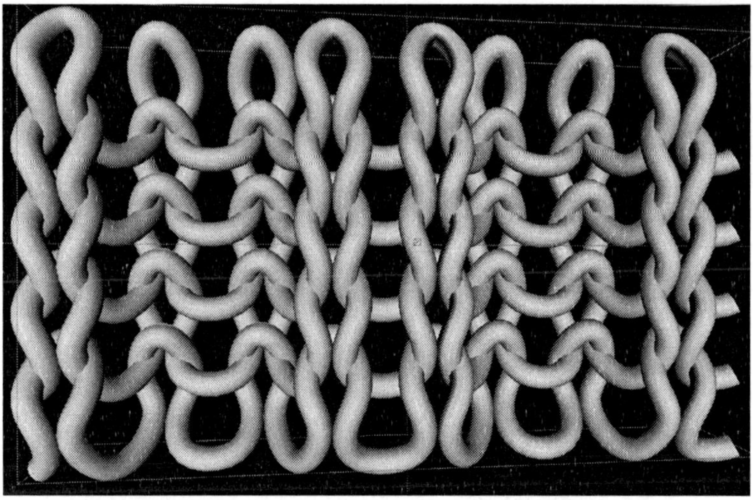

8.38 3D Geometry of weft knitted structure using WeftKnit (Moesen and Lomov, 2011).

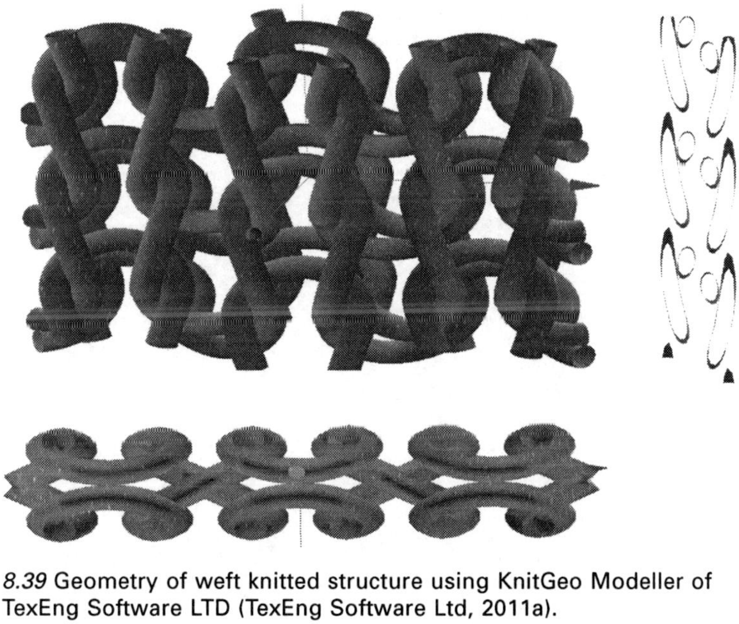

8.39 Geometry of weft knitted structure using KnitGeo Modeller of TexEng Software LTD (TexEng Software Ltd, 2011a).

8.6.2 Warp knitting examples

Göktepe and Harlock (2002) demonstrated a model of a simple warp knitted structures in 3D form, but this was insufficient for industrial use. Robatille *et al.* (2000) have described a method of modelling reinforced knitted fabrics

8.40 Photo realistic simulations of weft knitted fabrics with the software of Shima Seiki. Courtesy of Shima Seiki http://www.shimaseiki.com/.

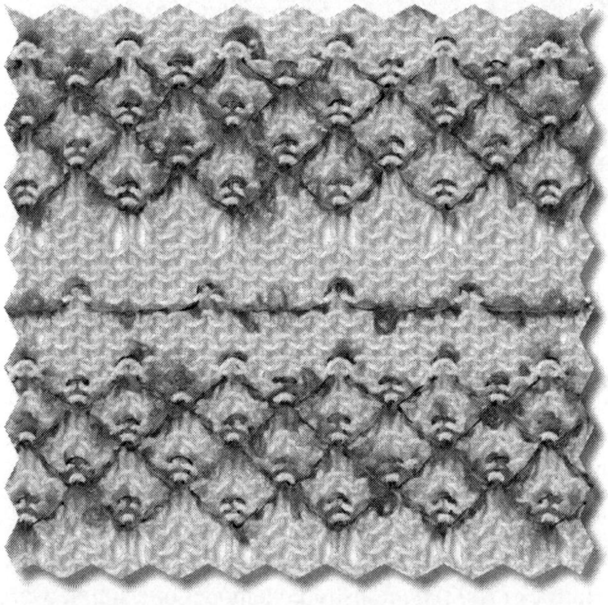

8.41 Photo realistic simulations of weft knitted fabrics with tuck stitches. Courtesy of Shima Seiki http://www.shimaseiki.com/.

with significantly thinner and more flexible looping threads than the warp, weft or diagonal yarns (rovings). Both models, however, were developed for a small group of structures only and are thus not suited for more general applications. Current studies of the mechanical properties of warp knitted structures, for instance bending properties by Jiang *et al.* (1999), are valid only for a certain group of fabrics, and are not generally applicable. Based on the basic method presented at the beginning of Section 8.6, warp knitting simulation tools were developed and reported in Renkens and Kyosev (2007) for single needle bed machines, and Renkens and Kyosev (2011) for double needle bed machines, including spacer and tubular fabrics. Such models are implemented in industrial CAD software packages developed by ALC Computertechnik, Renkens Consulting and TexMind UG (TexMind UG, 2011).

Figure 8.42 shows the difference between the geometrical model (a) and the model after performing a mechanical computation with it (b). The mechanical models in this case can be based on different assumptions, e.g. using trusses and joints or mass-spring systems with contact detection. The basic principles of these models are described in Chapter 6 and can also be found in Kyosev *et al.* (2005), Kyosev and Renkens (2010) and (de Araújo *et al.* (2004).

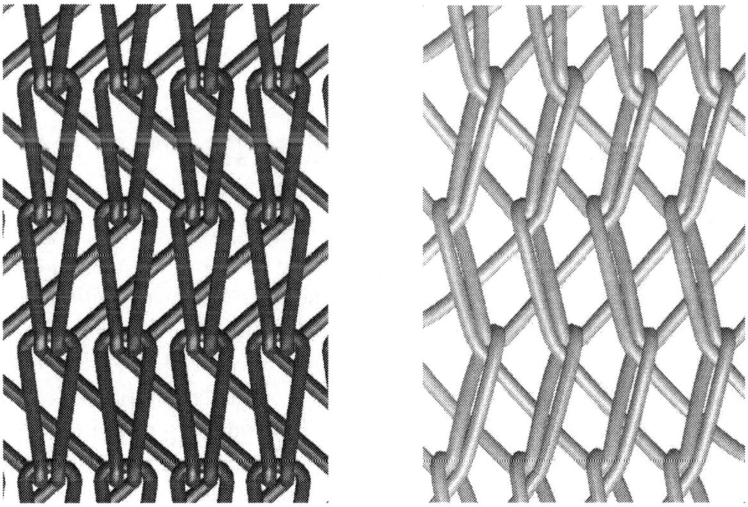

(a) (b)

8.42 Simulated warp knitted fabrics: (a) pure geometrical simulation and (b) after application of the mechanical model with force equilibrium.

8.7 Future trends

The simulation of the dynamic behaviour of textile fabrics is required for composites forming operations, for clothing simulation and other applications. The drape behaviour of fabrics can be simulated efficiently using mass-spring systems as shown in Han and Stylos (2010), Volino and Magnet-Thalmann (2000) and several other papers. This behaviour requires the material constants of the fabrics such as elastic, shear and bending modulus and area density, which depend on the material used and the woven structure.

Multi-scale approaches integrate the simulation on at least two levels – mezo-level – with the yarn geometry and pattern configuration and macro-level – with the effective fabric data, so that the complete fabric behaviour can be simulated. Such multiscale approaches require more complex computational models and special strategies which allow the material data at the macro level to be updated with the results from simulation at the meso-scale level. Interesting results are presented for instance in Vidal-Salle and Boisse (2010) for woven structures.

Another approach is to represent the single fibre or filaments in the yarns and model their complete behaviour. These models are extremely computational intensive, because several contact detections and calculations have to be performed at every time step; it is well known that contact calculations are time-consuming operations. With increasing computer power, such models are already proving useful, as reported for instance in Durville (2010). Another trend is the integrated simulation of the entire production process, including yarn and fabric deformation and the movement of machine parts which can lead to a simulation of the fabric structure (for instance using FEM; see Chapter 6).

8.8 Sources of further information

Books

The basics of the fabric mechanics can be found in several journal articles. A good overview of the classical methods is given in Postle *et al.* (1988). A more modern, computational oriented presentation of the theory is given in the *Theory Manual of WiseTex* (Lomov *et al.*, 2011). A new book, which summarizes the most known methods for the calculations of single layer woven structures is Behera and Hari (2010). The modelling of knitted structures is discussed in a separate chapter in Kyosev and Renkens (2010) and several other modelling aspects for the prediction of the textile behaviour can be found in Chen (2010).

Software for fabrics design and simulation

This is a slightly modified list of the software for weaving, taken from Apparel Research (2011), which gives only an overview about the most programs, and is not exhaustive. It must be noted that some companies have merged in recent years and their products hence appear under new names, while other companies have good products but their web pages are not listed on the top results when searching the WWW in some countries. There are as well several good simulation tools, developed for certain companies, which are reported in some papers, but are not available for a non-licensed user.

Arahne	http://www.arahne.si/
AVL Looms, inc	http://www.avlusa.com/
Canyon Art	http://www.weaveit.com/
The Design Scope Company (EAT)	http://www.designscopecompany.com/
Information Systems Consultants	http://www.isconindia.com/
JacqCAD International	http://www.jacqcad.com/
Mapple Hill Software	http://www.mhsoft.com/
MüCAD	http://www.mueller-frick.com
Point Carre USA	http://www.pointcarre.com/
Porini USA	http://www.porini.us/
Penelope®	www.infotex.es
Scot Weave	http://www.scotweave.com/
Startes Jaquard	http://www.startes.it/
TeckMen Systems LTD	http://www.teckmen.com/
Viable Systems Inc.	http://www.viacad.com/
Weave Point	http://www.weavepoint.com/
WiseTex	http://www.mtm.kuleuven.be/Onderzoek/Composites/software/wisetex
Yxendis	http://www.yxendis.com/

Weft knitting

FigPaint	http://www.fib.fr/gbsite/
Stoll	www.stoll.com
Shima Seiki	http://www.shimaseiki.com/
Steiger Textil	http://www.steiger-textil.ch/eng/profilo.htm
WeftKnit	http://liegebeest.studentenweb.org/

Warp knitting

MüCard	http://www.mueller-frick.com
TexMind	http://www.texmind.com
CADT	http://www.cadt.com/

ALC Computertechnik (now Texion)	http://www.texion.eu
EAT	http://www.designscopecompany.com

Braided fabrics

CAB Design	http://www.herzog-online.com/
TexMind	hww.texmind.com
WiseTex	http://www.mtm.kuleuven.be/Onderzoek/Composites/software/wisetex

8.9 References

A&P Technology, 2011. *Braiding calculator* [online]. Available from: http://www.braider.com/Resources/Braid-Calculator.aspx.

Akkerman, R. and Rodriguez, B.H.V., 2006. Braiding simulation for rtm preforms. *In: TexComp 8.*

Apparel Research, 2011. *Apparel Research* [online]. Available from: http://www.apparelsearch.com/software_knitting_and_weaving.htmte.

Arahne, 2011. *Arahne* [online]. Available from: http://www.arahne.si/.

Araújo, M. de, Fangueiro, R. and Hong, H., 2004. Modelling and simulation of the mechanical behaviour of weft-knitted fabrics for technical applications: Part III: 2D hexagonal FEA model with non-linear truss elements. *Autex Research Journal* [online], 4 (1), 24–32. Available from: http://www.autexrj.com/cms/zalaczone_pliki/5-04-1.pdf [Accessed July 2011].

Batra, S., Lee, D. and Backer, S., 1976. On the stress analysis of cylindrically wound packages: the case of a warp beam. *Textile Research Journal*, June, 453.

Beddoe, B., 1967. Anisotropy of yarn on a flanged tube. *The Journal of Strain Analysis*, 2, 207.

Behera, B. and Hari, B.K., 2010. *Woven Textile Structure: Theory and application.* Cambridge: Woodhead Publishing Ltd.

Chen, X., ed., 2010. *Modelling and Predicting Textile Behaviour*. Cambridge: Woodhead Publishing.

Durville, D., 2010. Simulation of the mechanical behaviour of woven fabrics at the scale of fibers. *International Journal of Material Forming*, *3* (S2), 1241–1251.

EAT, 2011. *Designscope* [online]. Available from: http://www.designscopecompany.com.

Efremov, E. and Efremov, B., 1982. *Basics of the theory of yarn winding*. In Russian. Original title: Ефремов,Е.Д., Ефремов,Б.Д. Основы теории наматывании нити на паковку. Moskau: Legkaia i pistevaia promislenost.

Ghosh, T.K. and Peng, H., 1995. Analysis of fabric deformation in a roll-making operation. *Textile Research Journal*, *65* (12), 739–747.

Ghosh, T.K., *et al.*, 1991a. Analysis of fabric deformation in a roll-making operation. *Textile Research Journal*, *61* (4), 185–192.

Ghosh, T., *et al.*, 1991b. Analysis of fabric deformation in a roll-making operation: Part I: A static Case. *Textile Research Journal*, *61* (3), 153–161.

Goktepe, O. and Harlock, S., 2002. Three-dimensional computer modeling of warp knitted structures. *Textile Research Journal*, *72*, 266–272.

Grishanov, S. and Lomov, S.E.A., 1997. The simulation of the geometry of two-component yarns. Part I: The mechanics of strange compression: Simulating yarn cross-section shape. *Journal of Textile Institute*, *88*, Part 1 (2).

Han, F. and Stylos, G.K., 2010. 3D modelling, simulation and visualisation techniques for drape textiles and garments. *In:* X. Chen, ed. *Modelling and predicting textile behaviour*. Cambridge: Woodhead Publishing, 388–421.

Herzog Maschinenfabrik, A., 2011. *CAB Design: Computer Aided braid design* [online]. Oldenburg. Available from: http://www.herzog-online.com/conpresso4/_data/CAB-Design_en.pdf.

Heuwinkel, F., 2011. *Analyse mechanischer Eigenschaften von Bandgeweben*. Hochschule Niederrhein.

Jiang, J., Hu, J. and Ko, F., 1999. Characterizing and modeling bending properties of multiaxial warp knitted fabrics. *Textile Research Journal*, *69* (9), 691–697.

Johansen, B., Lystrup, A., and Jensen, M., 1998. CADPATH a complete program for the CAD- CAE-and CAM-winding of advanced fibre composites. *Journal of Materials Processing Technology*, *77* (1–3), 194–200.

Kempner, E. and Hahn, H., 1995. Effects of radial stress relaxation on fiber stress filament winding of thick composites. *Composites Manufacturing*, *6* (2), 67–77.

Kessels, J. and Akkerman, R., 2002. Prediction of the yarn trajectories on complex braided preforms. *Composites Part A: Applied Science and Manufacturing*, *33* (8), 1073–1081.

Koltze, K., 2001. *Zur Optimierung des Zentrifugenspinnverfahrens durch robustes Prozeß-design*. PhD. Technische Universität Chemnitz.

Koussios, S., Bergsma, O. and Beukers, A., 2004a. Filament winding. Part 2: Generic kinematic model and its solutions. *Composites Part A: Applied Science and Manufacturing*, *35* (2), 197–212.

Koussios, S., Bergsma, O., and Beukers, A., 2004b. Filament winding. Part 1: Determination of the wound body related parameters'. *Composites Part A: Applied Science and Manufacturing*, *35* (2), 181–195.

Kurbak, A. and Ekmen, O., 2008. Basic studies for modeling complex weft knitted fabric structures Part I: A geometrical model for widthwise curlings of plain knitted fabrics. *Textile Research Journal*, *78* (3), 198–208.

Kurbak, A. and Soydan, A.S., 2008. Basic studies for modeling complex weft knitted fabric structures Part III: A geometrical model for 1 × 1 purl fabrics. *Textile Research Journal*, *78* (5), 377–381.

Kyosev, Y., 2007. Model generator for tubular braided fabrics. *In: Finite Element Modeling of Textiles and Textile Composites*. St Petersburg, Russia: Katholieke Universiteit Leuven & St Petersburg State University of Technology and Design, 1.

Kyosev, Y., 2011. *TexMind Braid Modeller* [online]. Available from: www.texmind.com.

Kyosev, Y. and Renkens, W., 2010. Modelling and visualization of knitted fabrics. *In:* X. Chen, ed. *Modelling and Predicting Textile Behaviour*. Cambridge: Woodhead Publishing, 225–262.

Kyosev, Y., Angelova, Y. and Kovar, R., 2005. 3D modelling of plain weft knitted structures from compressible yarn. *Research Journal of Textile and Apparel, Hong Kong*, *9*, 88–97.

Kyosev, Y., Reinbach, I. and Gries, T., 2006a. Stability problems of textile wound structures. *PAMM*, *6* (1), 235–236.

Kyosev, Y., Reinbach, R. and Gries, G., 2006b. Virtual bobbin: building; potential and problems. *Proceedings in Applied Mathematics and Mechanics*, *6*, 235–236.

Leaf, G.G.A., 1955. The geometry of a plain knitted loop. *Journal of the Textile Institute*, *45*, T587–605.

Li, H., Liang, Y. and Bao, H., 2005. CAM system for filament winding on elbows. *Journal of Materials Processing Technology*, *161* (3), 491–496.

Liao, T. and Adanur, S. (2000), '3D-structural simulation of tubular braided fabrics for net-shape composites', *Textile Research Journal*, *70* (4), 297–303.

Lomov, S., *et al.*, 2006. *Integrated textile preprocessor WiseTex, Version 2.5: Computational models, methods and algorithms*. Leuven: KU Leuven, Department MTM.

Lomov, S. *et al.*, 2011. *WiseTex* [online]. Available from: http://www.mtm.kuleuven.be/Onderzoek/Composites/Research/meso-macro/textile_composites_map/textile_modelling/textile_modelling_fe.

Moesen, M. and Lomov, S.V., 2011. *WeftKnit* [online]. Available from: http://liegebeest.studentenweb.org/weftknitEN.html.

Moesen, M., Lomov, S. and Verpoest, I., 2003. Modelling of the geometry of weft-knit fabrics. *In: Techtextil (Hg.) 2003 – TechTextil Symposium 7–10 April 2003.*

Neal, B., 1967. Behaviour of filament yarn wound on a flanged cylinder. *Journal of Strain Analysis*, *2*, 213–219.

Nishimoto, H. *et al.*, 2010. Prediction method for temporal change in fiber orientation on cylindrical braided preforms. *Textile Research Journal*, *80* (9), 814–821.

Peirce, F.T., 1937. 5 – The geometry of cloth structure. *Journal of the Textile Institute Transactions*, *28* (3), T45.

Pickett, A.K., Sirtautas, J. and Erber, A., 2009. Braiding simulation and prediction of mechanical properties. *Applied Composite Materials*, *16* (6), 345–364.

Postle, R., Carnaby, G. and Jong, S. de, 1988. *The Mechanics of Wool Structures* Ellis Horwood Limited.

Potluri, P., *et al.*, 2003. Geometrical modelling and control of a triaxial braiding machine for producing 3D preforms. *Composites Part A: Applied Science and Manufacturing*, *34* (6), 481–492.

Proshkov, A., 1986. *Mechanisms for yarn winding (topics of design)*. In Russian. Original title: Прошков, А.Ф. Механизмы раскладки нити (вопросы проектирования), Москва, Легпромбытиздат. Moscow: Legpromizdad.

Rawal, A., Potluri, P. and Steeleand, C., 2005. Geometrical modeling of the yarn paths in three-dimensional braided structures. *Journal of Industrial Textiles*, *35* (2), 115–135.

Rawal, A., Potluri, P. and Steele, C., 2007. Prediction of yarn paths in braided structures formed on a square pyramid. *Journal of Industrial Textiles*, *36* (3), 221–226.

Renkens, W. and Kyosev, Y., 2007. Geometrical modelling of warp knitted fabrics. *In: 'Finite Element Modelling of Textiles and Textile Composites', St Petersburg, 26–28 Sept. 2007.*

Renkens, W. and Kyosev, Y., 2011. Geometry modelling of warp knitted fabrics with 3D form. *Textile Research Journal*, *81* (4), 437–443.

Robitaille, F. *et al.*, 2000. Geometric modelling of industrial preforms: warp-knitted and multiple layer textiles. *Journal of Materials: Design and Applications, Proc. Institution of Mechanical Engineers (Part L), 214*, 71–90.

Sherburn, M., 2007. Geometric and mechanical modelling of textiles. PhD. University of Nottingham.

Shima Seiki, 2011. *SDS-ONE APEX* [online]. Available from: http://www.shimaseiki.com/product/design/sdsone_apex/.

Simon, L. and Hübner, M., 1983. *Vorbereitungsmaschinen für Weberei, Strickerei und Wirkerei.* Berlin, Heidelberg, New York, Tokyo: Springer Verlag.

Stueve, J. and Gries, T., 2009. Advances in the simulation of the overbraiding process using FEM. *In:* SAMPE, ed. *SEICO 09: Sampe Europe 30th International Jubilee Conference and Forum*, 618–625.

TexEng Software Ltd, 2011a. *KnitGeoModeller* [online]. Available from: http://www.texeng.co.uk/knitgeomodeller.html.

TexEng Software Ltd, 2011b. *Weave Engineer* [online]. Manchester. Available from: http://www.texeng.co.uk/2dfabricsl.html.

TexMind UG, 2011. *Modelling of Textiles with TexMind Software Products – an overview* [online]. Mönchengladbach. Available from: www.texmind.com.

University of Nottingham, 2011. *TexGen* [online]. Available from: http://texgen.sourceforge.net/index.php/Main_Page.

Ursiny, P., 1986. Neue Erkenntnisse über den Spannungsaufbau in Spulen. *Textiltechnik*, *36* (10), 535.

Ursiny, P., Hes, L. and Mägel. M., 1995. Vergleichmäßigung des Spannungsaufbaues in Spulen. *Meliand Textilberichte*, *76* (5), 314–315.

van Wyck, C., 1946. A study of the compressibility of wool, with special reference to South African merino wool. *Oderstepoort Journal of Veterinary Science and Animal Industry*, *21* (99).

Verpoest, I. and Lomov, S.V., 2005. Virtual textile composites software WiseTex: integration with micro-mechanical, permeability and structural analysis. *Composites Science and Technology*, *65* (15–16), 2563–2574.

Vicentini, V., 1969. Calculation of the elastic deformation of textile beams during winding. *Journal of Strain Analysis*, *4*, 48–56.

Vidal-Salle, E. and Boisse, P., 2010. Modelling the structures and properties of woven fabrics. *In:* X. Chen, ed. *Modelling and Predicting Textile Behaviour*. Cambridge, Woodhead Publishing, 144–179.

Volino, P. and Magnet-Thalmann, N., 2000. *Virtual Clothing – Theory and Practice*. Springer.

Wirjadi O., 2009. Models and algorithms for image-based analysis of microstructures. PhD Dissertation.

Wirjadi, O., *et al.*, 2009. Applications of anisotropic image filters for computing 2D and 3D-fiber orientations, stereology and image analysis. *In: Proceedings of the 10th European Conference of ISS 2009: The MIRIAM Project Series*, 107–112.

Zhang, W., Ding, X. and Li, Y., 2007. Microstructure of 3D braided preform for composites with complex rectangular cross-section. *Journal of Composite Materials*, *41* (25), 2975–2983.

9

Simulation of textile machinery

D. VEIT, RWTH Aachen University, Germany

Abstract: The simulation of textile machines is of increasing importance as trials of machines are often time consuming and expensive. In addition, many customers today demand quick production of small lots which in turn requires an optimized machine setting within a very short period of time. A wide range of methods is described in this chapter, covering processes from yarn production and processing through fabric manufacturing to finishing. The relevant systems of equations are explained where possible and simulation results are illustrated for selected examples of typical applications that have, in many cases, have been commercialized.

Key words: mathematical model, simulation, textile machine.

9.1 Introduction: simulation methods to optimize machine settings and product quality

The importance of developing mathematical models and simulation programs for textile machines becomes apparent when considering the following developments in recent years.

- Trials to run machines with new materials or new settings are increasingly time-consuming and expensive due to the ever-increasing complexity of textile machines.
- A lot of knowledge about proper machine settings is being lost due to many experienced textile workers retiring in developed countries where textile industries are longest established; workers in the developing world often lack the respective expertise and are thus not able to fill this gap.
- Customers expecting delivery of their orders within a very short period of time and ordering smaller lots in shorter intervals, thus not leaving enough time for time-consuming trials.
- Machines becoming faster which in turn increases wear on machine parts; this makes it ever more important to calculate optimum machine setting in order to reduce the stresses on the machine and its components to increase their economic life.

Developing simulation programs for textile machines requires both a profound mathematical and technical understanding of textile processes and textile materials as well as access to industrial-scale production facilities in order to verify the developed tools. Up until the turn of the millennium, these requirements were comparatively easy to meet in the industrialized world. Since then, the developing world has been gaining ground and it is to be expected that more and more computer simulations of textile machines will also be developed there in the years to come. So far, most simulations of textile machines that have been put into practice have been developed in Europe and this research thus predominates among the references at the end of this chapter. In the following sections, a range of mathematical models are presented, covering textile production from spinning through fabric production to finishing. Some selected examples, which have been put into practice or may in other ways be interesting for practical application, are described in detail.

Most simulation tools developed in recent years to improve machine settings for new materials, new products or new components use a numerical approach. They apply the methods described in the previous chapters such as artificial neural networks, evolutionary algorithms, fuzzy logic and computational fluid dynamics. The majority of models are based on data gathered through trials rather than on equations derived from the laws of physics. Only a very few published models apply the latter principle. Some of them are described in Section 9.2. In general, it is not difficult to optimize machine settings as the values by which the degree of optimization is assessed can normally be measured directly on the machine. This is not possible when it comes to improving product quality.

It has proven to be extremely difficult to relate product properties to machine settings via analytical equation systems for two reasons.

- Firstly, there are many parameters that have an influence on the final product quality that cannot be measured directly on a particular machine since several machines are normally employed to manufacture the final product.
- Secondly, and as a consequence of the first reason, there are a large number of parameters that influence each other or whose influence cannot be directly measured, e.g. yarn orientation, temperature, fineness and evenness of intermediate products. This makes it extremely complicated to include them all properly into a mathematical model.

Hence, when it comes to improve the quality of products by applying simulation techniques, the vast majority of models that have been put into industrial practice are based on numerical methods as described in Chapters 2–5. As Fig. 9.1 shows, these models can principally work in both directions: predicting product quality based on the machine settings and optimizing machine settings to achieve certain product properties. Since a certain machine setting should always lead to a certain product quality, whereas a

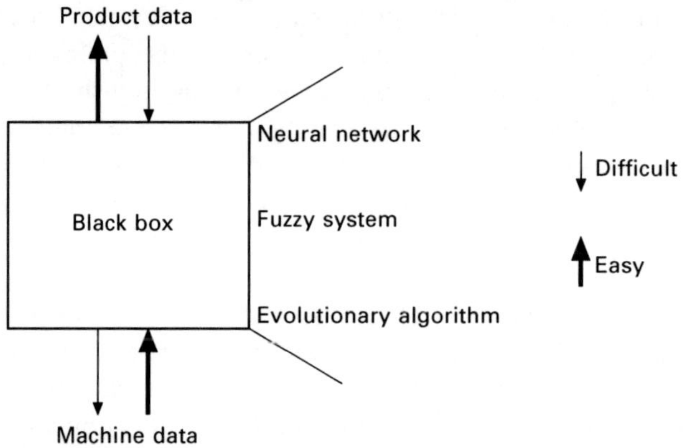

9.1 Principles of mathematical models to improve product quality.

certain product quality can be achieved potentially with different machine settings, the first approach is much easier to be put into practice than the latter, as has already been explained in Chapter 2. Nevertheless, there are a few noteworthy examples of analytical approaches to improve product quality which are explained in the next section.

In this chapter, typical examples for the simulation of textile machines are described. They cover yarn formation, fabric production, finishing and plant design. Since there is often a direct industrial application of the results of such simulations, there are a number of mathematical models that have been developed by researchers directly for use in the textile industry. Since these were not published, they are not cited here.

9.2 Simulating yarn production and processing: ring/rotor spinning and false-twist texturing

Basligil (1994) published an extensive literature review with regard to papers on machine interference, explaining the basic concepts of machine interference and naming possible solutions. This is not a true simulation but nevertheless provides a valuable source for further information about this topic which could be used to create a mathematical model. An early example of mathematically modelling staple-fibre spinning processes can be found in Crank and Whitmore (1953, 1954) and Crank (1953). In these studies the effect of balloon height, cap radius, bobbin radius and speed and yarn fineness on, for example, the size of the balloon and on the yarn tension within the bobbin are analysed. Certain simplifying assumptions are made, e.g. that the cross-section of the yarn is circular and the thread itself perfectly flexible and inextensible. With this model, it was possible to determine conditions under which a balloon can develop or the yarn breaks due to high tension.

The model itself is convincing, but experimental data to verify the results of the model were not given.

More recent papers about models to analyse balloon formation during ring-spinning include Batra *et al.* (1989a, 1989b). They showed that it is possible in principle to calculate, for example, the mass of the traveller required to achieve a certain yarn tension, the tension distribution along the yarn path and to determine the shape of the balloon. Data from actual trials to verify the model are not given.

Owing to increasing computer power, papers with more sophisticated mathematical models on ring spinning have been published. A typical example which is described below in more detail is Tang *et al.* (2010a). They designed a mathematical model and a numerical simulation to analyse dynamic behaviour and twist distribution in a modified ring spinning system which is shown in Fig. 9.2. In general, we have

$$F_{\text{yarn, acceleration}} + F_{\text{Coriolis}} + F_{\text{acceleration, centripedal}} = F_{\text{acceleration, yarn tension}} + F_{\text{air drag}} \qquad [9.1]$$

which is equivalent to

$$m\left[\frac{\mathrm{D}^2}{\mathrm{D}t^2} + 2\cdot\varpi\cdot e_z \times \frac{\mathrm{D}R}{\mathrm{D}t} + \varpi^2\cdot e_z \times (e_z \times R)\right] = \frac{\partial}{\partial s}\left(T\cdot\frac{\partial R}{\partial s}\right) + F \qquad [9.2]$$

where m stands for the yarn fineness, ω denotes the angular velocity of the pin, e_z coincides with the axis z in Fig. 9.3 and $R(s,t)$ is the yarn path with s representing the arc length and t the time.

The twist propagation equation according to Tang *et al.* (2010b) is given by

$$\frac{\partial^2 N(s,t)}{\partial t^2} + 2\cdot V\cdot\frac{\partial^2 N(s,t)}{\partial s\cdot\partial t} + (V^2 - V_0^2)\cdot\frac{\partial^2 N(s,t)}{\partial s^2} = 0 \qquad [9.3]$$

where $N(s,t)$ stands for the rotation angle of the cross-section of the yarn at position s at point-in-time t and V_0 represents the propagation velocity of the yarn twist, for which we have

$$V_0 = \sqrt{\frac{GI}{J}} \qquad [9.4]$$

with G, I and J representing the shear modulus, the polar moment of inertia and the unit moment of inertia.

In general, in ring spinning, it can be assumed that the twist propagation speed is much larger than the delivery speed of the yarn. The twist in each zone (see Fig. 9.2) then consists of three components: the original twist

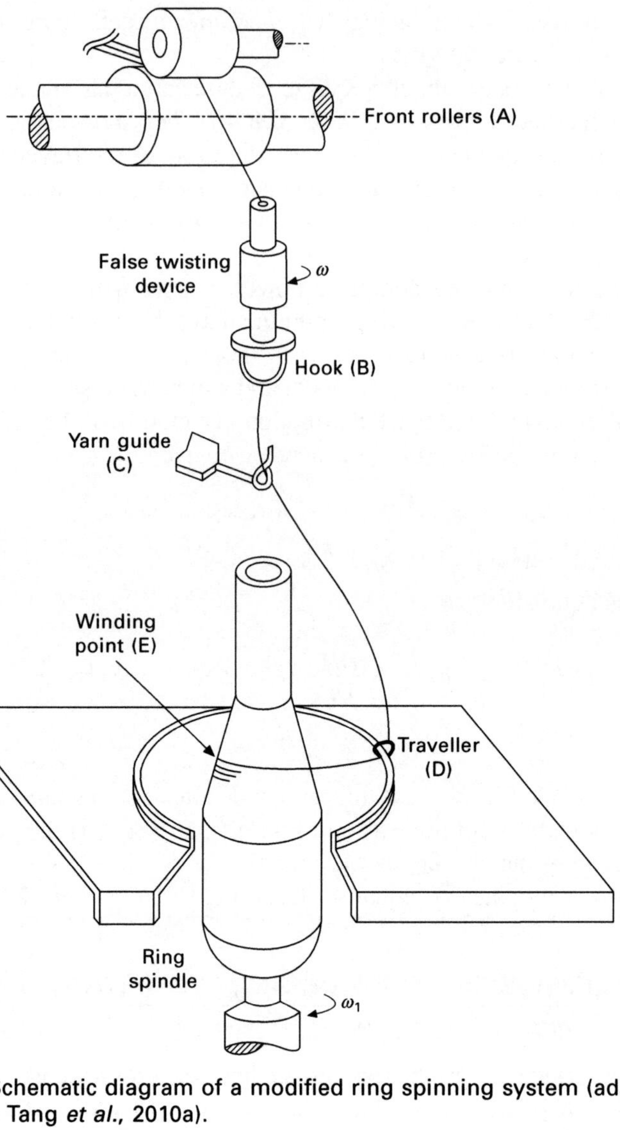

9.2 Schematic diagram of a modified ring spinning system (adapted from Tang *et al.*, 2010a).

wave and the two waves coming from each adjacent boundary zone. At the winding point, we have

$$\varpi_E = \varpi_D = \varpi_1 \qquad [9.5]$$

This model was used to create a computer simulation to predict the shape of the balloon and the tension along the yarn path from roller (Fig. 9.2, A) to the winding point at the bobbin (E). A typical bobbin curve is shown in Fig 9.3. Typical calculated results for the course of the yarn path are shown

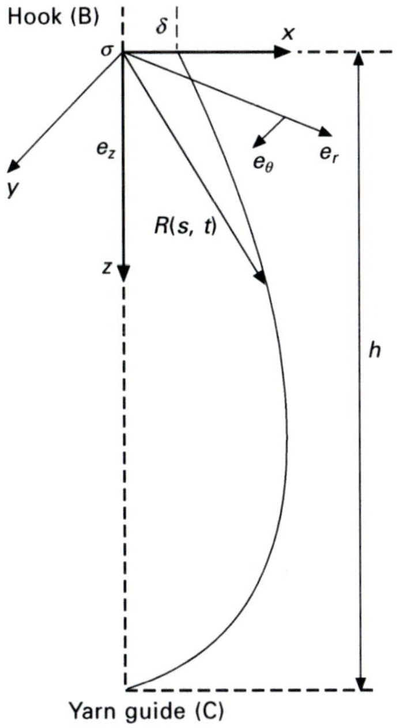

9.3 The balloon from the hook to the yarn guide (adapted from Tang *et al.*, 2010a).

in Fig. 9.4 for different yarn tensions T_0 at the yarn guide (C). These results were compared with yarn curves measured on an actual ring spinning machine and it was proven that the results coincide quantitatively. The calculated and the measured yarn twists were found to be very similar in all zones except CD. This was, according to the authors, due to the negligence of the yarn twist blockage of the yarn guide in the mathematical model. It remains to be seen whether a mathematical model of balloon formation will eventually also include this most interesting zone.

Zhu and Wang (2011) tried to detect the origins of faults caused by rollers in a spinning machine. They used an improved wavelet technique to first extract the relevant frequencies and then to determine the cause of the fault. Four different causes were identified using this approach.

An unusual approach to optimize the output of an open end (OE)-rotor spinning machine was published by Braccesi and Carfagni (1993). In automated OE-rotor spinning machines, a robot moves along the machine and removes full bobbins, reattaches broken yarns, cleans spin boxes, etc. In this paper, a mathematical model was designed that simulates the resulting output of the machine depending on different behaviours of the robot. Figure 9.5 shows the general approach. It was found that the factor with the highest influence on the machine output (e.g. number of full bobbins) was the number of attempts the robot was allowed to try to fix a problem before moving on and leaving it

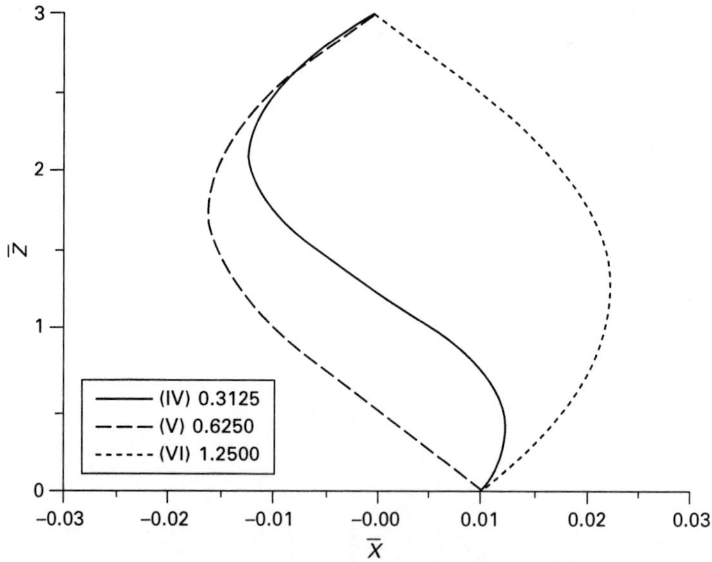

9.4 Typical bobbin curve (adapted from Tang *et al.*, 2010a).

to the operator to fix. Normally, a robot is allowed to try three times before resuming its movement along the machine with success rates in the range of 85%, 11% and 2% for the first, second and third attempts respectively. Allowing only one attempt does significantly increase the machine efficiency when the number of yarn breaks exceeds a certain limit, as can be seen in Fig. 9.6. The increase is more significant for coarse yarns than for thin yarns as the time needed to fix a problem relative to production time is directly correlated to the yarn fineness. Even if the success rate drops to 75% for the first attempt, this result persists.

The simulation model thus showed that in order to achieve optimized robot settings, the number of yarn breaks and the yarn fineness should be taken into account in order to increase the machine output. Whereas shorter yarn reattachment and bobbin removal times have a direct effect on machine efficiency, as was to be expected, the robot travelling speed showed no significant influence on machine output if the number of yarn breaks does not exceed typical values. This approach can also be used for any other type of textile machine when it comes to optimizing the automation process.

An interesting paper by Greenwood and Grigg (1985) explains the principle of texturing when using a false-twister with rotating discs. They developed an analytical mathematical model which could well be used to create a simulation of the false-twist texturing process. A similar model was developed by Olbrich (1995) which he used successfully to simulate the twist insertion in false-twist texturing to improve process stability.

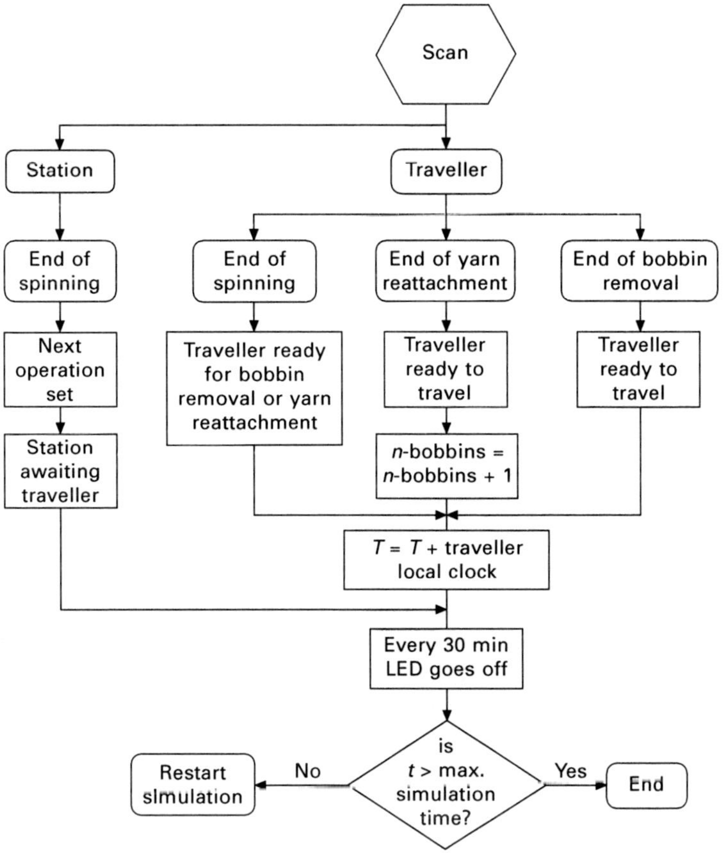

9.5 Block diagram of the simulation (adapted from Braccesi and Carfagni, 1993).

A paper by Baykara *et al.* (2009) describes modelling and predictive control of an air-jet texturing and twisting machine, based on a state-space model. The model shows the tension change in the twisting process. In designing the model predictive controller, two major objectives are considered:

- The first objective is that the yarn tension in the twisting process tracks a given reference tension.
- The second objective is that the yarn tension, with the texturing process assumed to be a step disturbance. has to track the same ideal tension value.

To verify the control algorithm performance, the proposed algorithm is compared with a PID controller using the Ziegler-Nichols method. All computer simulations were completed with the MIDSYS© toolbox and MATLAB© Simulink©. The simulation study shows that the proposed model predictive

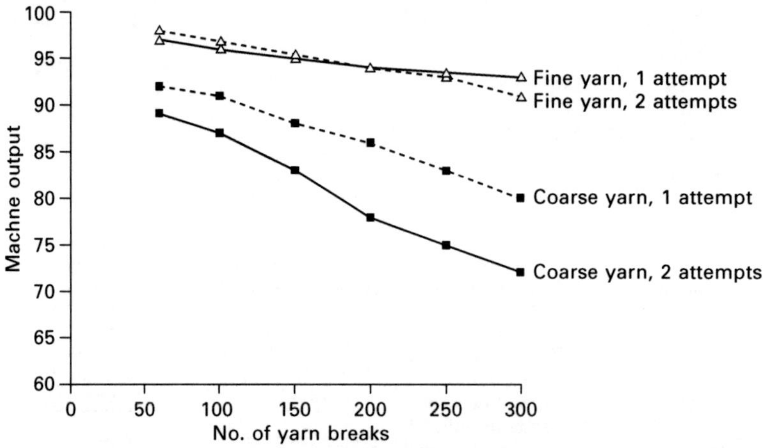

9.6 Machine output depending on no. of attempts to fix the problem (adapted from Braccesi and Carfagni, 1993).

control is an improvement over the proportional-integral-derivative (PID) controller for any control inputs.

Meier (1993) developed a model to describe heat transfer in the false-twist texturing process. This finally resulted in a new heater that set a new standard in texturing by increasing the texturing speed by about 50% and at the same time decreasing energy consumption. He proved that the transfer from heater to yarn takes place mainly through convection, hence we have

$$\frac{\dot{Q}}{A} = \alpha \cdot \Delta\vartheta \qquad [9.6]$$

with $\dot{Q}$ = heat flow, $\Delta\vartheta$ = temperature difference between heater surface and fluid and A = area through which the heat transfer takes place.

The heat transfer coefficient α represents, for example, the fluid velocity and other values that have an influence on the heat exchange. Assuming that the yarn is an ideal cylinder, the temperature inside the yarn is given by

$$\frac{\partial\vartheta}{\partial t} = \frac{a}{r} \cdot \frac{\partial}{\partial t}\left(r \cdot \frac{\partial\vartheta}{\partial r}\right) \qquad [9.7]$$

with r = radius of the cylinder and t = time.

All relevant parameters of the fluid and the yarn are taken into account by

$$a = \frac{\lambda}{\rho \cdot c} \qquad [9.8]$$

with λ = heat conductivity, ρ = density and c = heat capacity.

Equation 9.7 can be rewritten in non-dimensional form by introducing the normalized temperature by

$$\Theta = \frac{\vartheta - \vartheta_\infty}{\vartheta_A - \vartheta_\infty} \quad [9.9]$$

the normalized distance to the centre of the yarn according to

$$\xi = \frac{r}{R} \quad [9.10]$$

and the dimensionless time (Fourier number)

$$\tau = \frac{a \cdot t}{R^2} \quad [9.11]$$

This leads to a cylinder-symmetrical temperature field, which is given by

$$\frac{\partial \Theta}{\partial \tau} = \frac{1}{\xi} \cdot \frac{\partial}{\xi}\left(\xi \cdot \frac{\partial \Theta}{\partial \xi}\right) \quad [9.12]$$

which is equivalent to

$$\frac{\partial \Theta}{\partial \tau} = \frac{\partial^2 \Theta}{\partial \xi^2} + \frac{1}{\xi} \cdot \frac{\partial \Theta}{\partial \xi} \quad [9.13]$$

In order to solve this equation numerically, it must be transformed into a difference equation; hence we get

$$\Theta_{I+1,K} = \frac{\Delta\tau}{(\Delta\xi)^2} \cdot \left[\Theta_{I,K+1} \cdot \left(1 + \frac{\Delta\xi}{2 \cdot \xi_K}\right) + \Theta_{I,K-1} \cdot \left(1 - \frac{\Delta\xi}{2 \cdot \xi_K}\right)\right] + \left(1 - 2\frac{\Delta\tau}{(\Delta\xi)^2}\right) \cdot \Theta_{I,K} \quad [9.14]$$

The continuous time and location parameters τ and ξ are replaced by discrete values τ_I, τ_{I+1},... and ξ_K, ξ_{K+1},... Their values differ by $\Delta\tau$ and $\Delta\xi$ respectively. Using this equation, it is possible to calculate the temperature at any position ξ_K at point-in-time $\tau_{I+1} = \tau_I + \Delta\tau$ from the temperature at the adjacent positions ξ_{K+1} and ξ_{K-1} at point-in-time τ_I. This method works well if the increments meet the following requirement:

$$\frac{\Delta\tau}{(\Delta\xi)^2} \leq \frac{1}{2} \quad [9.15]$$

If $\Delta\tau/(\Delta\xi)^2 = \frac{1}{2}$, equation [9.14] can be rewritten, hence we get

$$\Theta_{I+1,K} = \frac{1}{2} \cdot \left[\Theta_{I,K+1} \cdot \left(1 + \frac{\Delta\xi}{2 \cdot \xi_K}\right) + \Theta_{I,K-1} \cdot \left(1 - \frac{\Delta\xi}{2 \cdot \xi_K}\right)\right] \quad [9.16]$$

The division of the position ξ has to be defined so that the temperature in the centre of the cylinder must not be calculated as this would lead to a indeterminate expression in equation [9.16]. Hence we have

$$\xi_K = \Delta\xi \cdot \left(K - \frac{1}{2}\right) \quad [9.17]$$

Dividing the yarn body into K_N sections where each has a thickness $\Delta\xi$, we have

$$\Delta\xi = \frac{1}{K_N} \quad [9.18]$$

In order to take into account the boundary conditions we need the temperatures at $K = 0$ ($\Delta\xi/2$ left of the yarn centre) and K_{N+1} ($\Delta\xi/2$ outside the surface). There, we have

$$\Theta_{I,0} = \Theta_{I,1} \text{ (symmetry)} \quad [9.19]$$

$$\Theta_{I,K+1} = \Theta_{I,K_N} \cdot \frac{1 - Bi \cdot \dfrac{\Delta\xi}{2}}{1 + Bi \cdot \dfrac{\Delta\xi}{2}} + \Theta_{I,\infty} \cdot \frac{Bi \cdot \Delta\xi}{1 + Bi \cdot \dfrac{\Delta\xi}{2}} \quad [9.20]$$

with Bi = Biot number which is the coefficient between the heat exchange between body and fluid and heat exchange inside the body according to

$$Bi = \frac{\alpha \cdot L}{\lambda}$$

with L = characteristic length.

For all temperatures within the interval $K = [1, K_N]$, we then get

$$\Theta_{I+1,K} = \frac{[(2K-1)+1] \cdot \Theta_{I,K+1} + [(2K-1)-1] \cdot \Theta_{I,K-1}}{2 \cdot (2K-1)} \quad [9.21]$$

With equations [9.19] to [9.21], it is possible to describe the heat transfer inside the yarn numerically.

The influence of the heat conductivity of the yarn is shown in Fig. 9.7. It is apparent that the influence of heat conductivity on the yarn temperature profile is small and can therefore be neglected. In contrast, the heat capacity of the yarn does have a significant influence on the absolute level of the yarn temperature profile as can be seen from Fig. 9.8. Applying the simulation programme, it was possible to determine the minimum time which is required to heat up the yarn to a certain average temperature. One boundary condition was that the difference between core and surface of the yarn must not exceed 15 °C as this has proven to pose no problems in subsequent processing (e.g. dyeing). Besides, due to the twist of the yarn, the yarn itself is shortened by approximately $E = 20\%$ which is also taken into account.

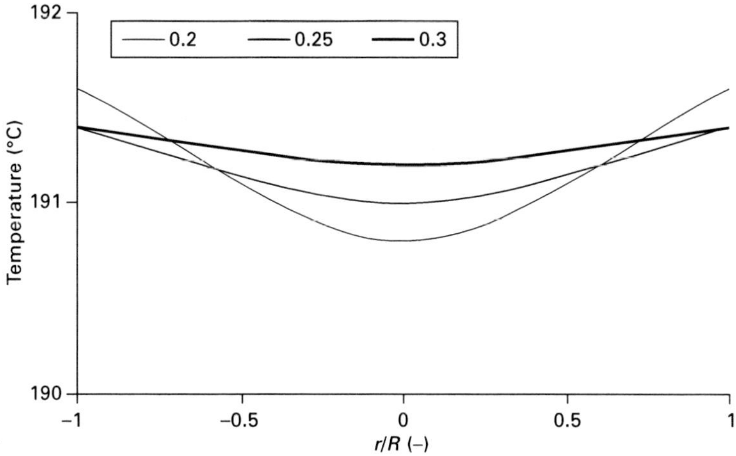

9.7 Influence of heat conductivity of the yarn on the temperature profile (adapted from Meier, 1993).

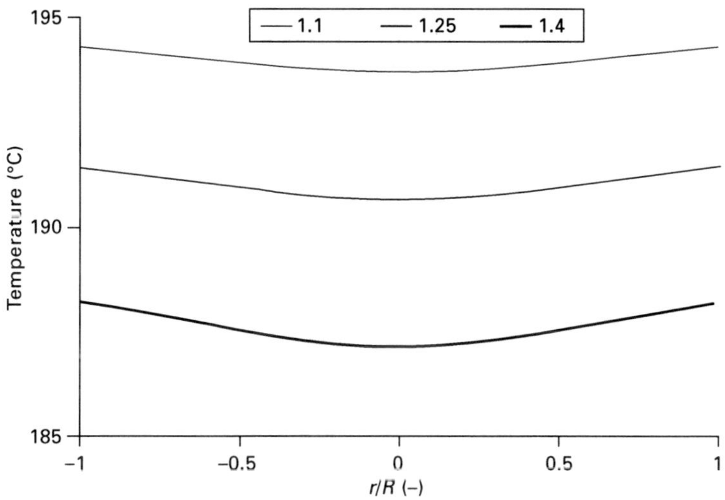

9.8 Influence of heat capacity on the yarn temperature profile (adapted from Meier, 1993).

From

$$t^* = \frac{l_{\text{heater}}}{V_{\text{texturing}}} \qquad [9.22]$$

with t^* = heating up time depending on the length of the heater l_{heater} and the texturing speed $v_{\text{texturing}}$, we hence get with

$$t = \frac{t^*}{(1 - E)} \quad [9.23]$$

the actual time t which the yarn has to spend inside the heater.

As Fig. 9.9 shows, the relation between minimum heating time and yarn fineness is linear for both, polyamide (PA) and polyester (PET). This leads to

$$t^*_{\min} = C \cdot Tt_{\text{tex}} \quad [9.24]$$

with C = material constant and Tt_{tex} = yarn fineness in the tex-system. The values for C can be calculated to $C_{\text{polyamide}} = 2.5\ 10^{-6}$ min/dtex and $C_{\text{polyester}} = 2.25 \times 10^{-6}$ min/dtex.

The time $t^*_{\min}$ in equation [9.24] therefore represents the physically shortest possible heating time for a yarn inside the heater. This makes it possible to calculate the efficiency of the heater according to

$$\eta_{\text{heater}} = \frac{t^*_{\min}}{t_{\text{residence}}} \quad [9.25]$$

where $t_{\text{residence}}$ stands for the residence time of the yarn inside the heater. This value is given by

$$t_{\text{residence}} = \frac{l_{\text{heater}}}{V_{\text{texturing}}} \quad [9.26]$$

Inserting equation [9.26] into equation [9.25] then leads to

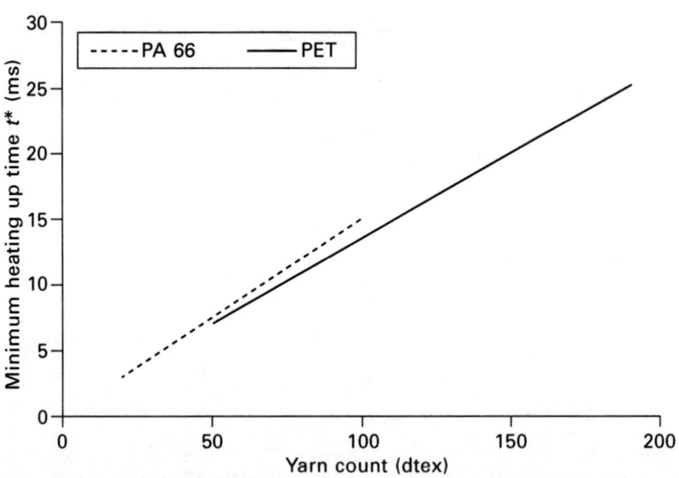

9.9 Shortest possible heating time $t_{\min}$ for polyester (PET) and polyamide (PA) yarns (adapted from Meier, 1993).

$$\eta_{heater} = \frac{C_{material} \cdot Tt_{tex} \cdot V_{texturing}}{l_{heater}} \cdot 100\%$$

Typical values for standard heaters are in the range of 10–15% which shows the great potential for improvement in this respect.

Figure 9.10 shows the average yarn temperature depending on the heating time for different heater temperatures. It is obvious that the relations are not linear, e.g. an increase of the heater temperature from 200 to 600 °C reduces the heating time t^* from 250 to 30 ms. In order to achieve a good heat-setting of the twist, experiments show that an average yarn temperature of 190 °C after the heater needs to be achieved. Figure 9.11 shows that the heater temperature directly affects the yarn temperature profile. The yarn temperature difference between core and yarn surface amounts to 30 °C for a heater temperature of 600 °C which needs to be equalized in the cooling zone in order to avoid differences in the dye take-up in later processes.

As Fig. 9.12 shows, the temperature of the core and the surface of the yarn can differ considerably at the beginning of the heater but are gradually equalized. In this example, the heater temperature was set to 200 °C and the fineness of the PET yarn was 167 dtex. Simulation results and measured values coincide very well as Fig. 9.13 shows. The slight deviations are due to a decreasing heat transfer coefficient with higher texturing speed in the practical trials whereas for the simulation a constant value was assumed.

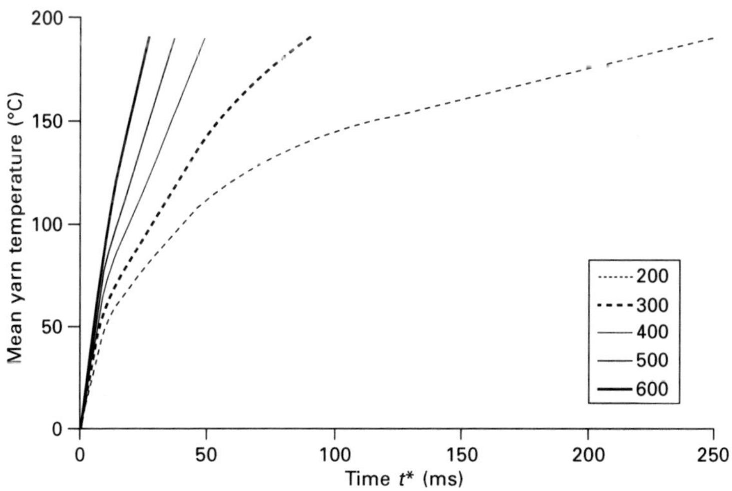

9.10 Average yarn temperature depending on heating up time for different heater temperatures (adapted from Meier, 1993).

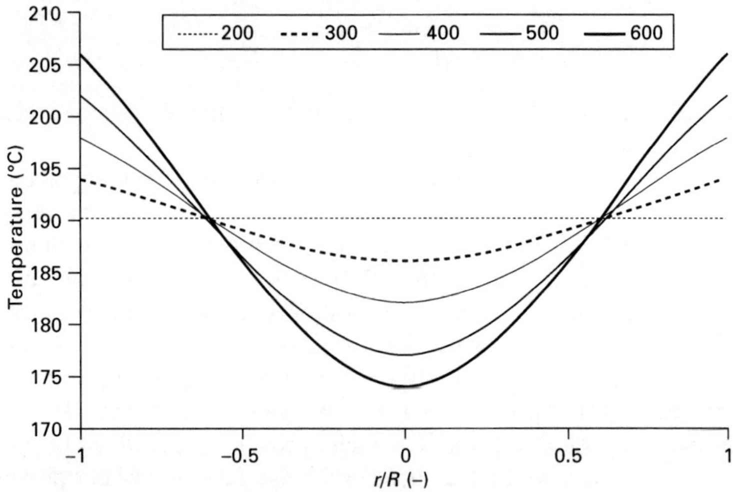

9.11 Influence of heater temperature on the yarn temperature profile (adapted from Meier, 1993).

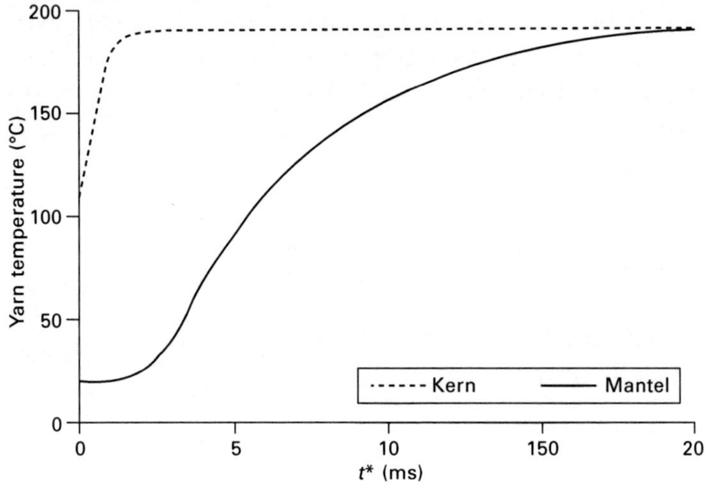

9.12 Temperature of core and surface of the yarn over time (adapted from Meier, 1993).

9.3 Simulating fabric production: weft insertion and warp tension

In this section, the focus is on fabric production through weaving as the number of published papers on knitting and warp-knitting machines in English is relatively small.

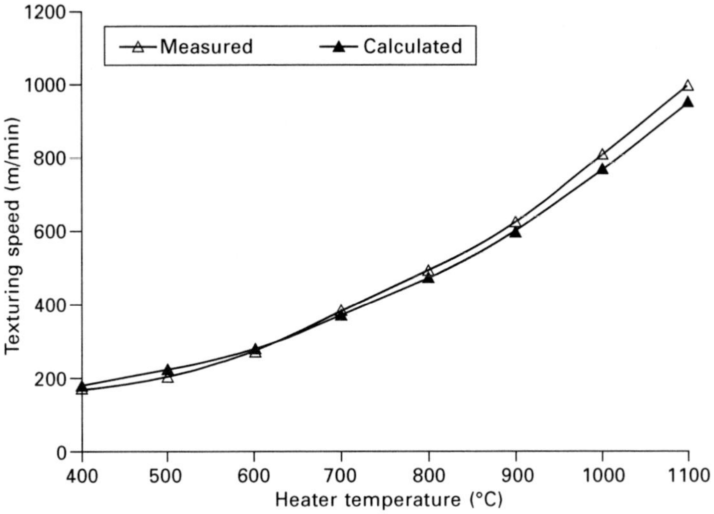

9.13 Comparison of calculated and measured values (adapted from Meier, 1993).

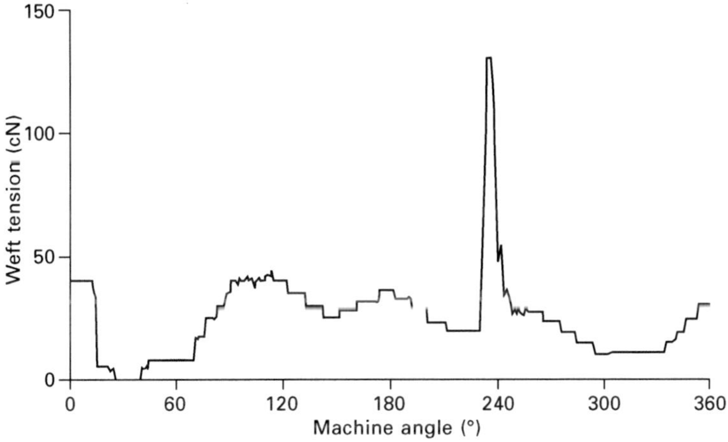

9.14 Weft yarn tension during weft insertion (adapted from Lehnert, 1994).

9.3.1 Weft insertion

Lehnert (1994) developed a simulation tool in order to increase the number of picks in modern weaving machines. The most important factor limiting the machine speed was determined to be the number of yarn breaks due to tension peaks. In a first step, Lehnert thus analysed the yarn tension during weft insertion (Fig. 9.14). It is apparent that the tension increases at the beginning due to accelerating forces and is nearly constant when the yarn passes through the shed. Then at a machine angle of approximately 240°, a tension peak occurs which is due to the strong deceleration of the weft when the yarn is stopped by the yarn tensioner. The aim of the simulation model was to model the behaviour of the yarn under different conditions during the weft insertion and thus design a yarn tensioner that would be able to reduce this peak by applying a constant braking force. Based on a model devised by Wulfhorst (1969), Lehnert develops a model for the stress–strain behaviour of yarns during weft insertion. Figure 9.15 shows typical shapes of stress–strain curves for yarns. The curves show areas of elastic and of plastic deformation.

The elastic deformation at the beginning of the stress–strain curve can be described by

$$F_{\text{linear}} = c \cdot \varepsilon \qquad [9.27]$$

with c representing the spring constant and ε as the elastic elongation. In many cases, yarns become stronger with increasing strain, which can be described by

$$F_{\text{elastic}} = F_{\text{str}} + F_{\text{linear}} \qquad [9.28]$$

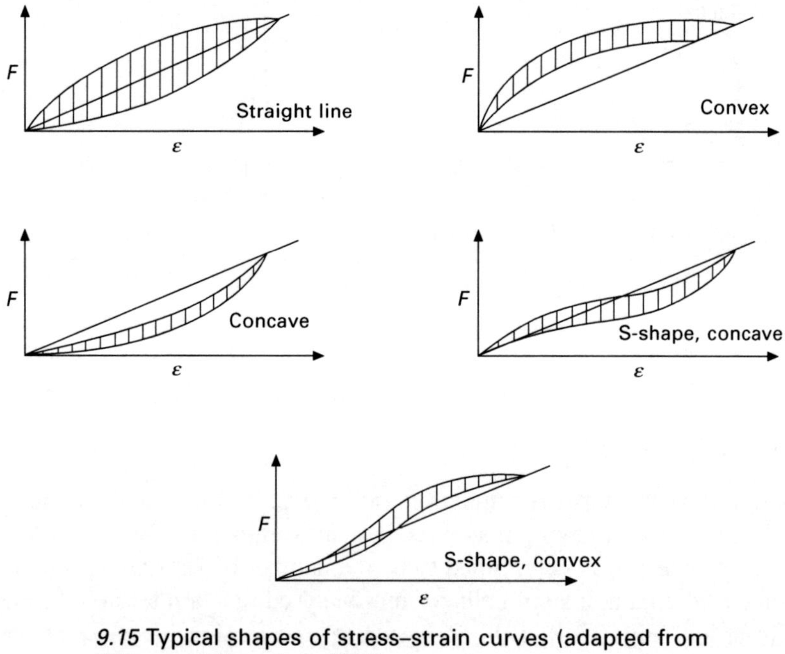

9.15 Typical shapes of stress–strain curves (adapted from Hättenschwiler *et al.*, 1984).

where

$$F_{str} = V_{str} \cdot \left(e^{-\left(\frac{\varepsilon}{V_{str}}\right)} + \frac{\varepsilon}{V_{str}} - 1 \right) \quad [9.29]$$

with V_{str} = additional force due to the yarn strain.

If the stress–strain curve deviates from a purely elastic behaviour, we also have plastic deformation according to

$$F_{plastic} = V_{plastic} \cdot \left(e^{-\left(\frac{\varepsilon,\,max}{V_{plastic}}\right)} + \frac{\varepsilon,\,max}{V_{plastic}} - 1 \right) \quad [9.30]$$

Hence we get for the dependence between stress and strain

$$F_{elongation} = F_{elastic} - F_{plastic} \quad [9.31]$$

It is possible to describe all curves as shown in Fig. 9.15 with this equation, provided that the deformation velocity is infinitely small.

In reality, with deformation velocities as they occur during weft insertion, the force due to plastic deformation increases the more plastic the yarn is stretched. Then we have

$$\dot{f}_t = \ln\left(\frac{\dot{F}_{plastic}}{V} + 1 \right) \quad [9.32]$$

with V = elongation parameter, $\dot{F}_{plastic}$ = velocity of viscoelastic deformation and $\dot{f}_t$ = initial value of the time-dependent additional force.

Hence we get for the force exerted on the weft during insertion

$$F_{overall} = F_{elastic} - F_{plastic} + F_{time\text{-}dependent} \quad [9.33]$$

This mathematical model is depicted in Fig. 9.16. The parameters in equations [9.27] to [9.33] were determined experimentally until the resulting stress–strain

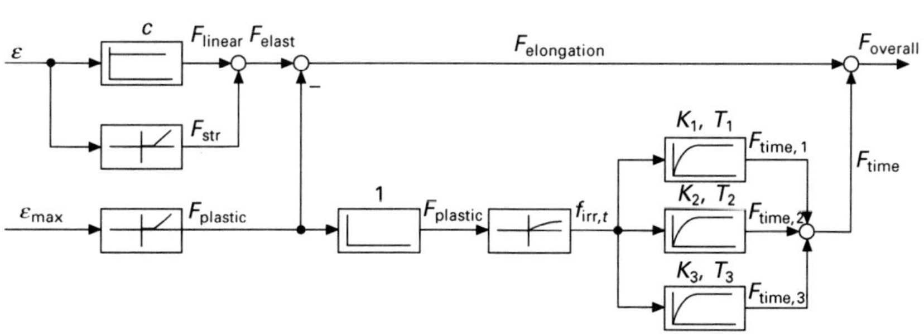

9.16 Signal flow chart of the yarn stress–strain model (adapted from Lehnert, 1994).

curves coincided sufficiently well with the actually measured curves for different materials (e.g. PET filament yarn, PET OE-rotor yarn).

The forces on the yarn during weft insertion are caused by frictional forces and forces of inertia. The latter depend on the yarn acceleration, its stress–strain behaviour and its mass. The forces of inertia can be calculated for a mass element representing the weft mass. This element exerts its force on the yarn at the end of the weft which is then considered without mass. Hence we have

$$F_{\text{inertia}} = F_{\text{yarn}} - F_{\text{friction}} \qquad [9.34]$$

Dividing the forces of inertia by the yarn mass, we get the weft acceleration. Integrating this equation twice, we get the distance which the weft has travelled for any given point in time during its insertion into the shed. This in turn affects the forces on the yarn according to its stress–strain behaviour and so forth. The respective signal flow chart is shown in Fig. 9.17.

Figure 9.18 shows that simulation and actually measured yarn tensions coincide very well. As is shown in Fig. 9.19, with this model it is possible to determine the tension components which exert an influence on the weft yarn. It is apparent that the plastic deformation increases throughout the weft insertion. At the end of the weft insertion, two scenarios can occur.

- Firstly, the braking force of the tensioner on the yarn is too low, thus resulting in a yarn loop and a possible fault in the fabric.
- Secondly, the braking force of the tensioner on the yarn is too high, thus causing unnecessary plastic deformation and hence possibly a yarn break.

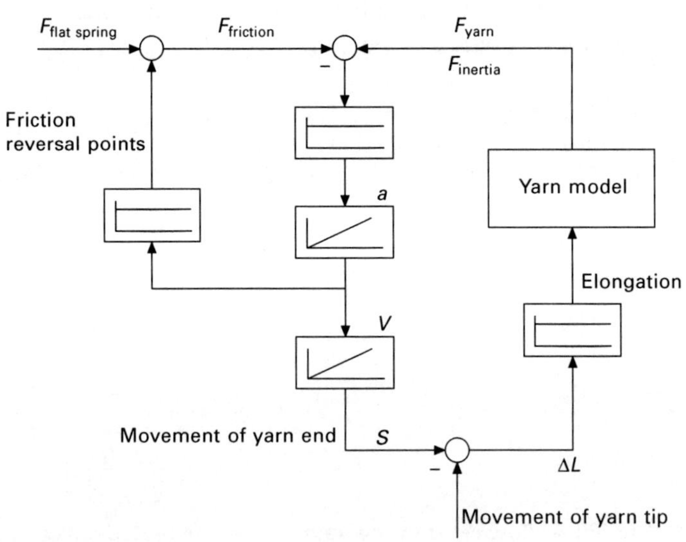

9.17 Signal flow chart of the simulation (adapted from Lehnert, 1994).

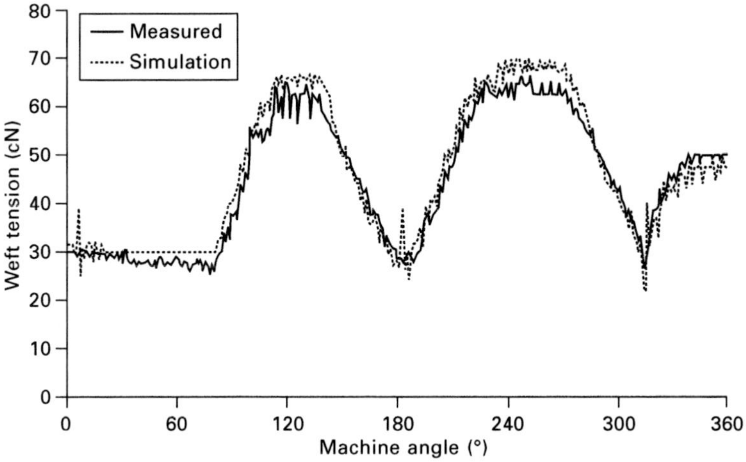

9.18 Calculated and actual tension values during weft tension (adapted from Lehnert, 1994).

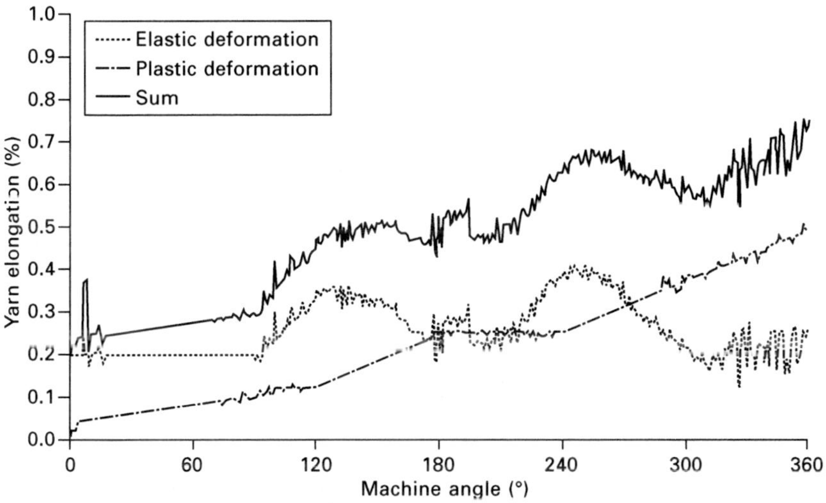

9.19 Deformation of the weft yarn during insertion (adapted from Lehnert, 1994).

In order to avoid both, it would be helpful to have the braking force exerted by the tensioner on the weft at as low a value as possible. The simulation model developed here can be used to achieve this by simulating the forces exerted on the weft yarn during insertion which allows the optimum braking force of the tensioner to be calculated at any given time. As can be seen in Fig. 9.20, the simulation model does indeed optimize the weft tension considerably.

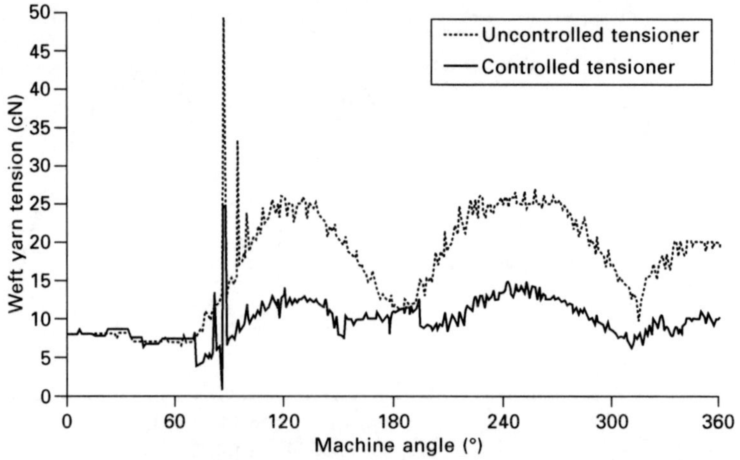

9.20 Comparison between yarn tension during weft insertion with and without a controlled yarn brake (adapted from Lehnert, 1994).

A more recent model was published by Celik *et al.* (2004). It is based on the following equation:

$$F_m + \Sigma_{n=1}^{6} F_{RJ,n} + F_{St} = F_B + F_X + F_A + F_R \qquad [9.35]$$

where F_m is the force on the yarn from the main jet, $F_{RJ,n}$ are the forces of the relay jets, F_{St} stands for the force from the stretching nozzle, F_B is the force of mass of inertia of the yarn from weft accumulator to the beginning of the reed, F_x represents the force of mass of inertia of the yarn through the reed, F_A is the force applied to the yarn to unwind it from the bobbin and F_R is the frictional force between yarn, machine components and air. According to their model, these forces are given by

$$F_m = C_R \cdot L_m \cdot 2 \cdot \sqrt{\frac{\pi \cdot Tt_{\text{tex}}}{\rho_{\text{yarn}}}} \cdot P_m \qquad [9.36]$$

with C_R = yarn force coefficient according to de Jager (1990), L_m = length of the main jet and P_m = main jet air pressure,

$$F_{RJ} = \frac{1}{2} \cdot C_f \cdot \rho_{\text{air}} \cdot (V_{\text{air}} - V_y)^2 \cdot \pi \cdot d_y^2 \cdot L_B \qquad [9.37]$$

with C_f = yarn force coefficient of the relay jets according to de Jager (1990), V_{air} = average air speed from the relay jets and V_y = weft speed. The yarn force coefficient was determined by measuring the force on the weft in the channel caused by the relay jets as a function of the air pressure.

The force by the stretching nozzle is regarded as constant:

$$F_{St} = \text{constant} \qquad [9.38]$$

The force of inertia on the yarn is given by

$$F_X = x \cdot Tt_{\text{tex}} \cdot \frac{d^2x}{dt^2} \qquad [9.39]$$

The force needed to unwind it from the bobbin is given by

$$F_A = \frac{1}{2} \cdot Tt_{\text{tex}} \cdot \left(\frac{dx}{dt}\right)^2 \qquad [9.40]$$

The frictional force F_R can be expressed by

$$F_R = C_{fd} \cdot f(t) \qquad [9.41]$$

with $f(t) = (1 - e^{[-(t-T_{sm})]/\tau})$ for $T_{sm} \le t \le T_{em}$ and $f(t) = e^{(t-T_{em})/2\tau}$ for $T_{em} \le t$, where C_{fd} stands for the frictional drag force coefficient for the weft, t is the running time, τ represents the pressure build-up time, T_{sm} the blowing start time and T_{em} the blowing stop time of the main jet. These equations lead to a nonlinear differential equation of the second order:

$$\frac{d_x^2}{dt^2} = \frac{1}{Tt_{\text{tex}} \cdot [x(t) + L_2]} \times \left(F_m + \sum_{n=1}^{6} F_{RJ} + F_{St} - F_A - F_R \right) \qquad [9.42]$$

where $x(t)$ stands for the position of the yarn tip while it travels through the reed and L_2 is the distance between weft accumulator and beginning of the reed.

For the simulation, the initial values of weft position, velocity and acceleration are set to zero. Figure 9.21 shows a comparison of the calculated

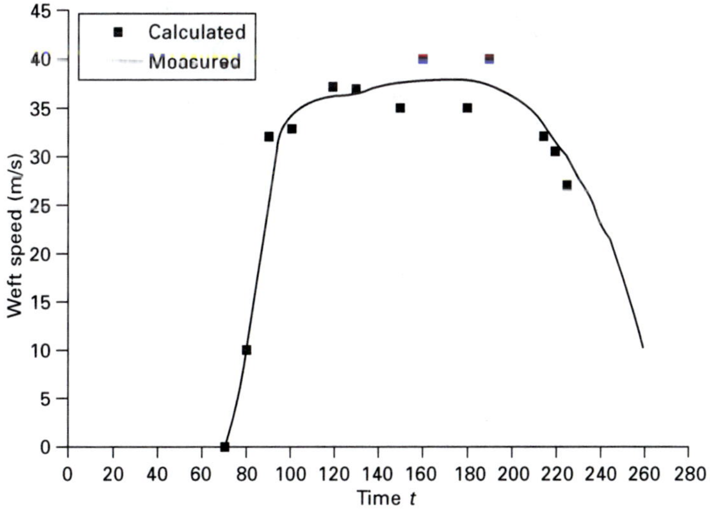

9.21 Comparison between simulation results and measured values (adapted from Celik *et al.*, 2004).

and experimental values for the weft speed for an OE rotor spun yarn. It is apparent that model and reality coincide well. This simulation method can be used to calculate and hence optimize, for example, the air consumption during weft insertion without having to resort to time-consuming trials.

9.3.2 Warp tension

In order to improve the machine settings without having to run expensive and time-consuming trials every time a new product is to be manufactured, a mathematical model describing the warp tensions was designed by de Weldige (1996). It describes the equilibrium of moments at the back rest of a weaving machine. The system and its boundaries are shown in Fig. 9.22. The boundaries are given by the point where the warp threads leave the warp beam and by the fabric beam. All weaving machine components that are taken into account are located above the grey plane.

The equilibrium of moments at the backrest is determined by taking into account:

- the forces of mass inertia (1 in Fig. 9.22);
- the forces of attenuation;

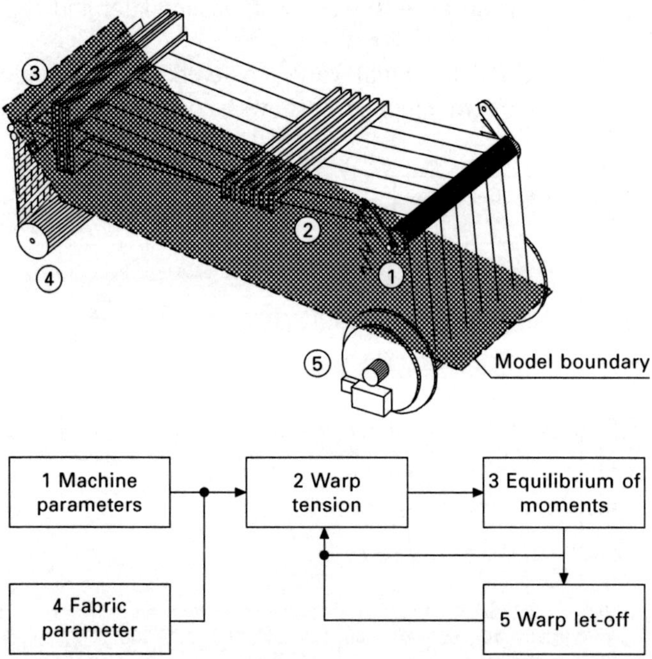

9.22 System boundaries for the mathematical model (adapted from de Weldige, 1996).

- the spring forces;
- the weight forces; and
- outer forces (2) causing the backrest to oscillate.

The warp tension is affected by the shed geometry, the weft beating up force, the fabric pattern and the weft density. The shed geometry and the weft beating up force are given by the machine parameters (3), fabric pattern and weft density result from the fabric parameters (4). Solving the equilibrium of moments leads to the movement over time of the backrest which in turn affects the warp tension. The control circuit is hence interconnected.

Several assumptions are made:

- The characteristics of all warp yarns are considered to be equal across the fabric width.
- Backrest, warp stop motion, sheds and reed are considered to be ideally stiff.
- The mass of the backrest is reduced to an equivalent mass; the springs and dampeners of the backrest are regarded as massless; their respective masses are taken into account in the equivalent mass of the backrest.
- The warp threads are considered to be spring elements with a linear stress–strain behaviour; this is allowed as the maximum elongation of the warp threads does not exceed 5%, which is well below the critical value of 10% where this assumption can no longer be made, as experiments show.
- The warp threads cannot take up pressure forces and are hence regarded as limp spring elements.
- The mass of the warp threads is several magnitudes smaller than the backrest mass and therefore neglected.
- The movement of heald frame and warp threads is the same, thus neglecting the heald allowance.
- The warp forces do not affect the movement of the heald frames.
- The influence of the warp stop motion drop wires on the warp threads is neglected.
- The frictional forces between thread guiding elements and the warp threads are not considered.
- The warp beam diameter is regarded as constant.

The warp forces are caused by the extension of the warp threads. Their absolute values for each heald frame depend on the geometric setting of the thread guiding elements, the shed geometry and the dynamic behaviour of all moving machine parts.

For a typical load situation, the stiffness of the warp threads is given by

$$E' = \frac{\Delta F}{\Delta \varepsilon}\left[\frac{N}{\%}\right] \qquad [9.43]$$

The resulting warp force is hence

$$F = E' \cdot \varepsilon \qquad [9.44]$$

The overall tension $F_{\text{ges},i}$ of the warp in heald frame i with m different warp yarns with j threads each and an elongation of $\varepsilon_{\text{ges},i}$ is then given by

$$F_{\text{ges},i} = \varepsilon_{\text{ges},i} \cdot E'_i = \varepsilon_{\text{ges},i} \cdot \sum_{k=1}^{m} (j \cdot E'_k) \qquad [9.45]$$

which leads to the overall warp tension

$$F_{\text{ges}} = \sum_{i=1}^{n} F_{\text{ges},i} \qquad [9.46]$$

The model and all relevant parameters are shown in Fig. 9.23. This model is valid for Dornier and Picanol weaving machines, for example. All values are measured relative to a Cartesian coordinate system (X, Y) which originates at the tip of the front shed. This point was chosen since it does not move during the weaving process relative to the machine frame. In order to formulate the equilibrium of moments at the backrest, a relative coordinate system (X', Y') is used. For the conversion of one coordinate system into the other, we have

$$\xi'_{x,y} = (\xi_{x,y} - X, Y_{\text{sbo}}) \qquad [9.47]$$

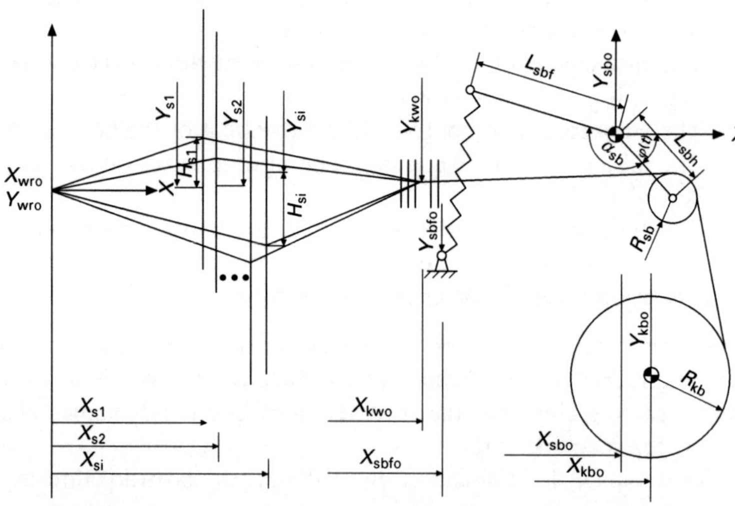

9.23 Schematic model of a weaving machine (adapted from de Weldige, 1996).

The position of the backrest with unstretched warp threads can be described by the angle φ_0. Its value can be calculated from the equilibrium of moments between the moment of equivalent mass of the backrest and the backrest spring forces. The latter can be determined using the overall spring constant c_{fsb} and the spring length l_{fsbo}. Hence, we have

$$\sum M_{sb} = 0 = -m_{sb} \cdot g \cdot L_{sbh} \cdot \cos(\varphi_0) + c_{fsb} \cdot (l_{fsb} - l_{fsbo}) \cdot L_{sbf}$$

$$\sqrt{1 - \{[L_{sbf} - A_{sbfsb} \cdot \cos(\beta_0 - \alpha_{sb} - \varphi_0)]/l_{fsb}\}^2} \qquad [9.48]$$

with

$$A_{sbfsb} = \sqrt{(X'_{sbfo} - X'_{sbo})^2 + (Y'_{sbfo} - Y'_{sbo})^2} = \sqrt{X'^2_{sbfo} + Y'^2_{sbfo}} \qquad [9.49]$$

$$\beta_o = \arctan\left(\frac{Y'_{sbfo}}{X'_{sbfo}}\right) + \pi, \text{ as } X'_{sbfo} < 0 \qquad [9.50]$$

and

$$l_{fsb}(\varphi) = \sqrt{l^2_{fsbo} + A^2_{sbfsb} - 2 \cdot l_{fsbo} \cdot A_{sbfsb} \cdot \cos(\beta_0 - \alpha_{sb} - \varphi_0)} \qquad [9.51]$$

from

$$A_{kbsb} = \sqrt{(L_{sbh} \cdot \cos(\varphi) - X'_{kbo})^2 + (L_{sbh} \cdot \sin(\varphi) - Y'_{kbo})^2} \qquad [9.52]$$

$$A_{sbkw} = \sqrt{(L_{sbh} \cdot \cos(\varphi) - X'_{kwo})^2 + (L_{sbh} \cdot \sin(\varphi) - Y'_{kwo})^2} \qquad [9.53]$$

and the angles

$$\beta_1 = \arcsin\left(\frac{R_{sb}}{A_{sbkw}}\right) \qquad [9.54]$$

$$\beta_2 = \arcsin\left(\frac{L_{sbh} \cdot \sin(\varphi) - Y'_{kwo}}{A_{sbkw}}\right) \qquad [9.55]$$

$$\beta_3 = \arcsin\left(\frac{R_{kb} - R_{sb}}{A_{kbsb}}\right) - \arctan\left(\frac{L_{sbh} \cdot \cos(\varphi) - X'_{kbo}}{L_{sbh} \cdot \sin(\varphi) - Y'_{kbo}}\right) \qquad [9.56]$$

we get

$$l_{kf,kb-sb} = \sqrt{A^2_{kbsb} - (R_{kb} - R_{sb})^2} \qquad [9.57]$$

$$l_{kf,sb} = R_{sb} \cdot \left(\beta_1 + \beta_2 - \beta_3 + \frac{\pi}{2}\right) \qquad [9.58]$$

$$l_{kf,sb-kw} = \sqrt{A_{sbkw}^2 - R_{sb}^2} \quad [9.59]$$

The coordinate Y_{si}^* depends on the fabric pattern and on the machine angle, hence we have

$$l_{kfi,kw-s} = \sqrt{(X'_{kwo} - Y'_{si})^2 + (Y'_{kwo} - Y_{si}^*)^2} \quad [9.60]$$

$$l_{kfi,s\text{-}wr} = \sqrt{(X'_{wro} - Y'_{si})^2 + (Y'_{wro} - Y_{si}^*)^2} \quad [9.61]$$

Y_{si}^* can then be calculated according to

$$Y_{si}^* = H_{si} \pm Y'_{si} \quad [9.62]$$

assuming that heald frame i is in the upper or lower shed during the rest stage or does not move at all. Hence

$$z = \frac{\alpha_{wm} - \frac{\varphi_{still}}{2}}{2\pi - \rho_{still}} \quad [9.63]$$

and thus

$$Y_{si}^* = c_{bindung} \cdot H_{si} \cdot \frac{1/2 - z + \sin(2\pi z)}{2\pi} + Y'_{si} \quad [9.64]$$

with φ_{still} = duration of heald frame resting stage in the open shed position [rad] and $c_{bindung}$ = indicator for the direction of the heald frame movement (+1: heald frame moves from upper to lower shed; –1: heald frame moves from lower to upper shed).

The thread length $l_{kf,kbab}$ is released by the warp beam according to

$$l_{kf,kbab} = R_{kb} \cdot i_{ka} \cdot \varphi_{kb} \quad [9.65]$$

where i_{ka} stands for the overall gear ratio of the warp beam drive and φ_{kb} represents the absolute angle of the warp beam drive [rad] since the machine start ($t = 0$).

It can be determined for a PI controller according to

$$\varphi_{kb}(t) = -\varphi_{kb,tensioning} \quad \text{for} \quad t \le T_{start} \quad [9.66a]$$

$$\phi_{kb}(t) = -\varphi_{kb,tensioning} + K_p(t) \cdot \Delta\phi(t) + K_i \cdot \int_{T_{ski}}^{t} \Delta\phi(t) - n_{start} \cdot t$$

$$\text{for } t > T_{start} \quad [9.66b]$$

with $\varphi_{kb,tensioning}$ = angle by which the warp beam was turned in a negative sense at machine start in order to achieve the given warp tension, $\Delta\varphi(t)$ = controller deviation with $\Delta\varphi(t) = \varphi(t) - \varphi_{soll}$

φ_{soll} = given value of backrest position,
T_{start} = starting point of warp let-off motion,
$K_p(t)$ = proportional controller amplification with $K_p(t) = K_{p1}$ for $t \leq T_{kp}$; $K_p(t) = K_{p2}$ for $t > T_{kp}$
T_{kp} = point in time at which the proportional controller amplification is switched from K_{p1} to K_{p2}
K_i = integrating controller amplification,
T_{ski} = point in time at which the integrating controller amplification is activated
n_{start} = rpm for warp let-off at machine start.

The warp load due to fabric take-up and reed beat-up is given by

$$s_{wr}(t) = l_{kf,riet} + l_{kf,wrab}$$

The fabric take-up can be calculated with

$$l_{kw,wrab} = t \cdot n_{wm} \cdot \text{density}_{weft}$$

where n_{wm} stands for the number of picks per minute and density$_{weft}$ represents the warp thread length per pick.

For the weft beating up, a sinusoidal warp thread extension according to

$$l_{kf,\,reed} = \frac{A_{reed}}{2} \cdot \left[\sin\left(2\pi \cdot \frac{\alpha_{wm} - \alpha_{reed-start}}{\alpha_{reed-duration}} - \frac{\pi}{2}\right) + 1\right] \qquad [9.67]$$

is assumed, where
A_{reed} = amplitude of the fabric movement owing to the weft beating up
$\alpha_{reed\text{-}start}$ = machine angle at which the fabric movement owing to the weft beating up commences
$\alpha_{reed\ duration}$ = duration of this fabric movement.
Hence

$$l_{kfo} = f(X_{kbo}, Y_{kbo}, R_{kb}, X_{sbo}, Y_{sbo}, K_{sbh}, L_{sbf}, \alpha_{sb}, m_{sb}, l_{fsbo}, c_{fsb}, X_{sbfo}, Y_{sbfo}, X_{kwo}, Y_{kwo}) \qquad [9.68]$$

The elongation of each group of warp threads assigned to a certain heald frame can then be calculated with

$$\varepsilon_{ges,i} = \frac{l_{kfi}(t) - l_{kfo}}{l_{kfo}} \qquad [9.69]$$

where

$$\varepsilon_{ges,i} > 0 \quad \text{for} \quad \frac{l_{kfi}(t) - l_{kfo}}{l_{kfo}} > 0 \qquad [9.69a]$$

$$\text{and} \quad \varepsilon_{ges,i} = 0 \quad \text{for} \quad \frac{l_{kfi}(t) - l_{kfo}}{l_{kfo}} \leq 0 \tag{9.69b}$$

The warp thread length $l_{kfi}(t)$ can be determined as the sum of the lengths of all warp thread segments between all components:

$$l_{kfi}(t) = l_{kb-sb} + l_{kf,sb} + l_{kf,sb-kw} + l_{kfi,kw-s} + l_{kfi,s-wr} + l_{kf,kbab} + l_{kf,wrab} + l_{kf,reed} \tag{9.70}$$

where
$l_{kf,,kb\text{-}sb}$ = warp thread length between warp beam and backrest
$l_{kf,sb}$ = warp thread length across backrest
$l_{kf,\ sb\text{-}kw}$ = warp thread length between backrest and warp stop motion
$l_{kfi,kw\text{-}s}$ = warp thread length between warp stop motion and heald frame
$l_{kfi,s\text{-}wr}$ = warp thread length between heald frame i and fabric
$l_{kf,kbab}$ = shortening or lengthening of warp thread due to warp let-off motion
$l_{kf,wrab}$ = lengthening of warp thread by the fabric take-up
$l_{kf,reed}$ = lengthening of warp thread due to weft beating up.

The starting point of the simulation of the warp thread tension is given by

$$F_{kf,soll} \overset{!}{=} F_{ges} = \sum_{i=}^{n} F_{ges,i}(\alpha_{WM}, \varphi, \varphi_{kb,spannen}) \tag{9.71}$$

which is equivalent to the overall warp tension F_{ges} reaching the given warp tension $F_{kf,soll}$.

This leads to

$$\begin{aligned} \Sigma M_{sb} \overset{!}{=} 0 = f(\varphi, F_{ges}) &= -m_{sb} \cdot g \cdot L_{sbh} \cdot \cos(\varphi) \\ &+ c_{fsb} \cdot (l_{fsb} - l_{fsbo}) \cdot L_{sbf} \\ &\sqrt{1 - \{[L_{sbf} - A_{sbfsb} \cdot \cos(\beta_0 - \alpha_{sb} - \varphi)]/l_{fsb}\}^2} \\ &- F_{ges,x'} \cdot L_{sbh} \cdot \sin(\varphi) + F_{ges,y'} \cdot L_{sbh} \cdot \cos(\varphi) \end{aligned} \tag{9.72}$$

where
$F_{ges,x'} = F_{ges} \cdot [-\cos(\beta_1 + \beta_2) + \sin(\beta_3)]$
$F_{ges,y'} = F_{ges} \cdot [-\sin(\beta_1 + \beta_2) + \cos(\beta_3)]$
$l_{fsb} = l_{fsbo}$ for $(l_{fsb} - l_{fsbo}) < 0$

The movement of the heald frames and the weft beating up leads to a high frequency excitation of the system. Hence, the forces of mass inertia and dampening can no longer be neglected.

The law of the angular momentum at the backrest pivotal point is then given by

$$J_{sbo} \cdot \frac{d^2\varphi}{dt^2} = \sum M_{sb}\left(t, \frac{d\varphi}{dt}, \varphi, \int_t \varphi \, dt\right) = -m_{sb} \cdot g \cdot L_{sbh} \cdot \cos(\varphi)$$

$$+ \, c_{fsb} \cdot (l_{fsb} - l_{fsbo}) \cdot L_{sbf}$$

$$\sqrt{1 - \{[L_{sbf} - A_{sbfsb} \cdot \cos(\beta_0 - \alpha_{sb} - \varphi)]/l_{fsb}\}^2}$$

$$- F_{ges,x'} \cdot L_{sbh} \cdot \sin(\varphi) + F_{ges,y'} \cdot L_{sbh} \cdot \cos(\varphi) - D_{sb} \cdot \frac{d\varphi}{dt} \qquad [9.73]$$

Apart from the constant parameters of the fabric and the machine, the solution of this equation depends on

- the time t since machine start;
- the angular velocity $\dot{\varphi} = d\varphi/dt$;
- the backrest angle φ; and
- the $\int_t \Delta\varphi \, dt$, which represents the integral of the deviation between backrest angle φ and given angle $\varphi_{reference}$.

The warp tension over time can then be calculated. The result for a typical example is shown in Fig. 9.24. The simulation of the warp tension makes it possible to determine an optimized machine setting (e.g. machine geometry) for any fabric prior to the actual trial thus saving time and hence money. This mathematical model was combined with evolutionary algorithms (see Chapter 3) to create the automatic weaving machine controller 'AUTOWARP' as described in de Weldige *et al.* (1997).

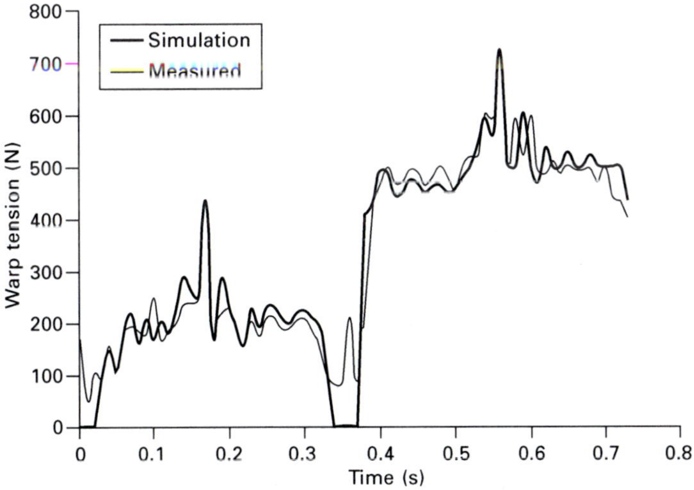

9.24 Comparison between measured and calculated warp tension (adapted from de Weldige, 1996).

A similar approach to this concept was published in Grossmann *et al.* (2010). They describe a concept of a new weaving and take-up technology to produce complete spacer fabrics without a separate assembly process. In the last part, a mathematical model for the simulation of the dynamic warp thread forces was developed. It was used to dimension the roller drives and to avoid unnecessarily high warp thread forces and thus fibre damage. The basic equations are similar to those derived by de Weldige (1996). The warp threads are regarded as spring-damper elements, the machine elements themselves as nodes with mass and hence a resulting inertia force. Velocity and displacements are calculated using MATLAB©. A schematic block diagram of the model is shown in Fig. 9.25. Predicted and measured values of the warp tension show a good correlation. The model can hence be used to predict and thus also to minimize the warp thread tension in order to reduce yarn damage.

Kuo *et al.* (2010) present a model for the dynamic behaviour and the control of a horizontal cross-lapper. Their analytical approach is an excellent example of a thorough analysis of such systems and could also help other researchers designing models for other applications.

9.4 Simulating finishing techniques: rolling and dyeing

An elaborate model for an improved roller system with regard to bending rigidity was developed by Kuo and Vu (2011). They developed a new model

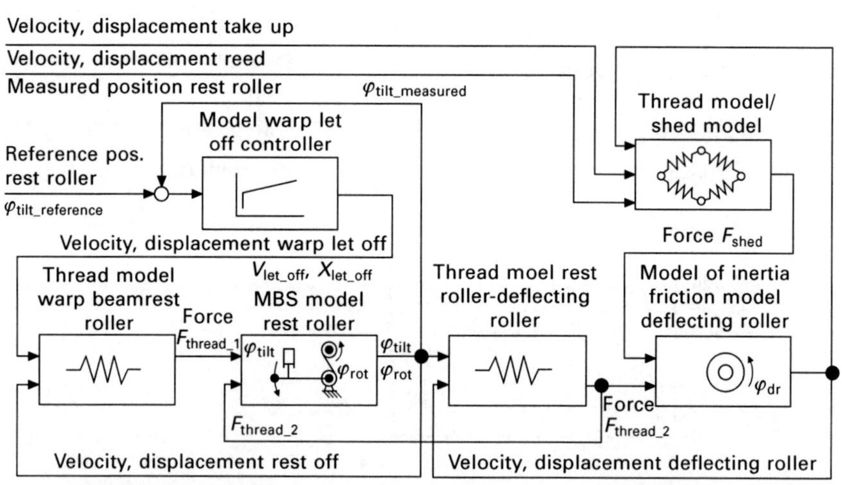

9.25 Block diagram of the simulation model (adapted from Grossmann *et al.*, 2010).

for a roller made of steel and nylon by applying beam theory according to Timoshenko. The results of the simulation indicate that the new roller has a high dynamic response performance and is easier to keep in a stable state than conventional rollers. A comparison between calculated and actually measured values for the new system is yet to be published. Other papers on the simulation of controllers for calendar roller systems worth reading include Kuo *et al.* (2011a, 2011b).

A mathematical description of a complete package or packed bed dyeing machine was published in Burley *et al.* (1985). The processes of convection, dispersion, absorption and desorption of dye to and from the packed bed of fibres were modelled, while account was also taken of the addition of dye to, and the possible removal of liquor from, the mixing tank. A numerical simulation of the mathematical model based on finite differences was developed and numerical results presented. Techniques for monitoring the accuracy of the results based on conservation properties of the system were discussed. Formulae for the expected theoretical steady states of the system were also given.

9.5 Simulating plant layout and production planning: finishing operations

Plant design and production planning are important issues in order to run a textile mill successfully. The main objective is to optimize the process flow by reducing queuing time, thus increasing the output of the plant. At the same time, the waiting time for customers to receive their ordered goods can also be decreased which is an often neglected but nevertheless important factor for the success of a textile enterprise.

An early noteworthy paper that deals with the optimized design of a textile mill is Livingston and Sommerfeld (1989). They apply the general purpose simulation system (GPSS) to perform a discrete-event simulation of a textile finishing mill that produces a variety of dyed and finished woven and knit fabrics. The GPSS is based on the queueing theory where the queue represents in this example work-in-process (WIP) inventory, batch chemicals or a custom order. The results of the simulation are, for example, machine utilization, queue length, WIP and idle time, production and residence time of one unit in the overall model. They can be used to calculate operator availability, possible bottlenecks, storage requirements and also to simulate stochastic behaviour of the operators, customers and the process itself. The aim of the simulation was to determine, among others, the impact of market demands, maintenance practices, quality control policies, capacity increase on equipment and manpower utilization, total production and manufacturing time on the overall performance of the mill. The flow diagram of the mill that was analysed is given in Fig. 9.26.

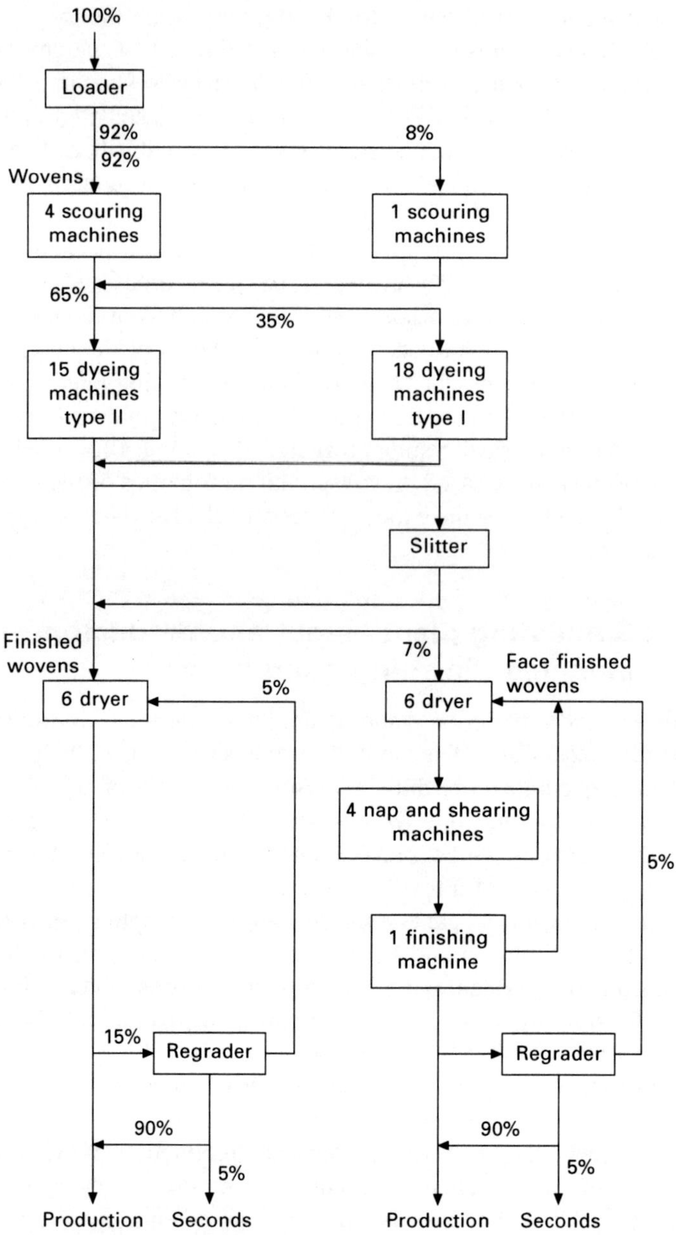

9.26 Flow diagram of textile finishing mill (adapted from Livingston and Sommerfeld, 1989).

After the loader unloads the wovens (92% of total production), they are cleaned in one of four scouring machines, the knits (8%) are cleaned in the fifth. All fabrics are then dyed in one of 33 dyeing machines. Some 15% of the wovens are checked after inspection by a regrader. Normally, 90% pass this test, 5% are sent back to the dryers and 5% are marked as seconds. In contrast, the knits are sent from the dyeing machines to the slitter, dried, napped, sheared and inspected. Again, 15% are checked by a regrader similar to the wovens. Besides this production process, also preventive maintenance procedures were taken into account.

The model is divided into segments which contain, for example, the machines and the queues before the machines. Figure 9.27 shows a typical example for the loading and scouring process: the loader distributes the fabrics to the knit scouring machine (8%) and the wovens scouring machines (92%). There, they are processed which is represented by queue–enter–depart–advance–leave operations. Each is assigned a certain time and capacity which can

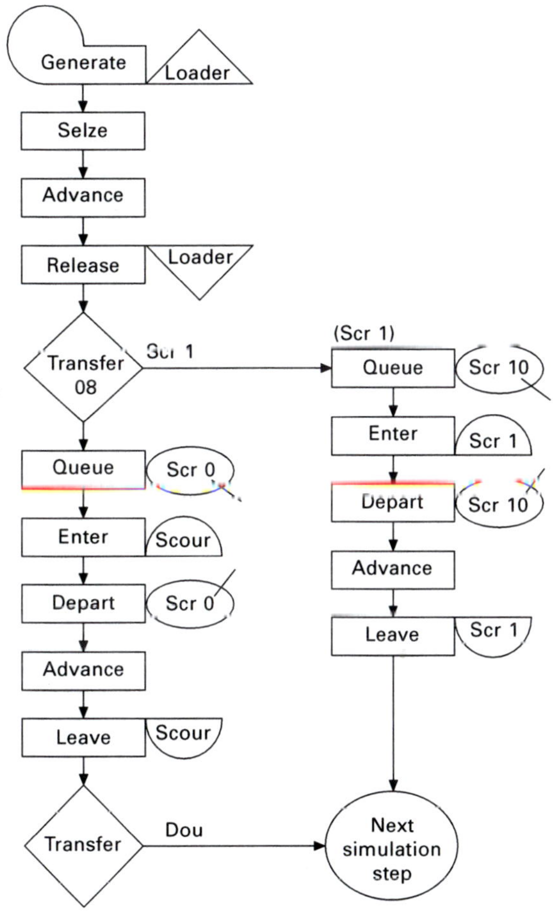

9.27 Model segment for loader and scouring machines (adapted from Livingston and Sommerfeld, 1989).

then be changed in the model. In a similar way, maintenance intervals are incorporated.

The authors then changed several parameters to determine their effect on the overall performance of the mill which led to the following results: an increase of the percentage of knits from 8% to 20% led to an increase of the one scouring machine utilization from 22% to 50% with no significant change in any of the other output results. Similarly, changes of other parameters such as the percentage of face-finished goods from 7% to 15%, the number of regraders and the percentage of re-inspected goods did not greatly affect the overall performance of the mill. In contrast, when the loader time was decreased which is equivalent to an overall production increase, it became apparent that the dyers were a bottleneck. A production increase by 17% could be managed with just one additional dyer whereas an increase by 33% required five additional dye machines to be added as can be seen in Fig. 9.28.

A recent paper by Azadeh *et al.* (2010) describes a more elaborate approach for a finishing mill which includes goal programming and a design of experiment (DoE) algorithm to optimize a mill with not just one but multiple objectives. It also gives references to a number of papers about the general approach of production planning. A schematic overview of the production process which was analysed is depicted in Fig. 9.29. Six different goods are produces using between one and three processes.

The main objectives were to minimize the completion time of each process, thus ensuring on-time delivery of the goods, and to decrease the processing times of all machines, hence lowering machine depreciation and

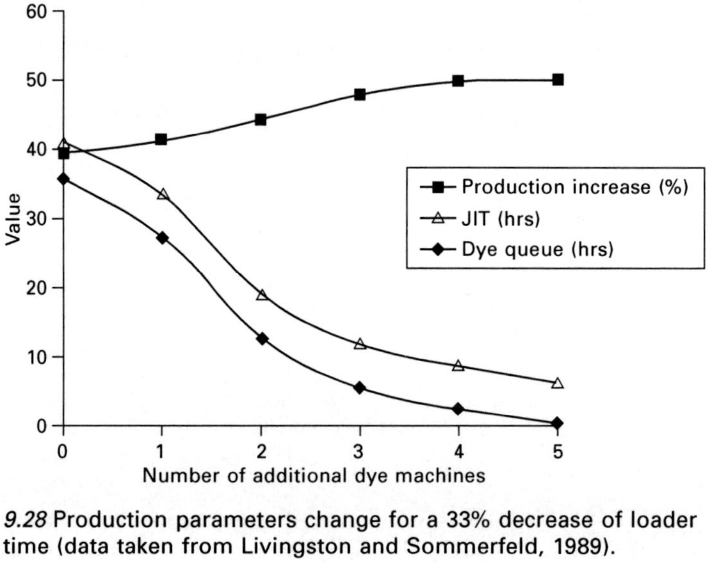

9.28 Production parameters change for a 33% decrease of loader time (data taken from Livingston and Sommerfeld, 1989).

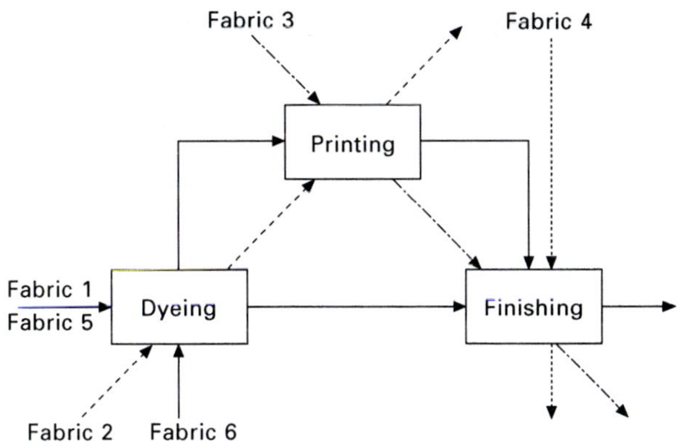

9.29 Schematic overview of the production process (adapted from Azadeh *et al.*, 2010).

other related costs. Typical assumptions such as 'all jobs are available at the starting time of the scheduling problem', 'one job cannot be allocated to more than one machine at the same time' and 'the processing time of each job on each machine includes setup times, transportation time', etc. were made.

With regard to the main objectives, the following five decision parameters were considered:

- specification of dyeing machines;
- working with either existing low technology machines or replacing them with high technology ones;
- temperatures of the printing and the stenter machine;
- the number of stenter machines; and
- five different workshop dispatching rules (e.g. first in first out, last in first out).

A different number of levels was assigned to each of the four factors which would have resulted in too large a number of possible combinations to be calculated. Hence, a complete factorial design using only the maximum and minimum values was employed which reduced the number of combinations to $5 \times 2^4 = 80$. By carrying out an analysis of variance (ANOVA), parameters which have no significant impact on the result are eliminated.

The simulation led to the following conclusions: set temperatures of printing and stenter machine to certain values, double the number of stenter machines, continue working with two low technology dyeing machines, the best dispatching rule is 'first in first out' and three extra workers are needed. This approach could also be applied to other kinds of textile mills and seems to be very promising with regard to an easy-to-use system for the optimization of plant design and production planning.

9.6 Practical advice in simulating textile machinery

Before trying to create an accurate model of reality, it is advisable to define the goal which is to be achieved by the simulation. This in turn requires a thorough analysis of the relevant parameters as an important and indispensable step. From there, designing a simplified model which still represents the real case with sufficient accuracy is the next step. In many cases, it is fully sufficient to consider not the machine as a whole but only certain crucial components which may separately be modelled much more easily than the entire system. In some cases, where one model does not fit reality, it is a sensible approach to create separate models for a certain parameter range. This often simplifies the simulation.

Whereas in the 1980s and 1990s most simulation programs were written by researchers from scratch, today there is a wide range of commercial tools available with which even complex models can be created. Engineers striving to describe textile machines and processes should hence concentrate on tailoring these programs to the problem in question rather than doing all the programming themselves.

Without the means to verify the results, a simulation of machines or components is pointless. Nevertheless, many papers that have been published in recent years do not present means to actually compare the calculated results with measurable values. Hence, a large number of papers describing such models have not been considered in this chapter although the simulation results appeared to be sensible.

9.7 References and further reading

Azadeh, A., Ghaderi, S.F., Dehghanbaghi, M., Dabbaghi, A. (2010), 'Integration of simulation, design of experiment and goal programming for minimization of makespan and tardiness', *International Journal of Advanced Manufacturing Technology 46*, 431–444.

Basligil, H. (1994), 'Determining machine interference through simulation method and its application to textile industry', *Modelling, Measurement & Control D: Manufacturing, Management, Human & Socio-Economic Problems 9, (1–3)*, 23–42.

Batra, S.K., Ghosh, T.K., Zeidman, M.I. (1989a), 'An integrated approach to dynamic analysis of the ring spinning process. Part I: without air drag and Coriolis acceleration', *Textile Research Journal 59 (6)*, 309–317.

Batra, S.K., Ghosh, T.K., Zeidman, M.I. (1989b), 'An integrated approach to dynamic analysis of the ring spinning process. Part II: with air drag', *Textile Research Journal 59 (7)*, 416–424.

Baykara, M., Öznergiz, E., Özsoy, C., Demir, A., Gülsen, S. (2009), 'The air-jet texturing and twisting machine's and model predictive control based on state-space', *Proceedings of 2nd IFAC International Conference on Intelligent Control Systems and Signal Processing, vol. 2, Part 1*, 73–78.

Braccesi, C., Carfagni, M. (1993), 'Using simulation to optimise output from an open-end rotor-spinning machine', *Journal of the Textile Institute 84*, 107–119.

Burley, R., Wai, P.-C., McGuire, G.R. (1985), 'Numerical simulation of an axial flow package dyeing machine', *Applied Mathematical Modelling 9 (1)*, 33–39.

Celik, N., Babaarslan, O., Bandara, M.P.U. (2004), 'A mathematical model for numerical simulation of weft insertion on an air-jet weaving machine', *Textile Research Journal 74*, 236–240.

Crank, J. (1953), 'A theoretical investigation of cap and ring spinning systems', *Textile Research Journal 23*, 266–276.

Crank, J., Whitmore, D.D. (1953), 'Balloon diameters and thread tensions calculated for different cap-spinning conditions', *Textile Research Journal 23*, 657–663.

Crank, J., Whitmore, D.D. (1954), 'The effect of cap-edge friction on spinning', *Textile Research Journal 24*, 434–437.

De Jager, G. (1990), 'Untersuchung und Simulation des Schusseintrags an Luftdüsenwebmaschinen', *Mitt. Inst. Textilmasch. Textilind.* (25).

De Weldige, E. (1996), 'Prozeßsimulation der Kettfadenzugkräfte in Webmaschinen', doctoral thesis, RWTH Aachen University.

De Weldige, E., Osthus, T., Wulfhorst, B. (1995), 'Reducing set-up times and optimizing processes by the automation of setting procedures on looms', *Mechatronics 5*, 147–163.

De Weldige, E., Osthus, T., Wulfhorst, B. (1997), 'Auto-warp process optimization and increased production through fully automatic loom setting', *Melliand Textilberichte 78*, 141–143.

Greenwood, K., Grigg, P.J. (1985), 'The development of twist in a false-twist texturing machine with a friction twister', *Journal of the Textile Institute 76*, 244–263.

Grossmann, K., Muehl, A., Loeser, M., Cherif, C., Hoffmann, G., Torun, A.R. (2010), 'New solutions for the manufacturing of spacer preforms for thermoplastic textile-reinforced lightweight structures', *Production Engineering Research Development 4*, 589–597.

Hättenschwiler, P., Pfeiffer, R., Schaufelberger, J. (1984), 'Die Zugfestigkeit von Garnen (Neue Erkenntnisse aus der Praxis)', *Melliand Textilberichte 65*, 23–26, 98–107.

Kuo, C.-F.J., Vu, H.Q. (2011), 'Dynamic modeling and control of a bending combined roller system using Timoshenko beam theory and constrained model predictive control', *Textile Research Journal 81*, 1448–1459.

Kuo, C.-F.J., Tu, H.-M., Liu, C.-H. (2010), 'Dynamic modeling and control of a current new horizontal type cross-lapper machine', *Textile Research Journal 80*, 2016–2027.

Kuo, C.-F.J., Chen, J.-S., Hiep, P.V., Tu, H.-M. (2011a), 'Theoretical control and experimental verification of a calendar roller system, Part 1: dynamic system modelling and model predictive controller design', *Textile Research Journal 81*, 10–21.

Kuo, C.-F.J., Chen, J.-S., Hiep, P.V., Tu, H.-M. (2011b), 'An entire strategy for control of a calendar roller system, part 2: settling time-optimal feedback control', *Textile Research Journal 81*, 542–549.

Lehnert, F. (1994), 'Simulation und Reduzierung der Schußfadenbelastung beim Weben', doctoral thesis, RWTH Aachen University.

Livingston, D.L., Sommerfeld, J.T. (1989), 'Discrete-event simulation in the design of textile finishing processes', *Textile Research Journal 59*, 589–596.

Meier, K. (1993), 'Shortening of the heating zone in false-twist texturing', *Proceedings of 32nd International Man-made Fiber Conference*, Dornbirn, 1–8.

Olbrich, A. (1995), 'Analyse und Simulation der Drallerteilung durch Friktion zur Optimierung der Falschdraht-Texturierung', doctoral thesis, RWTH Aachen University.

Osterloh, M., Wulfhorst, B. (1996), 'Opportunities for energy saving in the texturing process', *Chemical Fibers International 46*, 364–366.

Tang, H.B., Xu, B.G., Tao, X.M., Feng, J. (2010a), 'Mathematical modelling and numerical simulation of yarn behaviour in a modified ring spinning system', *Applied Mathematical Modelling 35*, 139–151.

Tang, H.B., Xu, B.G., Tao, X.M. (2010b), 'A new analytical solution of the twist wave propagation equation with its application in a modified ring spinning system', *Textile Research Journal 80 (7)*, 636–641.

Wulfhorst, B. (1969), 'Formänderungseigenschaften von Polyurethan- und Kautschuk-Elastomerfäden', doctoral thesis, RWTH Aachen University.

Zhu, M., Wang, Z. (2011), 'Textile spinning-frame roller fault diagnosis based on improved wavelet analysis and support vector machine', *Proceedings of International Conference on Electric Information and Control Engineering 2011*, 786–790.

Index